The Development of Security and Whole Care System for the Aged in China

Yanzhong Wang
Editor

The Development of Security and Whole Care System for the Aged in China

Springer

Editor
Yanzhong Wang
The Institute of Ethnology and Anthropology
Chinese Academy of Social Sciences
Beijing
China

ISBN 978-981-10-6480-7 ISBN 978-981-10-6481-4 (eBook)
https://doi.org/10.1007/978-981-10-6481-4

Library of Congress Control Number: 2017954475

Printed on acid-free paper

This Springer imprint is published by Springer Nature
The registered company is Springer Nature Singapore Pte Ltd.
The registered company address is: 152 Beach Road, #21-01/04 Gateway East, Singapore 189721, Singapore

Contents

Introduction

Population aging is one of the fundamental realities China faces at present and will face for a long period in the future. By the end of 2012, China's population over the age of 60 had reached 193.9 million, accounting for 14.3% of the total population, of which 127.14 million people were over the age of 65, accounting for 9.4% of the total population (National Bureau of Statistics, 2012). Population aging not only is an important manifestation of China's economic and social development and progress, but also poses a serious challenge to China's economic and social development. As population aging becomes increasingly serious, there is an urgent need to promote the pension security system to ensure that the elderly will be looked after properly and can spend their later years in comfort, and by doing so, they can have an equal share in the country's economic and social progress.

Old age is an inevitable part of everyone's life. Aging is not only a personal issue and a family issue but also a social issue. The elderly have different characteristics from other age-groups, such as a reduced physiological ability, decreased income from work, and a change in their social roles, which determines that they face a greater risk of diseases, poverty, and mental disturbances. Hypertension, heart disease, cancer, arthritis, Alzheimer's, and other chronic and critical diseases are common among the elderly. Most of the elderly face issues of daily care in their later years. A considerable number of elderly people may have to rely on others to take care of them to survive due to full or partial loss of the ability to work and look after themselves. As the income the elderly obtain through their own work gradually decreases, they begin to rely on their children or the pensions provided by the government and society to maintain their later lives. The risk to the elderly of psychiatric disorders is further increased because of a declining physiological function, decreased income, the "empty nest" phenomenon, accelerated population mobility, reduced social interaction, and other factors.

A rapidly aging population in China will lead to rapidly growing economic needs, service needs, and mental needs for the elderly. However, the current pension security system in China is still not fully developed, with an imperfect system, a narrow coverage, low level of security, poor service capability, management inefficiency, and many other practical issues. As a result, the contradiction between supply and demand in the pension security system becomes increasingly prominent. Getting old before getting rich, changes in the family structure, accelerated urbanization, and changes in the labor market pose new challenges to the development of the pension security system.

Promoting the development of the pension security system is an inevitable requirement for the realization of "providing for the elderly," an important part of the social security system, an important task of building a harmonious socialist society, and an important approach to the transformation of the economic development model and the building of a moderately prosperous society in a comprehensive manner and ensuring a "happy China." Promoting the establishment of the pension security system is not only an urgent necessity, but also feasible in reality: The Party and the government attach importance to people's livelihoods, which provides a strong guarantee for the pension security system. Fast economic growth provides a solid material basis for the development of the pension security system, and preliminary explorations in some areas have laid a good practical foundation for the development of the system.

In order to cope with the challenges of an aging population and effectively meet the elderly's various needs, a pension security system focusing on economic security, service security, and mental security for the elderly needs to be established. Furthermore, these factors will influence each other and interrelate.

Economic needs are the most basic needs the elderly face. Economic security should be the focus during the establishment of the pension security system. A scientific economic security system for the elderly needs to be established to ensure that the elderly can enjoy a moderate and stable source of income to sustain themselves in their later years. All forms of pension systems and benefit systems for the elderly must be established and improved as soon as possible. Economic security for the elderly needs to be adapted to the level of economic development. Certain differences between urban and rural areas as well as regional and occupational differences should be taken into considerations set against the fundamental principle that all the elderly can enjoy equal basic economic security for the realization of fairness and efficiency.

Service security is essential to help the elderly enjoy their later years and improve the quality of their lives. Led by the government via corresponding laws, regulations, and service organizations, the government, the family, the community, and many other agents should be involved to provide comprehensive, specialized, multi-level security services for the elderly. Daily care and medical care for the elderly should be given priority to. Paid and free methods should be combined to provide some basic free services to some elderly people with no families, in households that are entitled to the five guarantees and poor elderly people; those elderly people with relatively high incomes should be encouraged to realize security via market purchases, while the government enhances market regulation and provides some policy support.

Spiritual security is very important to the elderly as spiritual emptiness and loneliness are likely to cause senile mental illness and social issues. The presence and levels of spiritual security are directly related to the quality of the elderly's lives. Spiritual security comes from both formal channels and informal channels. All levels of government should play a role in spiritual security for the elderly, establish relevant laws and regulations, increase investment in spiritual security for the elderly and the training of professionals, and vigorously advocate attention to spiritual security for the elderly. Additionally, families, relatives, society, and other informal channels are also very important.

Establishing a pension security system requires not only capital investment and institution building, but also a scientific and efficient management system. The useful experience of other countries in pension security management must be fully absorbed and drawn on to establish a pension security management system for the elderly that is adapted to the existing political system and closely related to China's national conditions. Legislation on pension security needs to be actively and steadily pushed forward. The transition of the government's functions needs to be promoted toward supervision and regulation and the providing of basic pension security to build a moderately concentrated supervision and regulation system and coordination mechanism of pension security in which rights and responsibilities are consistent, and a scientific and reasonable management system for an endowment insurance fund needs to be established. An all-round and scientific management system for the service industry for the elderly needs to be established while the socialization of pension security's management and services being continuously pushed forward.

With economic and social development, pension security in all countries is undergoing constant change and improvement. Reform and development of the pension security system abroad provides useful experience for China to draw upon. Foreign experience in the economic security of the elderly is mainly reflected in the following areas: A multi-level and socialized endowment insurance system is established and improved. Investment in pension security is significantly higher than investment in other social security programs. A pension financing mechanism of shared responsibility is established with a focus on the accumulation of pension security funds. Economic security for the elderly is adapted to economic development. Foreign experience in the service security of the elderly is mainly reflected in the following areas: An elderly service security system that meets diverse and multi-level needs is established. A complete legal system of services for the elderly is established with attention paid to the specialization of services. In addition, a health-care insurance system is established and improved. Foreign experience in spiritual security for the elderly is mainly reflected in the following areas: The government attaches importance to the spiritual security of the elderly, promotes the role of the family in securing the spiritual security of the elderly, encourages the participation of nonprofit organizations and other social forces, and encourages community involvement for the elderly.

In terms of the pension security system's properties and the government's functions, the role of the government in the pension security system needs to be given full play to. Public goods have two characteristics, namely "non-excludability" and "non-rivalry." Most parts of the pension security system have a strong public nature and belong to pure public goods or quasi-public goods.

The government is the organizer or provider of these public goods and needs to give full play to its role in the pension security system. In terms of the functions of the government, the market needs to play a fundamental role in allocating resources. There is also a need to enhance the government's regulation of the market and macro-control to prevent "market failure." In the future, the leading role of the government needs to be given full play to in an appropriate way to provide good, efficient, and accessible pension security. There is a need to promote the government's guidance and overall planning in the development of the pension security system, to perfect regulations and policies related to the pension security system, to improve the financial input mechanism, and to give full play to the government's role in organization, implementation, supervision, and regulation.

Based on the current development of pension security, this study proceeds from population aging and the elderly's needs and establishes a pension security system integrating "economic security, service security, and spiritual security," on which systematical research and discussion are conducted respectively. In addition, in-depth special studies on how to promote the improvement of the pension security system are conducted in terms of management system, international experience, the role of the government, and other aspects. Theoretical research and empirical research are combined in this study in sociology, management science, economics, and other interdisciplinary and crossover research methods. Key and tough issues about pension security are fully discussed, and specific policy recommendations on the future development of China's pension security system are put forward.

The formulation and publication of this research report are due to the painstaking efforts of every member of the research team who referred to a lot of relevant material and data during work and in their spare time, conducted a number of seminars, and revised the report repeatedly before it was eventually completed. Experts in related fields who provided helpful advice and suggestions for the improvement of the research project during the discussion about and the formulation of this research report are gratefully acknowledged.

Pension security is a complex system. Although we have tried our best to study and analyze some of the theoretical and practical issues, it may not be satisfactory in some aspects due to limited time, energy, and capability. We look forward to seeing more academic studies on pension security to provide theoretical reference and policy recommendations for the improvement of the pension security system in China. We also look forward to cooperation between theoretical and practical fields as well as the combination of theoretical research and practical exploration to actively and steadily push forward the establishment of the pension security system. "Don't forget the elderly who are not related to you while caring for those who are." The development of the pension security system requires the efforts of the government, society, families, and individuals. We expect that all sectors of society care for the elderly and pay attention to their well-being and the establishment of the pension security system in China!

Chapter 1
Population Aging and the Construction of the Modern Pension Security System

Yanzhong Wang and Yuqi Long

Old age is an inevitable part of every normal person's life. Aging is not only a personal issue and a family issue but also a social issue. It's the right of elderly people to be looked after and to enjoy their later lives, and the responsibility of a country and society as well. With the population aging getting increasingly serious, it's an important task of the government to promote the building of the pension security system and ensure that the elderly can enjoy their later lives and that they can share the benefit of economic and social progress. The elderly have different characteristics from the groups at other ages, such as reduced physiological functions, decreased income from work and the change in social roles. They face the risks of diseases, poverty and mental illness, thus need securities on medical care, economy, service, mental care and other aspects. Currently, the building of the pension security system in China is both important and urgent. A complete pension security system focusing on economic security, service security and spiritual security of the elderly needs to be built in the future. The development of the pension security system requires funding, legal institutions, management, talents and the involvement of the government, families, society, communities and individuals. Effective measures need to be taken on the basis of right principles and a scientific understanding.

Y. Wang (✉)
Institute of Ethnology and Anthropology, Chinese Academy of Social Sciences,
Building 6, #27 Zhongguancunnan Street, Haidian District 100081, Beijing,
People's Republic of China
e-mail: wangyanzhong01@163.com

Y. Long
School of Management, Capital Normal University,
#83 West Third Ring Road North, Haidian District 100089, Beijing,
People's Republic of China
e-mail: lyqdy2008@163.com

Y. Wang (ed.), *The Development of Security and Whole Care System for the Aged in China*, https://doi.org/10.1007/978-981-10-6481-4_1

1.1 Characteristics of the Elderly and the Risks They Face

A clear understanding of elderly people includes their characteristics, the risks they face and the needs they have. On top of that, a scientific understanding of the pension security system is needed first in the research on the aging issue aimed at the building of a scientific and reasonable pension security system.

1.1.1 Characteristics of the Elderly

The elderly refer to the people over a certain age, and particularly, the group of people living in a certain mode of social production and time, and certain districts with a certain number and quality, at higher ages according to the sequence of age.[1] Definitions of the elderly vary by time and country. They also have different meanings from different perspectives including calendar age, physiological age, mental age, social age and others, among which the first two are most commonly adopted. In the late 19th century and the early 20th century, Swedish demographer Gustav Sundbary defined 50 as the beginning of old age on the ground that most people experience menopause and lose fertility at 50. France was the world's first aging country, where 60 was regarded as the beginning of old age.[2] Prolonged life expectancy has shifted definitions of the onset of old age in some countries. The age of 65 has been defined as the beginning of old age in many countries. Prior to the founding of the New China, life expectancy in China was less than 35 years old, while it's more than 70 now.

On the whole, the elderly refer to a group of people with clear distinctions from adolescents, adults and the middle-aged. This group has the following four main characteristics:

(1) Deterioration of physiological functions. This is the most obvious characteristic of the elderly and an objective law of human physiology as well. The vast majority of people go through this stage. With the increase in age, the physiological functions of the elderly gradually deteriorate with decreased immunity and increased age-related diseases,[3] such as arthritis, hypertension, diabetes and respiratory problems. There are some visible manifestations of deteriorated physiological functions of the elderly, such as white hair, baldness, kyphosis, slow movement and slow response. With the gradual deterioration of physiological functions, the elderly gradually lose the ability to look after themselves in daily life.

---

[1]Tian [1].

[2]Xiong [2].

[3]Yan and Xiaoyun[3].

(2) Decrease in income from work. With the increase in age, the elderly's ability to work declines continuously and they obtain less and less income from work. At this point, they usually rely on the care of their children or the support of the family and society, while some of them maintain their lives with their savings in the past. Particularly in the age of industrialization and information, the knowledge and skills the elderly have gradually lagged behind the needs of the times due to the rapid changes in various technologies and the emergence of new management ideas. Besides, as their ability to work declines, the income they get gradually decreases.
(3) Changes in social roles. After the elderly reach old age, their social roles also change. They shift from workers to complete consumers and from child and elderly people supporters to dependents. After reaching old age, they gradually withdraw themselves from work and eventually no longer engage in formal work. As the social relations established through work before old age gradually fade away, social interaction also decreases. Due to changes in social roles, individuals, families and society need to provide certain support for the elderly in order to promote smooth transition of their roles.
(4) Certain social experience and skills. The elderly have dealt with various people and matters in work and life before old age, thus being more experienced in handling problems. Some elderly people get both lessons and success through various hardships. The social experience of the elderly plays an important role in guiding young people during their growth. They have acquired a variety of skills through education at school and all types of in-service training, which are of much help to young people and are important manifestation of the elderly's functions.

1.1.2 Major Risks the Elderly Face

As a special group, the elderly face greater risks than other age groups. With an understanding of the risks, the pension security should have a definite object in view. The greater risks they face, the better security they should need. The elderly mainly face the following four risks:

(1) Risk of disease. A healthy body is the common pursuit of people throughout the world and also an important manifestation of the quality of life and the level of happiness. However, as the elderly's physiological functions deteriorate and their immunity declines, a variety of diseases show up. Compared with adults, elderly people face greater risk of disease. Once getting sick, they are subject to both physical and mental pains. Their ability to look after themselves in daily life gradually declines or even gets completely lost, so they need other people to look after them. With the acceleration of urbanization and industrialization as well as the increasingly serious environmental pollution, the risk of diseases of the elderly increases day by day. The older the age is, the higher the frailty

Table 1.1 Fortnight morbidity rate of residents in different age groups

Age	Total (‰)		Urban areas (‰)		Rural areas (‰)	
	2008	2003	2008	2003	2008	2003
0–4	174.2	133.0	146.7	104.2	179.8	139.5
5–14	76.9	72.2	63.9	60.9	79.8	74.5
15–24	49.7	49.8	50.6	40.4	49.5	52.4
25–34	74.9	82.5	63.2	59.5	79.6	90.4
35–44	136.0	126.2	101.6	100.0	147.6	135.9
45–54	227.2	191.5	213.8	163.1	232.8	202.6
55–64	322.7	251.8	355.1	258.1	310.0	249.0
Over 65	465.9	338.3	580.9	396.9	398.2	302.1

Source National Health Service Survey in 2003 and 2008

index is and the higher the risk of death is.[4] According to data of the National Health Service Survey, the fortnight morbidity rate of residents over the age of 65 was 338.3% in 2003, which increased to 465.9% in 2008; wherein the morbidity rate of elder residents in urban areas was higher, which was 396.9% in 2003 and increased to 580.9% in 2008 (see Table 1.1). From the age of patients discharged from the hospital nationwide in 2009, people over the age of 60 accounted for a quite high proportion in various types of diseases (see Table 1.2). According to previous National Health Service Surveys and related research, a high proportion of elderly people suffer from various chronic diseases, mainly including hypertension, heart disease, tumor, arthritis, Alzheimer's and others.

(2) Risk in daily care. With the decline in physical and mental functions and the increase of diseases, the majority of elderly people face the issue of daily care in their later years. A considerable number of elderly people have to rely on others to take care of them to survive due to full or partial loss of the ability to work and to look after themselves. However, without an effective security system and material, financial and personnel reserves, the elderly will have a very hard time in daily care during their later years. This risk is very common and is a very tough issue.

(3) Risk of poverty. As the elderly's ability to work declines and the income they obtain through their own work gradually decreases, they begin to rely on their children's maintenance or the pension provided by the government and society to maintain their later lives. With the centralization and empty-nest phenomenon, children's financial support to the elderly declines, and some are even NEETs (not in Education, Employment, or Training). Although the current pension security system continues to be improved, the pension security that the elderly get from the government and society is still limited, especially in the

[4] Gu et al. [4]. Frailty index refers to the proportion of deficits present in an individual out of the total number of age-related health variables considered.

Table 1.2 Age and disease of patients discharged from the hospital in 2009

Disease	Under 5 (%)	5–14 (%)	15–44 (%)	45–59 (%)	Over 60 (%)
Total	11.1	4.6	33.6	20.3	30.4
Infectious and parasitic diseases	32.4	12.2	27.4	13.5	14.5
Tumor	1.0	1.4	27.3	34.1	36.2
Hematological, hemopoietic organs and immune diseases	11.7	18.5	29.7	16.5	23.6
Endocrine, nutritional and metabolic diseases	1.9	0.9	20.9	34.6	41.7
Mental and behavioral disorders	3.0	3.0	52.4	24.6	17.0
Nervous system diseases	5.7	4.6	21.7	25.9	42.2
Eye and appendage diseases	1.5	2.9	15.2	20.5	59.9
Ear and mastoid diseases	2.8	5.8	39.1	28.8	23.4
Circulatory system diseases	0.7	0.7	9.6	24.0	65.1
Respiratory diseases	38.4	12.7	14.0	9.4	25.5
Skin and subcutaneous tissue diseases	12.8	9.3	35.3	19.5	23.0
Genitourinary system diseases	2.3	3.5	43.1	25.3	25.8

Source China Public Health Statistical Yearbook (2010)

vast rural areas. Scholars have different estimates about the number of elderly people in poverty in China according to different standards. Qiao Xiaochun, Zhang Kaiti and others estimate that elderly people in poverty accounted for 17.5% of all the elderly people in China according to the data of the fifth national population census. The number of elderly people in poverty was 22.748 million (The minimum living allowance in urban areas is regarded as the standard of absolute poverty in urban areas, while 30% of it is regarded as that in local rural areas).[5] Yu [6] once estimated that the number of China's poor elderly people in different poverty measure methods was 38.53 million (according to Engel coefficient), 44.86 million (according to the international poverty line, including 12.64 million in urban areas and 32.22 million in rural areas) and 42.86 million (according to the subjective measure, including 9.32 million in urban areas and 33.54 million in rural areas) respectively.[6] The risk of poverty is closely related with the risk of disease. Deteriorated health because of poverty and the lack of money for treatment increases the risk of disease. On the contrary, the risk of disease will further increase the risk of poverty of elderly people, leading to "poverty resulted from disease." Therefore, it is an important task of the country and society to release the risk of poverty elderly people are faced with. With rapid economic and social

[5]Qiao et al. [5].

[6]Yu [6].

Table 1.3 Suicidal tendencies of the elderly

Behavior	Urban and rural areas	
	Urban area (%)	Rural area (%)
Wanting to die in the past month	32.7	26.7
Thinking about suicide in the past month	19.3	18.9
Having committed suicide in the past month	21.3	6.5
Wanting to die in the past year	41.2	40.2
Wanting to die in the past five years	53.8	54.2

Source Zhang Kaiti (ed.) [36]

development, it is also an inevitable requirement to solve the poverty issue of the elderly so that they can share the results of the reform and development.

(4) Risk of mental illness. Germany's famous psychoanalyst Erik Erikson believed that people have tasks of self-growth and possible crises during every stage of their lives. The conflict between the sense of accomplishment and that of loss become the main issue in old age, which, if not handled properly, will produce a sense of loss, despair and self-abasement.[7] This feature in old age is also apt to cause psychological and mental problems, and elderly people are prone to the sense of loneliness and depression and may even conduct deviant or illegal behaviors, such as suicide or crime. With the decline of elderly people's physiological functions, the continuous decrease in their ability to adapt to changes in the environment and the reduction of income they get from work, they get increasingly dependent in their lives.

With changes in their social roles, elderly people have less social interactions and activities, and a stronger sense of loss, which increases their sense of loneliness to some extent. According to a survey of elderly population in urban and rural areas in 2006, the rate of elderly people in urban areas suffering from mild, moderate and severe depression was 30.9, 9.4 and 4.2% respectively. As the age increased, so did the rate of depression.[8] Suicidality among the elderly is quite obvious (see Table 1.3). Take a district in Nanjing for example, 23 unnatural deaths were reported to the police in 2009, including 13 cases of suicide of elderly people (over the age of 60), accounting for 57%.[9] Elderly people are attached to the past, fear the future and are confused about the meaning of life, thus being unable to bring their value into full play and falling into depression. They languish in age discrimination for their frailty and low competitiveness,[10] further increasing the risk of mental disorders in the elderly, which will increase the risk of mental illness to a certain extent. With declined physiological functions, reduced income, an empty nest, faster population flow, less social interactions and other factors, the risk of mental

[7]Chen and Liu [7].

[8]Zhang and Guo [8].

[9]Zhaohui [9].

[10]Song and Pu [10].

illness is further increased. Therefore, it is both the needs of the elderly and the task of pension security to reduce elderly people' sense of loneliness and enrich their spiritual life.

1.2 Theories on the Risks the Elderly Face and Their Needs

Population aging has also encouraged reflections on relevant theories for a comprehensive and in-depth understanding of aging, elderly people's needs and their behaviors. Typical theories about the elderly include:

(1) The disengagement theory. This theory was first formulated by E. Cumming and W.E. Henry in 1961 in the book *Growing Old*[11] and has been continuously improved afterwards. According to the disengagement theory, as elderly people get older, they have less social activities and interpersonal communication with other people. Besides, their role in interpersonal communication changes from a leading role into a supporting one and from a positive one into a negative one. Elderly people also believe that, as they get older, changes in their behavior are reasonable and that they should gradually withdraw from major roles in society and choose the life and behavior suitable for senior citizens. The disengagement theory is apt to lead to self-isolation and social exclusion of the elderly, and may even cause relevant social issues.
(2) The activity theory. According to this theory, the participation in social activities is the foundation of people's social lives, and the same goes for the elderly. There is no clear or significant difference between the elderly and adults in their participation in social activities. The only difference is that they slow down.[12] Elderly people also have the desire and demand for activities and it is very important for them to participate in a reasonable number of social activities. When they stop working, they need other activities instead. The activity theory is of important guiding significance for the promotion of elderly people's participation in social activities to live a positive life at old age.
(3) The continuity theory. This theory was brought forth by American scholars Richard, Leifson and Petersen.[13] According to the continuity theory, different stages in the life circle are highly continuous. Elderly people usually maintain the same life and activities as they did in their earlier years of life instead of changing everything all of a sudden. Living habits, personalities, hobbies and others formed during adulthood are difficult to change suddenly. For example, a person who liked activities in adulthood won't stand still at old age, and a

[11]Cumming and Henry [11].

[12]Tian [12].

[13]Zhang [13].

confident and optimistic person won't suddenly become passive or self-abased. Therefore, the lifestyle and habits in adulthood need to be connected with those at old age.

(4) The social exchange theory. Introduced by American sociologists Emerson and Blau,[14] this theory, based on the psychology of instrumental behavior and utilitarian economics, posits that social interaction is the exchange between both sides of the interaction. During the interaction, both sides think about their own interests and determine the form and extent of the interaction. Therefore, people need certain controllable resources, such as wealth, achievement, capacity and health, so do elderly people. For all kinds of reasons, the resources that elderly people own continue to decrease, thus having lower status in social interaction and exchanges. Therefore, the elderly need resources that can be used for exchanges to maintain normal social interaction. There are also exchanges between elderly people and their children even within the family, who provide for and take care of them as payback. "Bringing up a son to provide for one's old age" is the best example. The social exchange theory is far too rational and utilitarian in explaining the care for the elderly.

(5) The elderly subculture theory. Subculture is a non-mainstream, partial and frontier cultural group within a larger culture in a specific region, often having beliefs or lifestyles at variance with those of the larger culture. According to this theory, the elderly form a unique subculture, values and lifestyles with distinctive differences with those of other age groups in their lives and activities. The elderly subculture theory is favorable for mutual aid and interaction between the elderly, but regards them independent from other age groups. It pays too much attention on the specificity of the elderly, while ignoring their similarity to adults, thus being apt to cause rejection and isolation.

The relevant theories above interpret the elderly and their lifestyles from different perspectives and are of great value for a better understanding of the needs, risk analysis and the construction of the pension security system. Elderly people share the same needs with other age groups, while having their special needs, which mainly include:

(1) The need for financial support. Like other age groups, the elderly need clothing, food, shelter and transportation. Financial support is supposed to meet these needs. For various reasons, elderly people get less income from work, so they need the help of their families, government and society to get through old age. Only a small number of elderly people can maintain their lives with their savings before old age. It is safe to say that the availability of adequate financial support determines the quality of elderly people's lives.

(2) The need for medical security. As mentioned above, elderly people face great risks of disease and are especially prone to a variety of chronic diseases. Suffering physical and mental pains, elderly people need medical security more

[14]Emerson [14].

than adults do. As they lack sources of income, the government needs to establish institutions and funds to satisfy their needs for medical service. If elderly people without medical security get sick, their family will have more burdens, or their health conditions will get worse.

(3) The need for daily care. With the increase of age and risks of age-related diseases, elderly people's ability to take care of themselves declines. Therefore, they need help to live a normal life in terms of eating, dressing, bathing, indoor activities and others. According to the analysis of elderly people's ability to take care of themselves with data about the changes in national population in sampling survey of China's elderly people in 2004, some scholars found that 8.9% of the elderly in China could not take care of themselves that year. The number of elderly people who could not take care of themselves was more than 12 million.[15] By the end of 2009, 18.99 million of 167 million elderly people in China were over the age of 80, accounting for 11.4%, the majority of whom had gradually lost part of or all of the ability to take care of themselves. There were 10.36 million disabled elders and 21.23 million semi-disabled elders in China, accounting for 18.9% of the total population of elders, who needed daily care at different levels.[16]

(4) The need for mental support. The majority of the elderly, after reaching old age, no longer work, so the social relations they established in previous work gradually decline and the free time they have increases. During this period, they need more mental support for psychological satisfaction to improve the quality of their lives in their later years. At some point, the elderly need even more mental support than financial support. Especially when their needs in material life are met, spiritual needs become increasingly prominent. The lack of mental support often leads to spiritual emptiness and mental disorder of the elderly, resulting in a variety of relevant social issues.

(5) The need for social involvement. As social people, elderly people have the same need for social interaction and involvement with adults. Elderly people's social involvement includes economic involvement (such as certain economic activities), political involvement (such as democratic management and supervision and Party activities), recreation involvement (such as various recreational and sports activities) and others. Through social involvement, the elderly can show their personal values and get a sense of spiritual satisfaction.

Needs of the elderly have the following features: First, elderly people's needs are all-round. That is to say, the needs of the elderly concern all aspects of life, including the needs for financial support, medical security, daily care, mental support, social involvement and other aspects. Second, elderly people's needs are of a hierarchy. That is to say, the importance of their needs is not the same, featuring a certain hierarchy. In general, financial support, medical security and daily care are priorities, while mental support and social involvement are needs of higher levels

[15]Du and Wu [15].

[16]Xinhuanet, http://news.xinhuanet.com/politics/2010-11/07/c_13595080.htm.

on this basis. Third, elderly people's needs are relevant. These needs of the elderly are often interrelated and influence each other. For example, whether the need for financial support is met has an impact on the needs for medical security, daily care and spiritual support; whether the need for social involvement is met is directly related to the spiritual life of the elderly; while receiving medical security (when sick) and daily care, the elderly often get certain spiritual satisfaction. Fourth, elderly people's needs are different. Different elderly people have different types of needs. For example, needs of elderly people in urban areas are different from those in rural areas; the needs of those engaged in different professions (before old age) are different; the needs of those with different income are different; the needs of those with different education background are different. Fifth, elderly people's needs are evolving. With the development of economy and society, the focus of elderly people's needs is different. The needs of elderly people in agricultural society, industrial society and information society are different, showing continuous evolution on the whole.

1.3 Multiple Subjects of Pension Security and Their Evolution

1.3.1 The Supply System of Pension Security

The objects of pension security and their needs have been covered above, while the supply of the pension security system will be mainly talked about next, including its subjects, content and approaches.

(1) Subjects of pension security, namely its providers. Different types of elderly groups have different needs. Besides, the needs of elderly people are all-round and diversified, so the subjects of pension security should also be diversified. It is difficult for a single subject to provide good pension security for all the elderly people. The subjects of traditional pension security are families and individuals, while those of modern pension security are increasingly diversified, including the government, various social organizations, employers, businesses, families, individuals and others. With economic development and the government's better public services, it has an increasingly important role in the construction of the pension security system and has become the most important subject of pension security in most countries. The government provides various types of pension security for the elderly mainly by improving the security system, providing funds, training personnel, enhancing management and others. Families are still important subjects of pension security. In the vast rural areas in particular, families are the most important subject, from which the elderly get pension security. As members of society, elderly people should get

help from society; relevant organizations, non-profit and volunteers' organizations in particular, should provide help for the elderly; employers of elderly people should provide material and spiritual support for them. Of course, elderly people, especially the "young elders" with working capacities, should participate in certain economic activities or social welfare services, both to enrich their own lives at old age and to relieve the burden on the government and their families.

(2) Content of pension security. Security is aimed at demands and risks. As mentioned previously, the risks and major needs of elderly people are important part of pension security, including financial support, material support, medical security, daily care, mental support and others. According to the major risks and features of the needs of elderly people, the content of pension security should focus on its all-roundness, hierarchy, difference and evolution. The key points of pension security need to be grasped, with emphasis on the equality and efficiency of pension security.

(3) Approaches of pension security. Approaches of pension security refer to the specific ways to realize its objectives, mainly including family care, community care and institutional care. (1) Family care is the most important approach of pension security in agricultural society when, given undeveloped productivity, families played an important role in the economic society, while elderly people played a leading and dominant role in the family. After reaching old age, they mainly depended on the family. After a long process of development, the culture of respecting elderly people was formed and has been carried on till today when family care is still an important approach of providing care for the elderly. In modern society, it is probably difficult to achieve better security objectives without families. (2) Community care, also known as home care, means that the elderly live in their own homes and enjoy a variety of pension services provided by service agencies in the community, especially materials, daily care, mental support and other services provided for them. With a variety of services for the elderly, home care not only reduces the pressure on families, but also allows the elderly to enjoy the comfort of staying at home, thus becoming a more and more important approach of providing care for the elderly in modern society. Survey shows that 77.5% of elderly people in urban areas stated that their communities provided at least one type of pension service.[17] (3) Institutional care refers to that elderly people live in a variety of care institutions established by the government and society (such as nursing homes, rest homes, apartments for the elderly, day-care centers for the elderly and others) and purchase services with government subsidies or their own money. With the changes in family structure and accelerating urbanization, industrialization and labor mobility, institutional care is increasingly popular.

[17]Zhang and Guo [16].

1.3.2 From Home Care to the Pension Security System in Modern Society

Existing pension systems in the world were not available in ancient times. Instead, they came into being through continuous exploration in a long process of development. The modern pension system has a history of only more than 100 years. Pension systems in the world are not set. With economic and social development, they are also in constant changes and improvement.

It is an old topic to ensure that seniors are supported appropriately. Its coming into being from scratch, its development from being informal to being formal and from family care to social care is an important manifestation of the progress in human civilization. In primitive society, it was difficult to ensure elderly people's basic livelihood due to the low productivity and the lack of materials. In prehistoric times, elderly people who lost the ability to work as a result of old age were seen as burdens that competed with young people for living resources. Consequently, they were usually abandoned, abused or even killed. In ancient tribes on the east coast of Greenland, dying elders were sometimes drowned or drowned themselves. The practice of killing and abandoning elders also existed in many Indian tribes.[18] In some rural areas in ancient Japan, elderly people were taken into the mountains to let them die there. The practice of throwing elderly people over the age of 60 into the Tiber was followed in ancient Rome.[19] With the development of productivity and the progress in civilization, surplus products and means of subsistence got increasingly rich, and elderly people gained respect for their life experience and skills. Family care gradually became a tradition in the world and played a leading role in a long period in history and continues to play an important role today and even in the future.

Given the development of productivity, changes in the demographic structure, progress in science and technology and accelerating industrialization, it is more and more difficult for elderly people to bring their traditional skills into play. With the increase in the number of elderly people and mass production with machines, families have weaker and weaker economic functions and increasing burdens in terms of care for elderly people. The dominant role of family care began to falter, and countries began to explore new approaches of providing care for the elderly in order to meet the needs of future economic and social development. In the Middle Ages, pension insurance sprouted in fraternities, mutualités, friendly societies and other groups, which raised funds from their members to provide aid for elderly, ill, injured, unemployed and impoverished members. In 1601, the Elizabethan Poor Law (the Poor Relief Act for short) of the UK specified social relief activities in the country. Orphaned, elderly, ill, disabled and other people were accommodated mainly according to the poverty line set by the government, and the principles of

[18]Carr-Saunders [17].

[19]Hooyman [18].

removing "undeserving" poor, workhouse test and unified government management were put forward. It was important exploration before the social security system and the pension insurance system came into being in the world. However, the care measures stipulated in the *Elizabethan Poor Law* were only acts of charity and endowment from the ruling class in a capitalist country to maintain their rule and were rather random. Its objects lost their dignity and freedom. Besides, it only covered a very narrow range at a very low level. During accelerating industrialization, the poverty issue was getting worse, endangering social stability and the ruling class. The *Poor Law Amendment Act* was implemented in 1834, with some progress compared with the old poor law, but its essence, content and approaches remained basically unchanged, and it was far from realizing the objectives of social insurance.

The pension for employees in public sectors rather than ordinary workers was explored in a more standard manner, and the pension insurance system for them came into being before that for ordinary workers. England has a long history of paying pensions to retired employees (mainly government employees). In 1269, William, a servant of King Henry III, was awarded a pension of four pence per day because of his old age and weak body. The first cumulative occupational pension plan in the UK, the Chatham fund, established in 1590, was mainly aimed at the Royal Navy. At the end of their voyage, they were paid pensions according to their injury or disability rather than age, which were deducted from their salary.[20] In 1669, the French government enacted the *Code of Annuity*, which specified that the country paid a pension to seafarers who had lost the ability to work. Constrained by economic development, the practice was not implemented on a large scale. In 1862, the US established the pension system during the Civil War, which specified that disabled veterans could receive relief funds. In 1868, the Meiji Restoration was implemented in Japan, which started the country's rapid industrialization. During this period, Japan enforced superannuation on troops and government officials. From 1875 to 1884, Japan launched the *Navy Retirement Order*, the *Army Superannuation Order* and the *Officials Superannuation Order*. As seen above, the pension insurance system and the social insurance system mainly originated in the UK, where industrialization started first, followed by other countries that explored various forms of measures to provide care for the elderly and established the social pension insurance system.

The modern pension system in the world was set up during the Industrial Revolution and mass production with machines. In the 1880s, with the rapid progress of industrialization in Western countries, economic and social contradictions kept emerging and struggles of the working class arose one after another. Rulers of all countries began to maintain social stability and class rule through suppression, the development of social security and other means. The modern pension insurance system was essentially established in Germany in 1889, marked by the Old Age and Disability Insurance Bill promulgated by Bismarck's government, which stipulated

[20]Lewin [19].

Table 1.4 Time of the establishment of the pension insurance system in some countries

Country	Time of the establishment of the pension insurance system	Country	Time of the establishment of the pension insurance system
Denmark	1891	New Zealand	1898
Sweden	1903	Austria	1906
Australia	1908	UK	1908
France	1910	Netherlands	1913
Italy	1919	Russia	1922
Chile	1924	Canada	1927
Bulgaria	1924	Belgium	1924
Hungary	1928	Greece	1934
South Africa	1928	US	1935
Norway	1936	Finland	1937
Japan	1941	Switzerland	1946

Source Zou [37], Jiang and Geng [38]

that the old age and disability social insurance system, aimed at all workers and general officials in Germany, would be implemented, and that the country, employers and individuals contributed to the insurance funds together, in which the contributions of employers and individuals played a main role with certain subsidies from the government. After retirement, laborers received the pension according to their in-service salary. Those who reached the age of 71 and had paid the premiums for more than 30 years were entitled to the benefits after retirement.[21] Thereafter, some European and American countries also established social pension insurance systems, such as Denmark (1891), Norway (1894), Sweden (1903), Austria (1906), the UK (1908) and the US (1935). By the late 20th century, social pension insurance systems had been established in more than 166 countries in the world.[22] The time of their establishment in some countries is shown in Table 1.4.

After World War II, with rapid economic and social development, pension security systems and social security systems went through the golden age of rapid development around the world with the guidance of Keynesian economics. Government investment in the pension security system continued to increase, contributing to expanded coverage, better benefits, improved institutional systems, enhanced management systems and perfected operation mechanisms.

Take the UK for instance. The *National Insurance Act*, implemented in 1948, stipulated that pensions were contributed by three parties according to the contribution standard: Each man paid four shillings and seven pence per week and each employer 3 shillings and 10 pence, with a subsidy of 2 shillings and 1 penny from

[21]Jiang [20].

[22]Zheng [21].

the government. Each woman paid 3 shillings and 7 pence per week, each employer 3 shillings, with a subsidy of 1 shilling and 7 pence from the government. After 1951, the contribution standard was increased by 4 pence (per week) for each man and woman, of which employers and employees shared half. In terms of the national insurance allowance, each man or woman received 26 shillings per week, a married man 42 shillings, the insured person's wife or each of his/her adult relatives 16 shillings, and the first child 7 shillings and 6 pence. In 1950, the pension allowance was increased, and each married couple received 50 shillings per week.[23] In 1950, Prime Minister Clement Attlee was proud to announce that the foundation of the UK as a welfare state had been well and truly laid.[24] In 1959, the new *National Insurance Act* proposed to establish a pension insurance system associated with the earnings, complementary to the pension system of the same standard and based on the average salary in an individual's lifetime. Shortly after this, the national earnings-related pension was continuously increased. In 1963, it increased to 15 to 18 lb a week. The basic pension for a single elderly increased from 2 lb, 17 shillings and 6 pence to 3 lb, 12 shillings and 6 pence per week, and that for every couple increased from 4 lb, 12 shillings and 6 pence per week to 5 lb and 9 shillings.[25] In March, 1965, the national basic pension increased by 12 shillings and 6 pence per week, that for singles increased from 15% of the average earnings in 1948 to 21%.

The pension system of the US developed rapidly after its establishment. Especially during the rule of the Democratic Party, active social security policies were advocated, pension coverage continuously expanded, benefits improved, and investment increased, which could be found during most of the rule of the Democratic Party. Roosevelt's New Deal (1933–1953), Truman's Fair Deal (1945–1953), Kennedy's anti-poverty program (1961–1963), Johnson's Great Society (1963–1968) and others constantly promoted the development of the pension insurance system in the US. From 1947, Truman required several times in his messages to the Congress that pension coverage be expanded and that the pension insurance premium be increased. Old-age and survivors premiums were added in the Social Security Amendments passed in 1952, increased from $50 per month to $70. In 1954, the contribution base of old-age and survivor insurance payroll tax, from $3600 per year to $4200. During the time when Johnson was in office, the retirement allowance was increased twice, by 7% in 1965 and 13% in 1967. Since the establishment of the pension insurance system, the number of beneficiaries increased from 0.222 million in 1940 to 26.23 million in 1970 (118 times that in 1940).[26]

Over the same period, the pension insurance system in Germany also continued to be improved. In 1957, the German government adjusted the pension insurance

---

[23]Ding and Yang [22].

[24]Pelling [23].

[25]Ding and Yang [24].

[26]Zhang [25].

system, transferring the cumulative public retirement insurance into a compulsory pay-as-you-go public pension insurance system, in which people at working age paid pension premiums as required and received the pension after reaching the statutory retirement age (65). In 1957, Germany also established an index adjustment principle, according to which the pension standard changed simultaneously with the salary level of in-service employees. The pension level in Germany increased by almost 70% that year. Thereafter, the pension replacement rate in Germany kept increasing, and the number and rate of people relying on social aid decreased sharply.[27] In addition to the basic pension insurance, an occupational pension insurance system and a voluntary individual pension insurance system were established on the principle of voluntary participation. In 1972, the Germany government stipulated that employees with 35 or more working years could get the full pension in early retirement (at the age of 63).[28]

From the end of World War II to the 1970s, welfare systems were established in Sweden, Norway, Finland, Denmark and other Nordic countries except the UK. The entire social insurance system including the pension insurance system fully developed. Japan enacted the *National Pension Insurance Act* in 1959 and implemented the national annuity system that covered farmers and the self-employed in 1961, which was the first pension insurance system in East Asia covering all citizens. Afterwards, Japan continued to improve the level of pensions and established a benefit adjustment mechanism associated with the salary of in-service employees. As a result, the pension increased from 3000 yen in 1965 to 10,000 yen and even more than 20,000 yen.[29] After World War II, pension insurance systems developed rapidly around the world, mainly to cope with the needs of economic and social development after the war. In addition, stable domestic and international economic and social environment as well as rapid economic growth laid the material basis for the development of pension insurance systems.

After the mid-1970s, with the outbreak of the oil crisis, economies around the world got in dire predicament. High inflation rate, high unemployment rate and low economic growth appeared in some countries. The increase of social security and welfare expenditures after World War II became a heavy burden to some countries. Due to worsening economic conditions, accelerating population aging and rigid welfare expenditures, the original pension insurance system had to be reformed, adjusted to promote sustainable development and adapted to economic and social development.

China, as an ancient civilization, started the theoretical research on the elderly and the practice of pension security early. Compared with other countries, China's pension security had been focusing on family care in a very long time. Some other pension security measures had been adopted since the establishment of the country, aimed at certain groups of people, especially royal members and government

[27]Hinrichs [26].

[28]Lixin [27].

[29]Wang [28].

officials, and public funds were arranged to compensate for the shortcomings of family care. On the whole, families or expanded families—clans had always been the basic system in China in a very long time in history, according to which seniors were supported, and bringing up a son to provide for one's old age had been the main way for families to ensure pension security. The concept of not forgetting the elderly who are not related to you while caring for those who are advocated in Confucianism was only an ideal in pre-industrial society featuring low productivity and could only play a small supplementary role in parts. After industrialization began, huge progress in productivity and significant increase in public wealth provided the material conditions for the establishment of a modern pension security system. China had been working on the establishment of a social pension security system outside the family since modern times. With a growing range and security level of the social pension security system, China's pension security system is going through a transition from the focus on family care to less emphasis on it, and embracing a stronger social pension security system. Currently, the social pension security system is developing rapidly. Especially after the early 21st century, a basic pension security system aimed at various groups has been established with increasing coverage and constantly improved security. The pension security system in China is playing an increasingly important role in ensuring the elderly's basic livelihood, maintaining social stability and promoting economic development.

Meanwhile, China's pension security system, with many problems, is far from perfect and can't meet the needs of the aging population. Conflicts between the supply of pension security and various needs of the elderly are very prominent. The fact of getting old before getting rich, changes in family structure, accelerating urbanization and changes in the labor market have posed new challenges to the development of the pension security system.

1.4 Population Aging and the Construction of the Modern Pension Security System

Elderly people need not only financial support but also daily care, social involvement, mental support and others. With accelerating population aging, the focus of the pension security institutions also needs to transfer from the construction of the financial support system to that of the pension security system.

1.4.1 The Development of Population Aging

Population aging is the result of declining birth rate, declining mortality, prolonged life expectancy and other factors that work together. It is a symbol of the progress in human civilization, and also brings pressure to the economic and social

Table 1.5 Changes in world population 1950–2050

Year	1950–1955	1975–1980	2000–2005	2025–2030	2045–2050
Ratio of the population over the age of 60 (%)	8.2	8.6	10.0	15.0	21.1
Total fertility rate (%)	5.0	3.9	2.7	2.3	2.1
Life expectancy at birth (years)	46.5	59.8	66.0	72.4	76.0

Source UN: *World Population Aging (1950–2050)*, 2002

development of all countries, thus becoming a global issue. The reform and development of the pension insurance system, as a formal security system for the elderly, are directly affected by population aging. From a global perspective, since France became an aging society in 1865 with more than 10% of its population over the age of 60, the number of elderly people around the world has increased rapidly. According to data and forecast of the United Nations, the average ratio of the world's population over the age of 60 to the total population was 8.2% from 1950 to 1955, increasing to 8.6% from 1975 to 1980 and 10% from 2000 to 2005, and would rise to 21.1% from 2045 to 2050 (see Table 1.5). Besides, people are aging at an increasing speed in some countries. It would take 115 years for the population in France at and over the age of 65 to increase from 7% to 14%, 85 for Sweden, 66 for the US, 45 for the UK and only 26 for Japan.[30]

China is facing the fastest population aging in the world. In 2000, its population over the age of 65 reached 88.11 million, accounting for 6.96% of the total population[31] and marking that it has joined aging countries. The number of its population at and over the age of 65 was 118.83 million in 2010, accounting for 8.87% of its total population, an increase of 1.91% from 2000.[32] China's population aging has the following features:

(1) A large aging population. With a large population, declining birth rate, declining mortality and prolonged life expectancy, China has an increasing population, which has been the largest in the world so far and will remain so in the next 50 years.[33] At present, China's total aging population has exceeded the total population of most countries in the world. According to the forecast of the United Nations in 1999, the number of people at and over the age of 65 would be 235 million and 334 million in 2030 and 2050 respectively. By the mid-21st century, the number of Chinese people at and over the age of 65 would be 4.4 times the number of elderly people in the US, very close to the total population of the US. In 2050, China's total population would be 51 million less than that of India, but the number of Chinese people at and over the age of 65 would be

[30]Xiong [29].

[31]Data of the fifth national census in 2000.

[32]Bulletin on data of the sixth national census in 2010 (No. 1).

[33]Du and Yang [30].

Table 1.6 Changes in the age structure of China's population

Year	Ratio of population between the age of 0–14 (%)	Ratio of population at and over the age of 65 (%)	Ratio of the elderly and the young (%)	Old-age dependency ratio (%)	Median age (years)
1953	36.3	4.4	12.2	7.4	22.7
1964	40.7	3.6	8.8	6.4	20.2
1982	33.6	4.9	14.6	8.0	22.9
1990	27.7	5.6	20.1	8.3	25.3
2000	22.9	7.0	30.4	9.9	30.8
2010	16.6	8.87	53.4	12.0	–

Source Processed data of previous national censes. The ratio of the elderly to the young refers to the ratio of population at and over the age of 65 to that between 0 and 14. Old-age dependency ratio refers to the ratio of population at and over the age of 65 to that between 15 and 64

103 million more than that of India.[34] Such a large aging population would have a huge demand for pension security.

(2) Fast aging. China has a large number of elderly people who age fast. Since the founding of the New China, the number of elderly people in China has been increasing at an accelerating speed in general, and the ratio of elderly people to the whole population and old-age dependency ratio have been increasing rapidly (see Table 1.6). In 1953, the number of Chinese people over the age of 65 was 25.93 million, accounting for only 4.4% of the total population, 88.21 million in 2000, accounting for 7%,[35] 118.83 million in 2010, accounting for 8.87%.[36] On the supposition of medium fertility and mortality rates, the ratio of elderly people at and over the age of 65 to the total population would be increasing rapidly from less than 6% in 1990 to 15.8% in 2030 and 23.1% in 2050.[37] According to the statistics and forecast of the US Census Bureau, it would take 115 years for the ratio of the population at and over the age of 65 in France to rise from 7 to 14%, 85 for Sweden, 66 for the US, 45 for the UK and only 25 for China.[38] Fast population aging has brought great pressure to China's pension security system.

(3) Advanced age. With a large aging population that age fast, China has a large number of elderly people at advanced age. According to demographic categories of elderly people, those between the age of 60 and 69 are young elders, those between 70 and 79 middle-aged elders, and those over 80 venerable

[34]Zeng [31].

[35]According to internationally accepted standards, if a country's people over the age of 60 make up 10% of the population, or people over the age of 65 make up 7%, the country or region becomes an aging society.

[36]Data from the bulletin on data of the sixth national census in 2010 (No. 1).

[37]Zeng [31].

[38]Xiong [32].

elders. The average annual growth rate of China's elderly people at and over the age of 65 will be 2.3% in the future, while that of people over the age of 80 will be 4.2% over the same period. In the future, the average growth rate of venerable elders in China will be much higher than that of population aging. By 2050, the number of China's venerable elders over the age of 80 will be likely to be over 100 million, accounting for about 22% of the population at and over the age of 60.[39] The majority of venerable elders are complete consumers who need daily care, which further increases the demand for services security, mental support and medical security for the elderly.

1.4.2 Aging Population in Urgent Need for the Pension Security System

Population aging is a symbol of economic and social development and progress in human civilization. In addition, it poses severe challenges to economic and social development. Therefore, it is a difficult task for many countries during their economic and social development to ensure that seniors are supported appropriately, to constantly improve the quality of elderly people's lives and to allow them to share the fruits of economic and social development.

Old age means more than decreased ability to engage in economic activities. With reduced or suspended income resulted from the retirement system, elderly people need financial support from the pension security system. Meanwhile, with advancing age, declined physical functions and more diseases, elderly people lacking the ability to take care of themselves also need long-term care, which requires the establishment of a long-term care security system for them. In addition, as elderly people gradually withdraw themselves from social life with smaller families as more children work away from home, relieving elderly people's loneliness and meeting their needs for spiritual security are also paid attention to.

With rapid population aging, it is of great significance to strengthen the construction of the pension security system, which is mainly reflected in the following aspects:

(1) Strengthening the construction of the pension security system is the inexorable demand of the goal of ensuring that seniors are supported appropriately. Ensuring that elderly people spend their twilight years in comfort, continuously improving the quality of their lives and allowing them to share the fruits of economic and social development are essential to economic and social development. The goal of ensuring all the seniors are supported appropriately was proposed during the 17th National Congress of the CPC. To achieve this goal, there is an urgent need for the construction of the pension security system to be

[39] Lu [33].

strengthened in terms of economic security, medical security, service security, spiritual security and other aspects. A complete and perfect pension security system is directly related to the vital interests of elderly people and the civilization and progress of the whole society. At present, the pension security system in China is far from perfect. With many problems, it still has a long way to go till the goal of ensuring that seniors are supported appropriately is realized. Therefore, it needs to be improved with great efforts.

(2) Strengthening the construction of the pension security system is an important part of the improvement of the social security system. Social security is a system that includes pension security, medical security, employment security, basic living security and other content, and contains social insurance, social welfare, social relief, social preference and compensation and other programs. Pension security is an important part of the whole social security system, and is even the most important. The perfection of the pension security system is directly related to that of social security. China is going through a critical period of the improvement of the social security system. It needs to seize the opportunity and to proceed in terms of legislation, funds, personnel and other aspects to constantly improve the pension security system so as to realize the objectives of social security featuring fairness, sharing and maximum coverage.

(3) Strengthening the construction of the pension security system is an important task of constructing a harmonious socialist society. During the sixth Plenary Session of the 16th CPC Central Committee, the decision of constructing a harmonious socialist society was put forward, which requires that the government be people-oriented, make overall plans and take all factors into consideration, adhere to the Scientific Outlook on Development, solve the most direct and most practical interests-related issues that people are most concerned about, and focus on the development of social undertakings. The construction of the pension security system is directly related to people's practical interests and is an important part of the development of social undertakings. An imperfect pension security system may lead to a number of social issues concerning elderly people, affecting family harmony and damaging social stability. Therefore, we should improve the pension security system with great efforts while constructing a harmonious socialist society.

(4) Strengthening the construction of the pension security system is an important way to transform the economic development model. According to the 12th Five-year Plan, people's livelihood shall be ensured and improved with scientific development as the theme and accelerating transformation of the economic development model as the main line. At present, there are many unreasonable problems in China's economic development: the imbalance between savings and consumption, too high savings rate and too low

consumption rate.[40] The industrial structure is irrational: The proportion of the first and second industries is too high, while the third industry is underdeveloped. Besides, citizens' income disparity is too large.[41] These problems have seriously affected the sustainable and healthy development of China's economy and society, and must be addressed. Strengthening the construction of the pension security system is an important part of ensuring and improving people's livelihood and an important means of expanding domestic demand and promoting consumption. Through the development of services for elderly people, industrial restructuring can be promoted. A complete and perfect pension security system can also help to adjust income distribution and narrow the income disparity. As is shown, strengthening the construction of the pension security system is an important way to transform the economic development model.

(5) Strengthening the construction of the pension security system is an important support for the construction of a "happy China". With the continuous development of social production and the growth of per capita income, people have gradually got rid of the overall lack of materials in the pre-industrial era and entered an "affluent society" with abundant materials. However, economic growth and material prosperity didn't promote the sense of happiness while bringing more satisfaction to people, and even resulted in conflicts between modernization and the sense of happiness. It is the dream of many people and the goal of many countries to promote the sense of happiness during economic development and modernization. That is to say, economic development and income growth are not an end in themselves. The key is to promote the happiness and well-being of all the citizens during the development. Old age is a stage almost all the people go through. The happiness and well-being of elderly people are an important part of the total social well-being. Elderly people are treasures of families, whose happiness is related to both themselves and other family members. As the number of elderly people increases, promoting elderly people's sense of happiness is the objective need and important support of the construction of a "happy China".

---

[40]In 2008, China's savings rate was 51.4, 27.5% higher than the average savings rate in the world. From 1978 to 2008, China's savings rate increased by 13.5%, while the consumption rate dropped significantly over the same period. In 1978, China's consumption rate was 62.1%, dropping by 13.5 to 48.6% in 2008 [34].

[41]Long [35].

References

1. Tian Xueyuan. 1991. *China's Elderly People, 5.* China: China Economy Publishing House.
2. Xiong Bijun. 2009. *Economics of Aging*, 33. China: China Social Press.
3. Yan, Qin, and Mei Xiaoyun. 2007. An overview of the relations between physical characteristics and geriatric conditions. *Journal of Nanjing University of Traditional Chinese Medicine* 2: 131–134.
4. GU, Danan, Yi Zeng, Liu Yuzhi, and Xianxin Zeng. 2007. Study on Chinese elderly people's frailty index and its relationship with their painful death. *Population Research* 9: 35–41.
5. Qiao, Xiaochun, Kaiti Zhang, Sun Lujun, and Ling Zhang. 2005. The estimate of China's poor elderly people. *Population Research* 3: 8–16.
6. Yu Xuejun. 2003. Studies on poverty of elderly people. In *Survey data analysis with single sampling of China's elderly people*, ed. China Research Center on Aging. China: Standards Press of China.
7. Chen Qi and Rude Liu. 2007. *Contemporary educational psychology*, 42–44. Beijing: Beijing Normal University Press.
8. Zhang Kaiti and Ping Guo. 2010. *The blue book on population aging and the status of elderly people in China*, 215. China: China Social Press.
9. Zhaohui, Yin. 2010. Features, causes and countermeasures to suicide in the elderly population. *Journal of Jiangsu Police Officer College* 1: 138–142.
10. Song, Bao'an, and Xinwei Pu. 2005. Psychological features and values of elderly people in contemporary China. *Social Science Front* 5: 193–197.
11. Cumming, E., and W.E. Henry. 1961. *Growing old: The process of disengagement*. New York: Basic Books.
12. Tian Xueyuan. 1991. *China's Aging Society*, 50. China: China Economy Publishing House.
13. Zhang Kaiti (ed.). 2010. *Research on the Social Activities and Mental Health of China's Elderly People in Urban and Rural Areas*, 12. China: China Social Press.
14. Emerson R.M. 1964. *Power-dependence Relations, American Sociological Review*, 1962, 27 (1): 31–41. P.M. Blau. 1964. *Exchange and Power in Social Life*. New York: Wiley.
15. Du, Peng, and Chao Wu. 2006. China's elderly people's ability to take care of themselves and its changes. *Population Research* 1: 50–56.
17. Zhang Kaiti and Ping Guo. 2010. *The blue book on population aging and the status of elderly people in China*, 306. China: China Social Press.
18. Carr-Saunders A.M. 1983. *The population problem*, 150–151. China: The Commercial Press.
19. Hooyman N.R. 1992. *Social gerontology: A multidisciplinary perspective*, 23–43. China: Social Sciences Academic Press.
20. Lewin, Chris. 2005. Pension developments in the UK. *Economic Herald* 6: 78.
21. Jiang Xiangqun. 2005. *Pension security system—history and changes*, 23. China: China Renmin University Press.
22. Zheng Gongcheng. 2005. *Social security*, 301. China: China Labour and Social Security Publishing House.
23. Ding Jianding and Fengjuan Yang. 2004. *Development of the social security system in China,* 113–114. China: China Labour and Social Security Publishing House.
24. Pelling Henry. 1984. *The labor government, 1945–1951*, 117–118. London.
25. Ding Jianding and Fengjuan Yang. 2004. *Development of the social security system in the UK, 139*. China: China Labour and Social Security Publishing House.
26. Zhang Song. 2009. *Study on Pension Insurance in China's Population Aging*, a doctoral dissertation at Jilin University in 2009, pp. 101–102.
27. Hinrichs, Karl. 2005. *New century-new paradigm: Pension reforms in Germany*. London: Edward Elgar.
28. Lixin, Zhao. 2008. Development and reforms of Germany's state pension insurance. *Social Sciences Perspectives in Higher Education* 3: 42.

29. Wang Dewen. 2004. *Japanese under the setting sun and the deficit in its pension account.* http://www.finance.sina.com.cn.
30. Xiong Bijun. 2009. *Economics of Aging*, 15. China: China Social Press.
31. Du, Peng, and Hui Yang. 2009. Comparison of population aging of China and other Asian countries. *Population and Development* 2: 75–80.
32. Zeng, Yi. 2001. Features of China's aging population and countermeasures. *Population and Economics* 5: 3–10.
33. Xiong Bijun. 2009. *Economics of Aging*, 26. China: China Social Press.
34. Lu, Jiehua. 2007. Rapid population aging: Challenges and strategies. *Gansu Social Sciences* 6: 12–16.
35. Long, Yuqi. 2011. Evolution income disparity and reforms of China's income distribution system. *Southeast Academic Research* 1: 103–115.
36. Zhang Zhuoyuan. 2010. *Adjusting the economic structure and transforming the economic development model*, *People's Daily*, November 25, 2010, China News Network, http://www.chinanews.com/cj/2010/ll-25/2679722.shtml.
37. Kaiti Zhang(ed.). 2009. *A study of the social activities and psychological status of the elderly in urban and rural areas in China*, 195. China: China Social Press.
38. Genbao Zou. 2001. *Social security system—experience and reforms of EU countries*, 84. China: Shanghai University of Finance and Economics Press.
39. Shouming Jiang and Liang Geng. 2002. *Introduction to western social security systems*, 33. Science Publishing.

Chapter 2
Theoretical Bases for the Construction of the Modern Pension Security System

Zhenjie Liu

The issue of an aging population is one of the major issues that mankind must face in the 21st century. An aging of population is a natural and periodical process in history, but different measures and efforts to deal with this issue may often lead to different social effects. An expected positive effect should be that the aging level become positively correlated with the level of economic development, so that countries can successfully get through the aging period in a healthy, calm and harmonious way. However, if we fail to cope with this via sufficient preparations on multiple levels, then many problems will arise.

The period of an aging population arrived much earlier in Western countries with industrialization and modernization more developed than in China. Therefore, in terms of theoretical and practical research on the issue of pension security, there are a lot the country can learn from their experiences. In particular, theories on welfare economics, the principle of the equalization of public services, the modern social security theory, Pareto optimality, Keynesian theory of government intervention and the life cycle theory, developed in Western countries, depict the issue of care for the elderly citizens from different perspectives. Systematic and in-depth research on the above-mentioned theories will benefit China in accelerating the construction and improvement of a pension security system that covers all citizens.

Z. Liu (✉)
Policy Research Centre, Ministry of Civil Affairs of the People's Republic of China, #147 Beihe Street, Dongcheng District, Beijing 100721, People's Republic of China
e-mail: 547334498@qq.com

Y. Wang (ed.), *The Development of Security and Whole Care System for the Aged in China*, https://doi.org/10.1007/978-981-10-6481-4_2

2.1 Theories on Public Goods

1. Types and features of public goods

Public goods, a concept different from private goods, is product or service that is non-competitive, as it is consumed or used by the majority and is non-exclusively beneficial. This includes, for example, property and labor service in national defense, public security and judicial system, as well as compulsory education and public welfare service.

According to the principles of public economics, social products can be classified as public goods, private goods and quasi-public goods. In a broad sense, public goods refer to both material and spiritual public goods; in a narrow sense, public goods are goods in which used by one individual does not reduce availability to others. All goods that are competitive, excludable and dividable and that can be owned and used by individual consumers are private goods. Those that are not the above-mentioned two kinds are quasi-public goods, for example, pension, medical care, education and housing.

A good is public if it at least meets the following two points: First, it is for public welfare, rather than for a few consumers with privileges. Second, it is realized in the process of public management in a democratic, fair and sharing manner.

2. Supply of public goods

Generally, the government provides public goods in four ways: for free, after purchase, price intervention and production control. Those providing pure public goods are usually governments or social organizations. Characteristics of pure public goods are non-rivalry and non-excludability, that is, the consuming of these products by some will not affect others consuming them and some people using them do not exclude others from using them. In general cases, public goods, characterized by their large scale and high costs, are like free lunches to some extent, that is, they can only be provided by the government. The market and ordinary people have no obligation to provide free lunches, and even if they had, they could only provide a very limited amount of public goods. Education and reforms of the medical systems in China over these years have exposed many shortcomings that arose due to the market being the sole provider of public goods or services. As the legal environment is improved and the market economy standardized, some current functions of the government may be undertaken by the market or social organizations, but the function of the government as the investor in public products must be strengthened, rather than weakened.

3. Pension security as public goods

A fair and reasonable pension security system is a public good in nature. Here the basic pension insurance is defined as a public good because pension security can be classified into many types and modes, including social pension insurance that manifests the unity of rights and obligations as well as commercial services of private nature. Private goods are generally provided by the market, for example, commercial insurance. As a supplement to basic pension insurance, supplementary pension insurance (like enterprise annuity) offers a high-level guarantee, and, to be more accurate, it should be classified as a private good, and thus provided by the market. Theoretically speaking, quasi-public goods should be supplied by the government and the market. However, if both the government and the market are dysfunctional, a third-party department should play an active role. Rural social pension insurance is mainly a regional public good, which is a social product that individuals do not want to provide in rural areas. In the end, as a pure public good, rural social pension insurance should be provided by the country and be collective following the principle of the government taking the lead.

Even home-based elderly care services are not purely an individual or family issue, and need to be treated differently as conditions vary. For example, the welfare elderly care service is a pure public good that shall be provided by the government, public-good elderly care service is a quasi-public good that can be provided by the government and society, and profitable elderly care service is a private good that can be provided by the market.

Aside from commercial insurance, enterprise annuity also significantly has the nature of private good as it is an elderly care security measure that enterprises carry out to recruit, retain and incite employees, and therefore, it's obviously exclusive. However, it is a quasi-public good to some extent primarily, i.e., it is provided on the basis of the support of basic national policies (for example, tax policies) and a sound social basic pension insurance.

4. Public services as public goods should be equalized

Public services refer to the services that public or quasi-public sectors provide to meet public needs in society, including tangible material services, intangible non-material services, as well as reading, study, activities and social interaction, for example, building and opening museums, libraries and holding various salons.

Equalization literally means the process of adjusting and balancing to reach an equal state. However, here only a relatively, rather than definitely, equal state can be achieved. Regardless of which region they live in, the citizens of a country should enjoy equal rights to basic public services at the lowest standards. The equalization of public services should be realized between regions, rural and urban areas and different people.

The definition of public services dictates that they are pure public goods. Regardless of the region they live in or their status, any citizen should be able to enjoy equalized public services in terms of amount and range. Equalization is an ideal goal of the government to reach in the public services field. During this process, the government is tested in terms of its resource mobilizing and integrating abilities. Therefore, to some extent, equalization can promote the development of public services.

Equalization is a basic principle for public services. According to equalization, citizens should enjoy equal opportunities, just procedures and fair results as they use basic public services. Availability, fairness and sustainability are the major indicators for the appraisal of public services. Availability means the majority of citizens can receive public services timely and effectively, fair public service resources are distributed fairly and reasonably, and sustainability public service resources are supplied stably.

Currently the equalization of public services is emphasized primarily because of too many problems existing in the field of public services, among which the insufficiency of investment from public finance is most important. In addition, the huge gap in the allocation of public goods becomes one of the major issues that affect social fairness, justice and sharing. Therefore, the equalization of public services is of especially important significance now.

To achieve the equalization of public services, the government should invest more; and, at the same time, it should actively absorb private funds, expand investment channels of programs for public good, and concentrate its efforts on investing more in less developed areas and disadvantaged groups.

5. Sources of theories on the equalization of public services

Ideas on the equalization of basic public services primarily came from the economic field. Some traces of this thought can be found in the German new historical school, British Fabianism, Neoliberalism, welfare economics, Keynesian economics as well as many other ideological schools. With greater economic welfare in society as the goal, welfare economics marked the beginning of this field. From Pigou's theory to Amartya Sen's, then Pareto efficiency and the theory of the second best, each phase of new welfare economics inspired the formation of the theoretical system of the equalization of public services. According to Pigou, the higher the aggregate national income is, the greater economic welfare in society is; the more equalized the national income distribution is, the greater economic welfare in society is. For the first time in the history of western economics, he related social welfare to income distribution under national intervention and did research, which was of great significance in enlightening relevant research on the equalization of public services. The thought of equalization of national income, proposed by Pigou, was the inspiration for the equalization of public services. The government should try to maximize social welfare through equalization of public services.

There is a brilliant description in The Theory of Moral Sentiments by Adam Smith. According to him, if the fruits of a society's economic development cannot be shared by all the people, it is morally unsound and risky, because it is bound to jeopardize social stability.[1] Here we hold that fairness and justice should be the most important values of the security system in a socialist society.

Differing from old welfare economics, new welfare economics proposes the compensation principle, according to which any national policy change will lead to changes in market prices. As some people benefit from it, others' interests are damaged, thus the government needs to take measures to compensate these people, guaranteeing efficiency and fairness.

---

[1]Smith [1].

Amartya Sen, winner of the Nobel Prize in Economic Sciences in 1998, emphasized that the government should provide equalized basic public services, which should meet different standards in different stages, providing a guarantee for the improvement of citizens' abilities and welfare. According to him, the government cannot just focus on stimulating the economy when formulating an economic measure; the measure that the government makes should be equipped with a mechanism that guarantees citizens' basic living conditions and improves their living ability. To achieve this, the equalization of public services is an important method. These theories form a solid foundation for the idea of the equalization of public services.[2]

6. Equalization of elderly care services has become a priority

A country's social security system must go through the process from covering special groups to covering all citizens. Correspondingly, the allocation of public resources should also become equalized. However, public service resources are generally controlled by the government, and mainly allocated by it through public finance expenditure or other means. However, the government itself has to correct its mistakes if any improper allocation is made. Because of China's current situation and other factors, elderly care service resources in the country are obviously not evenly allocated, for example, there are huge gaps between different regions as well as rural and urban areas.

The existing elderly care facilities are far from enough in China: First, the existing elderly care service system is basically for the elderly in need. It mainly guarantees the basic living conditions of the elderly that have no labor capacity, income for life or support and the elderly that meet difficult situations. It only covers a small number of people, and provides only a few types of welfare. A systematic service system that can bring benefits to all the elderly has not been formed. Second, the supply of the beds for the elderly are seriously insufficient compared with demand. According to statistics, as of late 2008, there were 37,623 welfare institutions of various types and 2.45 million beds for the elderly in China. These accounted for enough for only 1.5% of all the elderly aged above 60, which was lower than the percentages in developed countries and some less-developed countries, 5 to 7% and 2 to 3% respectively. Third, the utilization rate of the existing facilities for the elderly is not high, and their functions and services need to be improved, as there is a large gap between the existing facilities and people's need. According to statistics, China invested 10.31 billion yuan in social welfare, including elderly care services in 2008, which only accounted for 0.16% of the national fiscal expenditure and 0.034% of GDP that year.[3] Because of the small amount of funding from national finance for development, the elderly care service is insufficient with few welfare functions.

---

[2]Chen [2].

[3]Website of the Ministry of Civil Affairs, http://Avw.mca.gov.cn.

Elderly care services are imperfectly distributed and experience a regional development imbalance. Based on the country's situation, the elderly care services in China should be provided at home, and in communities and institutions. However, currently, they are mostly provided in institutions, while the majority of the elderly do not have access to welfare services like community-based rehabilitation, cultural, entertainment and housekeeping services. At the same time, imbalanced economic and social development leads to the gap in the development of elderly care services between rural and urban areas as well as between different regions. The development of elderly care services in western areas lags far behind: There are no comprehensive welfare institutions in most counties, and the investment in nursing institutions for the elderly, personnel quality and service quality are much inferior to those in eastern areas.[4]

According to the part about promoting the construction of the financial system in *The 11th Five-Year Plan for National Economic and Social Development of the People's Republic China* (hereinafter referred to as *The 11th Five-Year Plan*) approved by the National People's Congress in March 2006, "accelerate the construction of the public financial system, clearly define the responsibilities for fiscal expenditure of governments of different levels and reasonably adjust the division of financial income between governments. Improve the financial transfer payment system of the central government and provincial governments, organize the financial management system under provincial governments, allow areas with conditions to implement the management system of the provincial government directly managing county governments, so as to gradually promote the equalization of basic public services". According to *The 11th Five-Year Plan*, the equalization of basic public services was proposed as a task in the construction of the financial system. It would be gradually realized on the condition that the financial transfer payment system of the central government and provincial governments is improved and the financial management system under provincial governments is organized. It is a result and goal of the construction of the financial system and the improvement of the financial transfer payment system. Efforts should be made to solve financial difficulties in counties and townships and improve the ability to provide public services of governments at the grassroots level. National financial investment should be gradually increased, and the ability to supply public goods and services be continuously improved. Generally, the equalization of basic public services can be realized through the equalization of financial resources. This was further emphasized in the 5th Plenary Session of the 17th Central Committee of the Communist Party of China and the 12th Five-Year Plan that efforts should be made to guarantee and improve people's livelihood, a quite comprehensive, sustainable basic public service system that is suitable for the national conditions and covers rural and urban areas be gradually improved, and the government's security ability be strengthened, so as to promote the equalization of basic public services.

[4]Zhang [3].

Currently, we should actively promote the transition of the government's functions, accelerate the construction of government providing public services, strengthen the awareness of governments of different levels to promote social development and stop them from blindly admiring and pursuing high GDP, so that they can provide various public service items that are needed in society. In terms of elderly care security, we need to do the following work:

(1) Further improve elderly care services. First, increase the supply of elderly care services through the government's leadership, community involvement and market operation, which is the prerequisite and basis for the availability of elderly care services. Second, make the distribution between rural and urban areas, different regions and groups reasonable through adjustments in the system.
(2) Establish an elderly care service allowance system. Provide necessary allowances, low-cost or free services for the elderly. The elderly that have no labor capacity, income for life or support can live together or in different places enjoying services, and it should be guaranteed that their basic living conditions be no worse than the average living standards. Funding should come from the financial budget, and the standards for the services provided to the elderly should be improved with the average living standards of the locals. Provide necessary allowances to the elderly that receive the minimum living allowance or meet financial difficulties to pay elderly care services, or provide low-cost or free services to them. The elderly that obtain stable incomes can pay elderly care services or their families can. Build more basic, demonstrative institutions for the elderly that mainly provide services and medical care, so that the elderly that have no labor capacity, income for life or support, that live in poverty or that are disabled can enjoy daily care, rehabilitation and all-day services.
(3) Establish a noncontributory pension system. Compared with the contributory pension insurance system, the noncontributory pension mode, called "zero pillar" by the World Bank in 2005, has many advantages, mainly including fairness and wide coverage. As the qualifications of the benefited groups have nothing to do with the applicants' income or employment history, it can cover the population with a low income or in poverty. Individuals do not need to pay, as the funds come completely from financial transfer payments. The only two conditions are that the individual is a national, and he or she has to reach a certain age. In comparison, the majority of the elderly are excluded from benefited groups of the contributory pension insurance system for their inability to pay the insurance. Therefore, in those less developed areas where residents obtain a relatively low income, the noncontributory pension system would enable the majority of the residents to benefit from this system.

2.2 Theories on Income Distribution

1. Basic theories of welfare economics on equalized income distribution

In the 1920s, Pigou, a British economist, established the whole system of welfare economics. In order to maximize welfare, Pigou took two issues into consideration: First, the increase of an individual's income would make him more satisfied; second, the transfer of the rich's monetary income to people living in poverty would make the whole society more satisfied. Based on these two points, he put forward two propositions: a higher aggregate national income would bring better economic welfare in society; and equalized national income distribution would bring better economic welfare in society. To a great extent, economic welfare is determined by the national income and the distribution of national income among social members. Therefore, to improve social welfare, the aggregate national income should be increased through production, and unequal distribution of national income should be eliminated.[5]

As a representative of supporters of new welfare economics, Jeremy Bentham took "the greatest interests of the majority" as the highest goal of utilitarian welfare. According to him, happiness doesn't refer to someone's happiness, but the happiness of the majority of all the social members. As he explicitly pointed out, "It is the greatest happiness of the greatest number that is the measure of right and wrong." In order to bring the greatest happiness to the majority, the relations between individuals' and all social members' interests shall be well managed; society shall care about individuals' interests, and individuals shall obey interests of society.[6] James Mill, who held the same view with Jeremy Bentham, also firmly believed that "the action that can promote the majority to obtain the greatest happiness is just."

Pareto, an Italian economist, proposed the well-known theory of Pareto optimality. According to this theory, increase some social members' well-being without reducing some others'; or significantly increase the well-being of the majority of social members when the well-being of a minority of social members is slightly reduced. In both cases, the well-being of the whole society is considered as having been improved, and this is the basic level that the social security system should reach. It complies with the principle of "utility maximization," and at the same time, it guarantees fairness. Social well-being is positively correlated to individuals' well-being, which is dependent on commodities and labor that the individuals consume. Individuals can most correctly judge their well-being, and make decisions according to their own preferences.[7]

---

[5]Pigou [4].

[6]Bentham [5].

[7]Pareto [6].

2. The gap in income distribution is large in China currently

The special natural and historical conditions and the structure consisting of rural and urban areas determine that the gap between different regions, rural and urban areas and different industries in China is even wider than other countries, and a series of social problems arise. The Gini coefficient, which is taken as a major index in China to reflect the income gap, reached 0.434 in China in 1994, which was higher than 0.4, a warning line recognized in the international community, and it exceeded 0.5 in 2010, higher than that in all developed countries and most of the less developed countries. Such a huge gap in income means economic development would be slowed down and the achievements of economic development are not shared among all the people; what's more, it has intensified social conflicts. If no effective measures are taken to narrow the income gap, social stability would be more seriously affected. Therefore, making the achievements of modernization benefit all the people through focusing on people's livelihood and equalized distribution will be another great challenge during the 12th Five Year Plan.

Currently the effect of labor income as a measure to adjust primary distribution gradually weakens. In this situation, social security with its unique function of redistribution shall take the important task of narrowing the income gap between social members. But redistribution shall be appropriate for the economic and social development. If wealth is too evenly redistributed, it may lead to absolute equalitarianism, reduce economic efficiency, damage the economic foundation for social development, and thus cause common poverty. Or if wealth is too lowly redistributed, income distribution can't be fair, and it may even enlarge the income gap after primary distribution.

3. Pension security can redistribute wealth

Whether adopting accounting on the cash basis or fund accumulation system, pension security can realize transfer payment and redistribution, through which personal life and stable development of society and economy can be guaranteed. Pension has an impact on intergenerational redistribution. If it's set to be too low, retirees may face a low-quality life, which is unfair to them who have paid for pension insurance; or if it's set to be too high, it would be a heavy burden to employees that are paying for pension insurance. On the one hand, retirees' welfare should be improved to let them share fruits of social development; on the other hand, employees should be in favor properly during intergenerational redistribution, so as to reduce the burden of paying for pension on them and encourage them to create wealth for society. Therefore, the pension security system designed by us should achieve these results: to guarantee that the elderly can enjoy their late life decently and redistribute social wealth through adjusting social and economic development.

At present, all expenditures of social security equal about 6% of GDP in China, significantly lower than those in most developed countries, which were 20% and even 30%. Against the background of national economy and financial resources continuously developing rapidly, we should especially re-consider its important role in adjusting the distribution of social wealth. China should remarkably increase the proportion of all expenditures of social security in GDP, that of financial expenditures of social security in national financial revenue and significantly improve the level of social security to rural and urban residents, especially low-income groups and ordinary laborers. Primary distribution, secondary distribution and third distribution should be reformed simultaneously, especially secondary distribution that includes social security.

According to the general data of national pension insurance, the income outnumbers the expenditure. But a problem we face now is the gap between regions, as some regions suffer deficits while some have surpluses. But overall in China, the income of pension insurance outnumbers the expenditure. Therefore, the current situation, the gap between regions, shall be improved to guarantee the fairness, financial stability and sustainability of this system.[8]

In the primitive society with extreme shortages of material wealth, “the elderly and the young were lonely and not well treated.” There was no resource for elderly care. But as society and economy developed to a certain level, the ruler would implement the pension security system differently to different groups. Generally, those who have been employed are put before those who haven’t been employed, those who have worked in public offices are before those who have worked not in public offices, and those in urban areas are before those in rural areas. After the pension security system for some social members is gradually established, other social members and even all citizens would demand pension.

According to the conditions and background of the establishment of social pension system in rural areas in foreign countries and after over 30 years of development since the implementation of the reform and opening up policy, national finance has basically been able to subsidize basic pension insurance. Over these years, China has enjoyed the high economic growth rate of about 10%, and its economic aggregate in 2010 exceeded 40 trillion yuan. China has surpassed Japan and is only second to the US in terms of economic aggregate. The average GDP per capita of China is expected to reach 10,000 dollars in 2015, and it’ll become a de facto developed country. Against this background, the pension security system that we design should be an important tool to adjust the distribution of social wealth, through which the majority of nationals can gain a decent life, rather than just be enough to allow the nationals to have adequate food and clothing or become a privilege exclusive to some people. Therefore, we have the ability and responsibility to establish a social pension security system that covers all the people as soon as possible. Firstly, include those marginalized groups that are out of the security system into it; secondly, improve the security level of the existing system in the scope that national finance allows. The system is going to transform from guaranteeing a basic life and seemingly covering all the people to guaranteeing a high-quality life and being fair to all the people, and become a pillar system that can promote and maintain the reasonable distribution of wealth.[9]

---

[8]Zheng [7].

[9]Zheng [7].

4. The elderly care service system realizes redistribution through driving social employment

The elderly care service industry is labor intensive. The building of the elderly care service system can not only provide society with corresponding elderly care service, but also drive social employment. As the aging situation intensifies, the elderly care service industry is rising with market demand, including apartments for the elderly, health and medical care, fitness cultivation, entertainment and tourism, daily care, intermediary service and various products for special use. The huge employment demand requires us to make a summary and popularize the successful experience of some regions in closely combining the development of elderly care service and the driving of employment and reemployment, and making them promote each other; and to develop posts for public welfare and establish a large team providing elderly care service, so as to offer labor support for the development of social welfare cause and drive social employment.[10]

Through practice, Shengzhou City, Zhejiang Province realized that it could play the basic function of market and let market allocate elderly care service resources. The government guided social forces to invest and build elderly service institutions according to market demand, which, classified into different levels, could meet varied, multi-level demand of the elderly based on the management concept of "all for the elderly, for the elderly's all demand, for all the elderly." In terms of distribution, the elderly service institutions are located on the same street, town or village, so as to highlight the scale and the agglomeration effect.[11] As the elderly service industry thrived, the elderly transferred the wealth they'd accumulated to the young through receiving services, which showed that the development of the elderly care service drove the employment in related industries and reasonably and effectively allocated social resources and wealth.

2.3 Theories on Social Mixing

With a meaning opposed to the meaning of separation, mixing together is a kind of status related to integration, that is, the process of various relevant factors getting involved with and affected by each other. Social mixing is a process that relevant social subjects destroy separation and get involved with each other through the integration system.

1. Theories on social mixing provide theoretical basis for the building of the harmonious society

---

[10]Zhang [8].

[11]Dong [9].

The building of the harmonious socialist society is a process of practice from social exclusion to social mixing. Social mixing, a concept on social policies to eliminate social exclusion, originated from European scholars' study on social exclusion, and later it was introduced into the field of social policies in China and became an important part in Chinese study on social policies. Amartya Sen held that a mixed society should be a community in which everyone enjoys fairness in life, actively and meaningfully participate in activities, share social experiences and obtain basic social welfare.

But as industrialization rapidly developed over the past century, stratification and even disintegration of individuals, families and even society occurred. Davis and More held that social stratification is a universal and inevitable social phenomenon, and impetus to promote social and individuals' development. Parsons explained the process of social and cultural changes based on the theory of evolution, and pointed out that society realizes higher-level mixing through changes and stratification, thus changing from being simple to being complicated. Here family is taken as a basic structure in society that plays a critical role. Then, through social reconstruction, small families are changed into large ones, and the young need to support the elderly. During this process, social exclusion is pushed to become social mixing. In addition, when there is social exclusion, social mixing is needed.[12]

Considering the current national conditions, the groups that are not covered by social security, like basic pension security, because of social exclusion, mainly include unemployed population in cities and towns, migrant population with rural migrant workers as the major part and people in difficulties, like those with physical problems, they are often excluded because of economic, political, cultural or psychological reasons, and can't be integrated into the mainstream society. As the concepts of putting people first and scientific development are gradually implemented in the 21st century, social mixing has appeared, and a social security system that allows the integration of rural and urban areas has come into being. As the income gap between rural and urban areas as well as groups with different positioning according to the household registration system widens, favorable policies to farmers and policies on the building of the harmonious society are formed. An era in which urbanization and continuous economic growth are promoted through social mixing is coming. Problems related to urban development and social disintegration will become prominent, and social mixing and economic growth will become new focuses of theories on development.[13]

The report of the 17th National Congress of the CPC points out that all people share the harmonious society, which is consistent with the value orientation of social mixing. Social mixing is to create a society shared by all people. Today as social stratification intensifies in China, we urgently need to establish a scientific, reasonable and effective mechanism of social mixing, timely coordinate the

[12]Ga [10].

[13]Liu [11].

relations between different social interest groups and form organic and dynamic atmosphere that advocates social solidarity and cohesion, so as to prevent social conflicts, which may hinder social development, and make sure the smooth operation of the social mechanism. Therefore, the concept of and theories on social mixing are of important practical significance to China and the building of a harmonious society.[14]

2. A perfect social security system is conducive to social mixing

Currently, promoting and accelerating social mixing is consistent with the Scientific Outlook on Development and conducive to the building of the harmonious society. Especially the establishment and perfection of a social security system and mechanism that allows integration of urban and rural areas will narrow down the gap of welfare enjoyed by residents with different identities and alleviate social conflicts, help realize the equalization and maximization of welfare enjoyed by residents in rural and urban areas, and promote the urbanization process and growth of national economy.

Social security with pension security as the major part helps resolve social conflicts, adjust social relations, realization of social fairness and share fruits of social development. Accelerating the perfection of the social security system is an important part in the maintaining of social fairness, coordinating of social interest groups and building of the harmonious society. The nature of social security is to adjust the gap between the rich and the poor, guarantee the people's vital interests and realize fair and reasonable social distribution through law and economic means. Its basic functions include: guaranteeing laborer's basic life, maintaining social stability, promoting economic development, keeping social fairness and improving nationals' welfare. Social security is an important means to promote the building of the harmonious society and a critical pillar that supports the harmonious society. Therefore, under the guidance of fairness, take the building of social security as an aspect related to national economy and the people's livelihood, and try to strike a balance between fairness and efficiency during practice. Promoting economic development shouldn't be an excuse for the negligence and even the hindering of the building of social security.

At present, the various unstable social factors are directly or indirectly related to social security. For example, during land requisition and demolition, elderly care, medical care, employment and other problems related to the parties involved can't be solved properly; the conflicts between capital and labor often break out because employees' rights of social security are infringed; many elderly people complain that their children don't support them, which lead to tension in family relations. All these social conflicts can be resolved by virtue of a scientific and reasonable social security system.

[14]Ga [10].

3. Stop the separation of pension security between organs, public institutions and enterprises

The pension insurance system treats enterprise workers, workers of organs and public institutions and civil servants differently. The two pension plans are different in terms of post-retirement welfare, rate of return and substitution rate, which not only leads to different welfare to retirees, but also prevents the flow of personnel between these two kinds of posts to some degree and increases management costs. If an enterprise worker is transferred to an organ or public institution, the money that he has paid for pension insurance can't be used in the pension system for organs and public institutions; if a worker of an organ or public institution is transferred to an enterprise, the money for his pension insurance is nowhere to be found.

Officials of organs and civil servants design and implement reforms, but they set themselves out of the reforms to protect their interests. Reform designers are not willing to set themselves as a model of reform, which makes reform harder, slows down development and hinders social harmony and stability. Objectively speaking, this leads to and heightens the occupational barrier between enterprises and administrative public institutions, and seriously hinders the reasonable and orderly flow of talents. Therefore, the current reform in the pension insurance system of public institutions should be taken as an opportunity to destroy the boundary between rural and urban areas and the occupational barrier, reconstruct the pension insurance system in China, and establish a single basic pension insurance system for all nationals.

4. Establish a pension insurance system that covers rural and urban areas and all occupations

The existing pension insurance system in China is obviously unconnected, in which the flow of personnel between different occupations is hindered and the social function of providing mutual aid can't be effectively played. Moreover, it can't solve the financial problems of pension insurance.

As the financial resources in China overly favored urban areas over these years, two different social systems are formed in rural and urban areas. In terms of pension security, four policies are made for urban residents that live on national finance, urban residents that live on enterprise finance, freelancers and rural residents. The most direct result of these policies is the latter three groups have to bear market risks that are extended level by level, and thus the gap between them and the first group is widened continuously. At present, firstly, since the government has weak supervision over the enterprises paying for employees' pension insurance, the employees' pension is regulated by market and the balance of their personal accounts keeps low; secondly, since the government doesn't take control of organs and public institutions in setting payment base, their employees can receive generous pension, which is paid by national finance, and the balance of their personal accounts keeps high. To solve this problem, the key isn't making the pension even for all these groups, but destroying the multi-track pension insurance in the macroscopic level, and establish a dynamic management mechanism, according to which the market risks shall be shared by all people.[15]

In addition, the current basic pension insurance system for city and town employees has low ability of incorporation, as it poses great difficulties for city and town residents, who are not employed, and rural social pension system to be incorporated in it. For one thing, the high premium stops rural migrant workers from paying for the basic pension insurance; for another, the inability of using the basic pension insurance for city and town employees in different places hinders rural migrant workers to pay for the basic pension insurance. Because of these, the system gradually becomes a "castle" that excludes rural migrant workers from the system these years.

The pension of enterprise retirees has been adjusted for several times over these years, and it's rising on the whole. But still, it's much lower than the pension of retirees of organs and public institutions. Moreover, these adjustments may easily lead to new unfair situations in society, for example, the pension problem of enterprise employees and the elderly that were unemployed when they were young, and other groups like the elderly in rural areas may compare their pension with that of other groups. In addition, the pension of enterprises in monopolized industries and in non-monopolized industries is also different. Because of the existing large gap in the pension of different systems and industries, the social security system is questioned, that is, whether it's fair or not. Therefore, to improve the social security system, it's an unavoidable topic for discussion that we should accelerate social mixing and eliminate the gap in social security, so that all nationals can fairly share the "cake" of social security.

Establishing a pension insurance system that covers the rural and urban areas is necessary to eliminate the imbalanced pension insurance development between rural and urban areas. For a long time, because of the separation of rural and urban areas, social security resources are unequally allocated. According to statistics, from 1991 to 2005, the average expenditure of social security per capita in urban areas accounted for 15% of GDP per capita, over 90 times higher than that in rural areas, which only accounted for 0.18%. This kind of dualistic social security system seriously affects the fairness of the social security system. It destroys social fairness, aggravates the gap between rural and urban areas, and goes against the requirement of fairness of the social security system.[16]

Therefore, we should lower the premium of pension insurance for cities and towns, so that groups like rural migrant workers may be able to pay for it. Cancel the limitations of the basic social pension insurance system, which classifies people into groups according to the household registration system, so as to connect rural and urban areas. Study and promote reform experimentation of pension insurance system for public institutions, draft relevant supporting policies for pension insurance system for public institutions, and focus on the study on the methods of

[15]Ma [12].

[16]Yan [13].

basic pension accumulation and payment and those of trial implementation of occupational annuity. As conditions permit, consider adjusting and reforming civil servants' pension insurance system. Establish an elderly service security system that allows a reasonable gap between different groups of people in five years, and build a free, open elderly service security system that covers all people and allows the integration of rural and urban areas with five more years. As the state increases investment in the old northwestern industrial base and middle and western areas in recent years, especially as the rural social security system that mainly includes rural social pension insurance and rural cooperative medical service is established and perfected, this kind of situation, that is, the dividing of the rural and urban areas, would be changed, and it would be a trend for these two areas to mix.

5. Push the mixing of elderly care cause and modern service industry[17]

Generally speaking, people may easily see the pressure caused by the aging issue on social development, but neglect the changes it brings about. In fact, the changes of the elderly care mode resulted from the aging issue would lead to the emergence of many industries, and drive the development of society and economy in China.

Currently, China is undergoing transformation along the path from family elderly care to institutional elderly care and finally to community elderly care. The changes of the elderly care mode will have great impacts on social and economic development, requiring the development of the elderly care service industry and pushing the mixing of elderly care cause and modern service industry. In the service industry, aside from providing regular items like daily care, housekeeping service, medical care, nutritious catering, rehabilitation, body building and commercial supporting facilities, surveillance systems can be gradually installed in the elderly's communities, for example, electronic surveillance system, alarm system, patrol system and e-card system. Expertise on different aspects can be applied in the development, building, operation and management of the elderly's communities, for example, community planning, environment protection, safety facilities, material science, the elderly's physical and mental status.

The mixing of elderly care cause and modern service industry is the result of innovation in the elderly care mode and also a trend of social restructuring. During the process to solve the elderly care issue, we should and can introduce in concepts and thoughts from modern service industry, through which we can drive the innovation in the elderly care mode and development of social undertakings in China.

6. Accelerate the organic mixing of ways of elderly care security

An appropriate way of elderly care would be the best. Since China has been in the agricultural society for a long time with bad natural conditions, family elderly care has become a long-exiting consensus and mainstream way of elderly care over the

---

[17]Xi [14].

past thousands of years, and it's also appropriate for the agricultural society after long-time selection.

But as the industrialization and urbanization processes accelerate in China, individuals' identity changes among a member of a family, unity or society. Because of this, the traditional family elderly care coexists with institutional, social elderly care and other elderly care security modes. Moreover, the above-mentioned modes often mix together. Among them, home-based elderly care service is suitable for China's conditions with Chinese characteristics. It refers to a form of socialized elderly care service, according to which communities and society help family members provide daily care, medical care, entertainment, mental comfort and other services to the elderly that live at home. It's an innovative form based on the traditional family elderly care service.[18]

Vienna International Plan of Action on Aging, produced by the first World Assembly on Aging in 1982, emphasizes: "They should be community-based and provide a broad range of preventive, remedial and developmental services for the aging, to enable them to lead as independent a life as possible in their own home and in their community." United Nations Principles for Older Persons, which was published in 1991, stresses once again: "Older persons should be able to reside at home for as long as possible" and "Older persons should benefit from family and community care and protection in accordance with each society's system of cultural values." From then on, the way of elderly care with home care as the major part, community care as supporter and institution care as supplement becomes a trend in the international community.

In terms of the content of security, elderly care security includes several levels: providing materials, guaranteeing basic life, offering daily care and mental comfort, which are consistent with Maslow's hierarchy of needs. The first two levels belong to basic security, and the latter two are to improve the elderly's life. If we can properly mix these levels, we can positively deal with the aging issue.

In addition, all institutions providing elderly care service, whether in cities, towns or rural areas, public or private ones, should be equally treated by the government and reformed so as to provide standardized services. Only through encouraging and supporting all institutions, providing elderly care service fairly and offering them equal development space can we accelerate their development.

2.4 Theory of Government Functions

All the ruling classes at all times and in all over the world treat the issue of supporting old citizens as a social problem. However in the early days, it was basically in a laissez-faire situation. With the social development, governments began to supply different degrees of relief to the old group, especially the elderly in

[18]Dong [15].

a jam, though with quite limited relief scopes and levels, which also led to unflattering implementing effects. After the Middle Ages, especially with the establishment of modern social security systems, whether the Church or governments all elevated the supporting for the old citizens to a very important position; and conformed it by forms of systems and even laws. As entering the modern society, the supporting for old citizens has already been the embodiment of basic human rights, which has also been implemented with positive intervening measures by governments. Therefore, as it were, the human pension securities experienced a historical developing process from laissez-faire to overall intervening, from sole and superficial to multilayer and thorough. As to the theories of government functions, liberalism and state intervention are two mainstream theories in western thinking public.

1. Theory of government functions of liberalism

The origin of the liberalistic thought of government functions can date back to European classical liberalism in the 17th and 18th century. The ideologists of the day advocated policies of nonintervention to give maximum laissez-faire policies to individuals and market subjects, and considered that, "the government intervening the least was the best government", while the main functions of the state and the government should concentrate on the aspects of safeguarding maximum freedom of individuals and their private property rights, etc. Some liberalists in the early days, such as Adam Smith, Locke, Montesquieu and others all had discourses on the systems of reducing intervention, advocating free trade.

Adam Smith creatively absorbed the viewpoints of physiocrats, and demonstrated the rationality of laissez-faire policy from the perspective of economics. His whole theory was based on the hypothesis of "economic man", and considered that, human was rational, and always planned for himself in economic activities to pursue the maximum of individual economic benefit; and the market with free competition was precisely the natural order conforming to the self-interest nature of human. Therefore, under the condition of market free competition, the economic behaviors of individuals can automatically agree on the realization of social benefit, with "an invisible hand" playing important functions in this process of benefit coordination. Smith wrote, "(he) always neither planned to promote public benefit, nor knew he himself promoted that kind of benefit to what degree… he just only calculated his own benefit. In this occasion, just like in many other occasions, he was guided by 'an invisible hand' to try his best to reach to a target that was not his real intention".[19] As by Adam Smith, the market is efficient with the function of self-regulation; the government should not destroy the operation of free-market mechanism by its own intervening actions. Hereby, he definitely laid out specific limitations to government functions: protecting national security from foreign invasion; protecting personal safety in the society from infringement and oppression by others; constructing and maintaining certain public facilities and public utilities

[19]Smith [16].

that the private sector was unable or unwilling to manage. In short, the government only played the role of "night-watchman" and "policeman".

The classical economic liberalism established by Adam Smith has gradually evolved into new classical economic liberalism after continuous development by economists of later generations. However, from the view of theory of government functions, they all laid emphasis on the laissez-faire system against excessive intervention to the economy by the state and the government. After the 1930s, with the impact of Keynesian economics, economic liberalism had been once depressed. In the 1960s and 1970s, the theory of liberalist government functions rose in the academic circle after "WWII" once again. They were generally called as "new conservatives", and Friedman, the representative of monetarism and American economist was considered as one of the most conservative economists in the West. The intervening activities by the state exactly impeded the healthy development of the market.[20] For another instance, the economists represented by James Buchanan proposed the concept of "government failure" and took this as the basis to analyze the intervening behaviors of the state.[21]

Other than economists, some scholars also strongly attacked the government functions of the state interventionism from the view of political philosophy, stressing that the continuous expansion of government authorities would damage individual rights and freedom of citizens. As its representative, Robert Nozick considered that individual rights were inviolable. On account of the disadvantages of government intervention, he advocated to control the government functions in the minimum scope and fully play the functions of market mechanism.[22]

In *Two Treatises of Government* wrote by Locke from the year 1680 to 1690, the main responsibility of the government was to protect individual freedom and property, and it could let them free without managing anything beyond that. In the monumental work *Wealth of Nations* published in 1776 by Adam Smith, the founder of the classical economic liberalism in the UK, which was consecrated as Bible in the economics field, it defined the responsibilities of the government from the perspective of economy, and considered that the government should grant maximum freedom to the people and things in the economic category. The theory of "an invisible hand" was the core viewpoint proposed by Smith in his classics of Wealth of Nations, and under the support of this viewpoint, western capitalism was widely adopted. From 1970s to 1980s, free market principles revived in the East and the West at the same time, and Hayek-type liberalism was delivered from oppression. Meanwhile, a great revolution quietly began in the field of government management. Mrs. Thatcher was elected as Prime Minister of Britain in 1979, and Reagan was elected as President of the USA in 1980, while a New Public Management Movement aiming at shrinking the scope of government functions and expanding the market field launched in Britain and the USA therewith, and then

[20]Milton [17].

[21]Buchanan [18].

[22]Nozick [19].

triggered global governance revolution. Under this background, the pension security responsibility undertaken by the government obviously weakened.

2. Theory of government functions of the state interventionism

The state interventionism generated mainly basing on market failure, personal nearsightedness and other circumstances. In the late 19th century, British ideologists represented by Green and Hobhouse had already perceived the necessity of intervening social life by the government. They pointed out that, the intervention in the aspects involving public benefit by the state could provide more social conditions and more favorable social environment for the free development of every citizen, which was not in conflict with personal freedom. Ideologists of Fabian socialism considered that it was necessary to economically guarantee the minimum standard for people so as to realize the fundamental values of freedom, philanthropism and equality of the bourgeoisie. The government should not only undertake the function of maintaining overall income, but also widely intervene in social life to solve the problems of education, medical treatment and so on of citizens. Some viewpoints of Fabian socialism became the theoretical basis of modern welfare state afterwards. As entering the 1970s, John Rawls, a famous American philosopher, published the work of *A Theory of Justice*, in which he had discussed relevant issues on government functions in an enormous ideological system. Rawls pointed out that, "justice as fairness" pursued a principle of improving the status of the least beneficiary to a maximum degree.[23]

During the period of "Great Depression" from 1929 to 1933, governments of principal capitalist countries including the USA, Britain and so on intervened in the field of public service in the round. Roosevelt Administration formulated *Social Insurance Law*, which stipulated that those wage-earners retired over the age of 65 could receive pensions varying from USD 10 to 85 dollars per month according to different wage levels. The unemployed could receive insurance benefits, and mothers with young children as well as the disabled could receive subsidies.

Keynes, a British economist, published *The General Theory of Employment, Interest and Money* in 1936, in which he referred back to the theory of government functions of the state interventionism *Social Security Law,* marking the establishment of modern social security systems emerged at the right moment with an epoch-marking significance by the midwifery of a series of elements. Keynes pointed out that the market with perfect competition assumed by the economic theory of classical liberalism did not exist in real life. Since people's economic behaviors were always influenced by the three basic psychological factors of "psychological consumption tendency", "psychological liquidity preference" and "marginal efficiency of capital", it was impossible to lead to the balance of social supply and demand only by the capitalism with market regulation, thus may trigger the deficiency of effective social demand, and the economic crisis may also arise subsequently. He considered that only sensible and positive policies by the

[23]John [20].

government could transform laissez-faire automatic balance into initiative balance of intervention and regulation by the government, thus made capitalism to be "adjustable capitalism" to prevent crisis. The functions of the government transformed from the role of "night-watchman" in the past to the dominant power of economy in a leap, playing decisive and indispensable functions.

In fact, in the modern west, people have already gradually realized that there were self-insurmountable imperfection and deficiency in the society, market and government. Thereupon, the two main theories of government functions above also began to emerge the tendency of mutual absorption and mutual fusion between them: the theory of government functions of the state interventionism still emphasized the maintenance of free competition and free trade; the theory of government functions of liberalism also no longer absolutely "rejected the state intervention", but required to confine this kind of intervening activities in a certain scope.[24] This was also the eclectic proposition of "the third road" developed afterwards.

3. The transformation of modern government functions

Since the so-called "omnipotent government" does not exist, which is also impossible, it has become a prevailing theory tendency in the academic circle in recent years to demand for the reduction of government intervention, and some scholars even take "small government, big society" in the West as the targeted model of the transformation of government functions. In fact, as confined by national conditions and the influence of traditional culture for thousands of years, this kind of seemingly reasonable model also has many restrictive factors. Thereupon, how to seek for the best joint point between government intervention and social autonomy then becomes a theoretic subject worthy of in-depth exploration for the aspect of the transformation of government functions in China. At present, the service-oriented government has already become the orientation of the transformation of government functions that has been widely accepted by all sectors of the society. The public service by the government also becomes the overriding joint point of government behaviors with citizen demands, which is also the main dimension to evaluate the government and its leaders by citizens. The service-oriented government is that, the government introduces market and society forces by effective institutional arrangement to form effective supply mechanism of public services to realize the target of supplying high-quality public products and services fairly, transparently, efficiently to better satisfy diversified demands of public services by the masses. Professor Zhang Kangzhi considered that, the service-oriented government was a government that "set public service as the fundamental purpose of the existence, operation and development of the government".[25] Therefore, the government must conduct governance together with

[24]He [21].

[25]Zhang [22].

non-governmental organizations, communities, the populace and private sectors to carry out the marketization and socialization of public services.

The research on government theory, the definition of government function by the economists in the early stage mostly only laid emphasis on the economic field, especially the aspect of protecting personal property regardless of whether they advocated weak government of "night-watchman"-type, or strong government with overall intervention, but they still paid less attention to the aspects of the government functions in social administration and public services, and even no attention on those. As a part of social affairs, the issue of supporting for old citizens also naturally had not entered their views too much. After entering the industrial society from the agricultural society, individual risks, family risks had evolved into social risks. Social risks naturally should be defended by social force, thus the state naturally become the leading force to defense risks for the elderly. In spite of different ethics and concepts of responsibility in various countries, the state is the main supplier of basic pension, and it has become consensus that public sectors should provide more pensions for all the elderly.

According to the principle of "justice as fairness" by Rawls, the government function is to improve the status of the least beneficiary to a maximum degree. At present, as the core of the transformation of government functions, public services of the government should be strengthened and expanded, and the target of the transformation of government functions should be the establishment of public service-oriented government. The government should provide basic and guaranteed pension, education, medical treatment, employment and other social public services, and put effort into solving the problem of social re-allocation to realize the target of fairness in the process of development.

As to the influence by China's economic reform to its pension security, an American scholar Halperin (2002) pointed out that China's economic reform had smashed people's "iron rice bowls", and the pension system also fell into crisis at the same time. Since it is just impossible for people to spend their remaining years in comfort in the fluctuating market, it is unrealistic for the government to fully pass the buck of pension responsibility. The Chinese government should further provide basic pension security for all the people. The 16th CPC National Congress also had a statement about the orientation of the transformation of government functions in China under the condition of socialist market economy, namely "economic regulation, market supervision, social administration and public service".

4. The government should undertake the entity responsibility for the pension of citizens

It can be roughly seen from the developing process of social security that, the role played by the government gradually changes from weak to strong, from the margin to the center, from dominance to subject. It is worth stressing that during the development of basic pension, the government undertakes the entity responsibility, not the dominant responsibility. The dominance is to initiate and promote; the subject is to directly participate and more deeply intervene.[26] No matter from the social insurance system (government-subject-individual-subject-type), minimum living allowance system (government-subject-family-subject-type), or welfare service system for the elderly (government-subject-family-supplement-type), the government undertakes the role of subject. The security types with the government playing dominant role include service system of home-based care for the aged (government initiating-family subject-type), preference system for the elderly (government initiating-society subject-type) and so on.[27]

The government undertakes entity responsibility in participating in subject investment and strengthening industrial management which is more important. The management of pension service is a new issue. The government should guide on a macro level by formulating industry planning of pension service, industry access standard, fundamental managing norms of pension institutions. Civil administration and other relevant functional departments should strengthen the management on the approval process, as well as support and specify it after the approval to strengthen supervision and inspection. It should also rely on the industry association of pension service to strengthen industry self-regulation.[28] The general trend is that, the government and society intervene more and more in the pension field to gradually establish corresponding pension security system and pension service system with home-based care for the aged as the subject, the community as the base, and institution pension as the auxiliary.

2.5 Life Cycle Theory

Life Cycle Theory is also called as Life Cycle Hypothesis (LCH) of consumption and savings, which was proposed by the American economists of F. Modigliani and R. Brumberg, A. Anton together. Its basic meaning can be vulgarly understood as the whole process of "Cradle-to-Grave". Rational consumers will arrange the consumption and savings throughout their life according to the principle of utility maximization to maintain the income and consumption in a lifetime in general balance. Life Cycle Hypothesis links consumption with the income and wealth in a lifetime, which is of important significance and status, for this reason, Modigliani won the Nobel Economics Prize in 1985.

1. Rational pension system can smooth the life course

The grasp of own life course by the human beings should firstly seek for existence and prolong the life as long as possible. The next is to live well with respect and quality, and be able to keep from the torment by hunger, cold and disease when

[26]Dong [23].
[27]Dong [24].
[28]Dong [25].

getting old and infirm, and then spend the remaining years in comfort. And the social security is absolutely necessary in pursuit of the two points.

From the perspective of individual life, the main function of old-age pension is income smoothing-old-age pension is a kind of measure to re-allocate the income of the middle-age period with production capacity to the period after retirement.[29] That is to say, individuals accumulate pension capital for the senectitude in the future by labor in the young period. Laborers prepare for the quitting of the labor market when they lose capacities to work by payment and accumulation of old-age pension when young, thus smooth the whole life cycle course to reduce fluctuation as much as possible.

According to the life-cycle hypothesis, people save money during their working lives and withdraw it during their post-retirement lives, in order to even out their consumption over their entire lifetimes by accumulating when their income is high and dis-saving when their income is low. This hypothesis was established on strict assumed conditions, of which, one is the assumption that future income is predictable. However, in real life, due to information asymmetry and lack of foresight (such as inflation, interest rate variations and other factors), it is difficult for people to estimate how much they should save. In other words, it is difficult for consumers to guarantee that they can achieve their retirement objectives with their voluntary savings and they, thus, often save too much or too little. This issue requires society to provide a form of institutional arrangement that can satisfy the needs of people engaging in this form of cross-period saving, which is a condition for the existence and development of a country's pension system (Sun Hongyu 2004). The pension system stimulated people' saving activities and allowed them to replace previous saving behavior with pension payments and withdrawals. Thus, pension payments and benefits reflect a savings trend whereby people maintain an appropriate income during their old age. Therefore, the contributions that people make toward pension funds can be seen as a form of saving that guarantees individual income during retirement, playing the comparative role of a piggy bank.

2. The family inter-generational support reflects Theory of Life Cycle

Theory of Life Cycle not only reflects in the aspect of individual life, but also reflects in the aspect of family as the base for individuals. In a family, the inter-generation fostering and supporting reflect obvious life cycle. From the view of modern social security, the costs of giving birth to and fostering children can be considered as the payment of security funds by parents at labor age for future provision when they get old. This fund is paid and accumulated year by year with the age growth of children, and its value will be preserved and increased with the gradual growing of children. When parents get old and lose labor capacity, and children grow up to labor age, the pension security funds paid in the past begin to pay until parents pass away. From the perspective of traditional families, this kind of payment, accumulation, appreciation and grant of pension fund is a process full

---

[29]Nicholas [26].

of ethical kindred, and is very natural.[30] The problem at present is that, this kind of inter-generational supporting model with the constraint type of non-institutionalized ethic is more and more impacted by external forces. For example, the advanced age with low birth rate leads to the decrease in competence of fostering children, or being too busy to nurse the elderly, which call for support from the government and social forces.

3. Postponing retirement age conforms to Theory of Life Cycle

The stipulation of retirement age in force directly leads to the sharp decrease of the proportion of individual working hours in the whole life cycle, which seriously injures inter-generational fairness, and challenges the financial sustainability of pension insurance system. Thus, postponing retirement age and adjusting life cycle become one of the measures adopted by most countries to deal with the pressure brought by population aging. Therefore, postponing retirement age can function double effects of increasing income and decreasing expenditure for pension funds. In the whole life cycle, the education investment by urban residents is higher than that of rural citizens, and those in the job is higher than those out of job. Researches show that, the years of working of professionals and administrators are mostly approximately 40 years, and those of low-end workers in agriculture, service and other industries can reach to over 50 years. It intensively performs in the aspects of the levels of personal income and pension security, and it is an inexorable trend that the former is always higher than the latter. According to this theory, combining the stipulation on retirement age at present and life cycle characteristics of the male and the female, the retirement age can be properly prolonged, and the scale of investment costs for individual human capital can be reflected by moderate distinction of the levels of old-age pension.

4. Theory of Life Cycle provides theoretical foundation for the implementation of the house-for-pension scheme

The house is the important fixed asset of a family, with the function of value preservation and increment. However, because of its enormous value content together with other factors, it is comparatively difficult for realization. Influenced by this, the living quality of the elderly with all housing property will also be greatly reduced correspondingly. According to Theory of Life Cycle, and the analysis on income and consumption cycle in a lifetime, the house-for-pension scheme can be taken into account for trial implementation in China. The elderly mortgage their own legal house properties to financial institutions or individuals, and the latter pay old-age pension to the elderly in due course, then the house properties will belong to the latter by a proper way after the elderly pass away. The tasks of provision for the aged are heavy with high pressure, and this kind of "reverse mortgage" model will not only raise the living quality of the elderly with economic independence, but also relieve the pressure of provision for the aged, exorbitant medical expenditure

[30]Liu [27].

and housing demand at present, thus make immovables to flow in fact, with rational allocation and optical utilization, which is also the enrichment and development of Theory of Life Cycle. It is also a kind of financial innovation for financial institutions, which can expand business scope by this to find new profit point.

Though there are comparatively successful precedents of the "house-for-pension" scheme in developed countries, it is still only in a starting stage in China, which is also faced with the influence by population longevity risk, the value assessment of real estate, price fluctuation in the real estate market and other uncertainties, therefore, the promotion of this work is still faced with many difficulties. We may select part of developed regions to pilot first, and then gradually popularize it after conditions get ripe. During this process, the government should strengthen guidance, supervision and risk control to make it become one of the important complementary measures to cope with population aging in China.

Commercial insurance companies have inborn advantages in operating "house-for-pension" business, for example, they can avoid market risks to a maximum degree, and design proper products according to the life expectancy of property owners and so on.

2.6 Theory of Civic Rights

According to the Constitution of China, the basic rights of citizens mainly include the right to elect and to be elected, cultural education right, labor and rest right, life security right and other rights. At present, the basic public services linked with the basic rights of citizens in China should include pension security, medical security, social relief, education, employment training and other rights at least. Certainly, basic public spiritual culture service should also be included. Besides, all citizens are free and equal when enjoying various rights, with the protection of the Constitution and laws.

1. Providing for the elderly is the basic right of citizens

The value of social security is to realize social equity and social justice. Welfare economists considered that social security is the public product provided by the government to all citizens, and should be equally enjoyed by all citizens regardless of their social situations, status and positions. As an important component of human rights, social security lays emphasis on the responsibility of the state, which reflects the dependence of civic rights on the state power, attaching extensive social responsibility to the government, and also creates conditions for citizens to enjoy other basic rights and freedom.[31]

The right of provision for old citizens is not only the core issue of pension security system, but also the logical starting point to solve the problem of pension

[31]Liu [28].

security. Under the circumstances that citizens have reached the age boundary to relieve labor obligations stipulated by the state or have lost the capacity to work because of old age, they are legally entitled to enjoy material assistance provided by the state and society, and be supported and helped by their families. Like subsistence right and labor right, pension right is also the basic right of citizens in the society. With the constant improvement of pension service system, the pension services acquired by citizens include not only material life, but also higher-level spiritual culture service.

Providing for the elderly is not only the right of citizens, but also the responsibility of the government. The state guarantees legal interest of the elderly by the form of legislation to let the extensive elderly live with dignity and security, which is determined by the objective law of social development, and is also the principal function of the state and government. Therefore, as one of the most important livelihood issues, providing for the elderly has been written into the Report of the 17th National Congress of CPC.

As one of the most important five types of insurance of social insurance, the pension insurance is an important component of social security system, which is a kind of social insurance system established by the state and society to solve the basic lives of labors under the circumstances that they have reached the age boundary to relieve labor obligations stipulated by the state or have lost the capacity to work because of old age and quit labor posts according to certain laws and regulations.

2. Relevant stipulations on the right of provision for old citizens by domestic and foreign constitutions and laws

The earliest *Old-age Pension Act* in Britain was passed in 1908. The Act stipulated that each of the elderly over 70 years old could receive 5 shillings as old-age pension per week. This made it become a kind of right of citizens to receive old-age pension for the first time in human history to confirm it in the form of law.

The four Constitutions issued by New China successively all had stipulations on the life guarantee right of retired workers and the right of material assistance of old citizens. Chapter 2 of the Constitution of the People's Republic of China passed in 1954 stipulated in "Basic rights and obligations of citizens" that "citizens of the People's Republic of China enjoy the rights and obligations of labor" (Article 42). "The state implemented retirement system on staffs of enterprises and institutions, and civil servants according to the stipulations by laws." "The lives of retirees are secured by the state and society" (Article 44). "Citizens of the People's Republic of China are entitled to acquire material assistance from the state and government under the circumstances of oldness, disease or losing labor capacity. The state develops social insurance, social relief and medical and health services for citizens to enjoy these rights." "The state and society secure the lives of disabled soldiers, comfort and compensate families of martyrs, give special treatment to families of soldiers." "The state and society help to arrange labor, lives and education for the blind, deaf, dumb and other disabled citizens" (Article 45). Obviously, the citizens mentioned in the Articles should include all urban and rural residents. Based on the Constitution, there are also stipulations on securing legal rights and interests of the elderly in Labor Law, Marriage Law, Law on Protection of Rights and Interests of the Elderly in China. Among them, Law on Protection of Rights and Interests of the Elderly explicitly stipulates that "the state establishes pension insurance system to secure basic lives of the elderly", "the state increases old-age pension according to the improvement of economic development, the living standards of people and the increase of staff salaries", and stipulates to implement five-guarantee system of "securing food, clothes, residence, medical treatment and burial" for the elderly in rural areas without labor capacity, source of income, supporters and dependents. Therefore, the right of provision for old citizens with subsistence right as the core has been basically notarized by the whole society and secured by laws.

Social Insurance Law of China issued in October, 2010 stipulates that the state establishes basic pension insurance, basic medical insurance, employment injury insurance, unemployment insurance, maternity insurance and other social insurance systems to secure citizens to legally acquire material assistance from the state and society under the circumstances of oldness, disease, occupational injury, unemployment, maternity and other situations. The payment ages of basic pension insurance relations of individuals employed in cross-planning areas are calculated by accumulation.

3. The draining of interests and rights of pension security by the marginal population in China

Section 2, Article 33 of the *Constitution of China* in 1982 explicitly stipulated that "all citizens of the People's Republic of China are equal before the law", and specific to the aspect of pension and social security, no matter the elderly in urban areas or rural areas, no matter old men or old women, they are generally equal to receive the pension services. However, the social pension issue of citizens in China is always in an incomplete condition for a long time, let alone equal and sharing. Although *Social Insurance Law* has been issued, its implementation effect still waits for strengthening. The salvation law, welfare law and other laws involving living quality of the elderly are still being brewed.

(1) The floating population with migrant workers as the subjects. In general, the features of flexible employment are informal labor relations, comparatively low remuneration, no institutional link with social insurance, low-threshold employment, easy to enter and quit. The flexible-employment population is always called as "edge population" or "vulnerable group" and so on by some newspapers, who are mainly temporary workers, seasonal workers, hourly workers, laid-off workers, migrant workers, individual households and free-lances, etc employed by formal sectors. The rate of joining social insurance is comparatively low by migrant workers, owing to their weak insurance awareness and lack of initiative, and the unwillingness of employers to participate insurance motivated by profit. Besides, it is also an important reason that the current system is not quite fit to the real employment situation of

migrant workers, which is not smooth in transferring and renewing and is with weak enforcement. The problem of transferring and renewing is the chronic illness perplexing the current pension system all along. Since the overall planning level is low, and migrant workers are in comparatively high liquidity during the process of moving from one place to another, the pension money paid previously cannot be transferred to the new working place, so migrant workers always choose to "cancel the insurance" or "clean up accounts" to retrieve the part of old-age pension paid by individuals, with the social overall planning paid by enterprises for them still leaving at the original place, which is like to make "contribution" to local finance. Because the current pension insurance system is designed for staffs in state-owned enterprises at the very start, not to taking the participation of migrant workers in account.
Researches show that, it has become a tradition by some local governments in Guangdong province to actively provide favorable assistance and services to migrant workers to "cancel the insurance". Therefore, migrant workers should be encouraged to gradually integrate into a new urban pension security system at current. It is an inevitable process of urbanization, while it is also an inevitable process that more and more labors come into cities. It is not difficult to imagine that there will be 300 million laborers constantly floating between the urban areas and the rural areas 10–20 years later.

(2) Pension security problems of land-requisitioned farmers. At present, the pension security problems of land-requisitioned farmers mainly lie in the following aspects: the constantly weakening of pension function of the land, the disunity of institutional model, too low compensation standard for land requisition, poorer sustainability of system and so on. As the source of food and clothing and the basis for survival, the pension security function of the land is the important basis for family pension in rural areas, which is also an important difference between the pension in rural areas and in urban areas. With the rapid development of market economy, and the acceleration of industrialization and urbanization, there comes out crisis in the quantity and quality of the land resources that can be occupied by farmers, with the pension function of the land emerging with weakening tendency. Since new Land Management Law issued in 1997, there have formed 40.27 million land-requisitioned farmers all over the country. With the acceleration of the process of urbanization, there are about 4 million land-requisitioned farmers newly increased every year in recent years. With the arrival of aging society, this part of "three-without" marginal population in the city will bring great impact to social harmony and stability, while the vast majority of unstable factors come from land requisition and demolishing at present. Therefore, it appears to be particularly necessary to properly solve the pension security problems of this part of population.

(3) College graduates that have failed to be employed in time. Since there is increase but no decrease in the employment pressure, a multitude of youthful students have been relentlessly pushed to the society from the campus. They cannot adapt to the complex social environment since they have not acquired

suitable job posts in time. Worse still, this group cannot be covered by social security network in time, which makes them increasingly in a marginal situation. In addition, this group always evolves into unstable factors of the society due to their active thoughts. Therefore, the security problems of this group should be focused by all sectors of society.

4. Empowerment to rural residents by accelerating the process of new-type rural social endowment insurance system

New Rural Co-operative Medical System with institutionally full coverage at present only relieves the problem of medical care for rural residents, and the system of subsistence allowances is just designed with the target of securing the most basic livelihood of the most disadvantageous group for the time being, with rather limited coverage and security level, tending to be unstable, unreliable and even non-institutional. In addition, compared to enjoying pension security, receiving minimum living allowances is a right notwithstanding, but it is not a fully dignified and honorable living style. In the long run, it still depends on the pilot and generalized new-type rural social endowment insurance system at present to solve the problem of provision for the elderly in rural areas to secure their basic livelihood, let them acquire various living rights more honorably, as well as improve and raise their living and survival quality. Such kind of institutional arrangement can not only fully improve the family status and social status of the elderly in rural areas, and help the elderly find back the lost dignity again, but also secure their basically quality daily life. Therefore, it is imperative to accelerate the new-type rural social endowment insurance system.

References

1. Smith, Adam. 1997. *The Theory of Moral Sentiments*. China: The Commercial Press.
2. Chen, Biao. 2009. *Study on Equalization of Basic Public Cultural Services in Zhejiang Province*, a master's thesis in 2009.
3. Zhang, Xiaofeng. *Socialist Core Value System and Elderly Care Service System Construction*. http://www.zgg.org.cn/pub/tgw/qnshl/qnsk/20100804_12482.html.
4. Pigou. 2007. *Welfare Economics*. China: Huaxia Publishing House.
5. Bentham, Jeremy. 1995. *A Fragment on Government*. India: The Commercial Printing Press.
6. Pareto. 2010. *The Rise and Fall of Elites*. China: Beijing Publishing House.
7. Zheng, Gongcheng. 2010. *Let Social Security Adjust Wealth Distribution*, *China Social Security Newspaper*, November 10.
8. Zhang, Xiaofeng. *Socialist Core Value System and Pension Service System Construction*. http://www.zgg.org.cn/pub/tgw/qnshl/qnsk/201008/t20100804_12482.html.
9. Dong, Hongya. 2010. *Shengzhou Mode of Elderly Care Service*, *Observing and Thinking*, June 1.
10. Ga, Rida., Huang, Kuangshi. 2008. *Analysis and Enlightenment of the Definition of Fusion in the Western Society*, *Modern Thoughts*, 1st issue of 2008.
11. Liu, Xiaofeng, Chen, Zhao and Lu, Ming. 2010. *Social Integration and Economic Growth*, *World Economy*, 6th issue of 2010.

12. Longsheng, Ma. 2009. *Pension System Should Be Designed According to Principle of 'All People Sharing Risks', Liaoshen Evening News*, February 1.
13. Yan, Dongbin, et al. 2008. *Study on Reform of Social Security System from the Perspective of Fairness, Teaching and Study of Finance*, 3rd issue of 2008.
14. Xi, Zhiyong. 2007. *Promoting the Mixing of Elderly Care Cause and Modern Service Industry, Jiefang Daily*, October 29.
15. Dong, Hongya. 2009. An Explanation of the Government's and Family's Function of Looking After the Old in the 'Shengzhou Model', *Journal of Zhejiang Education Institute*, 6th issue of 2009.
16. Smith, Adam. 1974. *An Inquiry into the Nature and Causes of Wealth of Nations*. China: The Commercial Press.
17. Milton, Friedman, et al. 1982. *Free to Choose: A Personal Statement*. China: The Commercial Press.
18. Buchanan. 1998. *Freedom, Market and State*, China: Beijing Institute of Economic Publishing House.
19. Nozick, Robert. 1991. *Anarchy*. State and Utopia: China Social Sciences Publishing House.
20. John, Rawls. 1998. *A Theory of Justice*. China: China Social Sciences Publishing House.
21. He, Wei. 1999. *Analysis on the Origin of the Theory of Government Functions in the West, Social Sciences in Nanjing*, the 7th issue, 1999.
22. Zhang, Kangzhi. 2009. *Grasping the Theory Orientation of the Research on Service-oriented Government, People's Tribune*, the 3rd issue, 2006.
23. Dong, Hongya. 2009. *Build Basic Pension System Focusing on Nursing to Pull off Providing for the Elderly, Northwest Population Journal*, the 3rd issue, 2009.
24. Dong, Hongya. 2010. *Research on Sharing, Coordinating, Complementary New Pension Security System-Taking Zhejiang Province for Example, Journal of Party School of Zhejiang Provincial Committee of CPC*, the 3rd issue.
25. Dong, Hongya. 2010. *Shengzhou Model of Pension Service, Observation and Consideration*, June 1.
26. Nicholas, Barr. 2003. *Economics of Welfare State*. Trans. *Zheng Bingwen*. China: China Labor & Social Security Publishing House.
27. Liu, Zhenjie. 2005. *Brief Analysis on the Function and Status of Family Security in Modern Social Security System, Inquiry into Economic Issues*, the 7th issue.
28. Liu, Lingzhi. 2010. *Statement on Right of Provision for Old Citizens in Pedigree of Basic Rights, People's Tribune*, the 26th issue.

Chapter 3
Study on the Modern Senior Care Service and Security System

Lianxing Wang

The 17th National Congress of the CPC established a strategic target of "providing senior citizens with a sense of security". In March 2010, when making a report on the work of the government, the then Premier Wen Jiabao put forward the following: "We will accelerate the establishment of a sound social senior care service system". In October 2010, the 5th Plenary Session of the 17th Central Committee of the CPC put forward the task of "paying attention to family and community functions, giving priority to the development of social senior care services, developing and expanding the business and industry of services for senior citizens" and taking "giving priority to the development of social senior care services" and "developing and expanding the business and industry of services for senior citizens" as an important part of "the 12th Five-Year Plan", revealing important practical significance and long-term strategic significance. With the development of the social economy, constant growth of overall national strength, great improvement in urban and rural residents' living conditions, an acceleration of an aging society, and senior citizens' diverse demands during retirement, the target of "providing senior citizens with a sense of security" not only meets low-level basic survival and the demand for senior care services is increasing. According to statistics from the Civil Affairs Department, as of the end of 2009, the number of disabled senior citizens in China was as high as 10.36 million, with the number of semi-disabled senior citizens as high as 21.23 million, accounting for 18.9% of the total number of senior citizens[1]; the 6th National Population Census showed that the percentage of China's population aged 60 and above accounted for 13.26%,

[1]Office of the National Working Commission on Aging, *2009 Statistical Bulletin on Cause for the Aging Population in China*, issued in June 2010.

L. Wang (✉)
Department of Finance, China Women's University, #1 Yu Hui Road, Chaoyang District 100101, Beijing, People's Republic of China
e-mail: lianxinabc@126.com

Y. Wang (ed.), *The Development of Security and Whole Care System for the Aged in China*, https://doi.org/10.1007/978-981-10-6481-4_3

an increase of 2.93 percentage points from the 2000 census, of which the percentage of the population aged 65 and above accounted for 8.87%, an increase of 1.91 percentage points from the 2000 census.[2] China's dependency ratio for senior citizens aged 65 and above reached 11.6%, a considerable burden on family and social senior care. China's GDP per capita just exceeded 4000 US dollars: there is a relatively large gap compared with the level of economic development of the developed countries that entered an aging population period during the same period. In a rapidly aging society, demands for social senior care services are growing. However, there is a relatively huge gap in senior care services among the gross supplies, financial capital investment, service quality and level and quality of employees, as well as policy implementation, supervision and management. How to realize the development target of "providing senior citizens with a sense of security, medical treatment, education, a sense of worth and happiness"[3] is a problem we must face in reality. Service security is an increasingly important part of senior security and also the emphasis on and difficulty in constructing China's senior care security system against a background of an aging population. This chapter starts with the connotations and urgency of senior care services and security, discusses senior care services and security in some developed countries, learns from the practical experience of senior care services and security in some areas of China, analyzes the present situation and existing problems in China's senior care services and security and puts forward policies and proposals with regards to the senior care service and security system.

3.1 Significance and Urgency of Senior Care Services and Security

Senior care services have a long history in China. *Guanzi* in the Western Zhou Dynasty said: "All countries should provide for senior citizens. For those aged over 70, one of their sons is exempt from military service and they will be provided with meat every three days; for those aged over 80, two of their sons are exempt from military service and they will be provided with meat every day; for those aged over 90, all their family are exempt from military service and they will be provided with meat and wine every day. The government should persuade younger generations to cook fine meals, and ask senior citizens what they want, hope and like. This is how senior citizens should be respected." In the Tang Dynasty, the system of giving special treatment to and serving senior citizens was established. Emperor Taizong of the Tang issued a decree providing five servants for those aged over 100 (*Book of Tang*). Many ancient dynasties punished unfilial people to ensure that children

[2]Website of the National Bureau of Statistics: *Data Dissemination of the Sixth National Population Census*.

[3]See Footnote 1.

provided for senior citizens. "Rearing children for old age" is a traditional concept of senior care in China. In 2000, with the acceleration of China's aging population, a number of experts and scholars studied senior services. Professor held that social welfare services for senior citizens in China refer to support, medical care, rehabilitation, entertainment, education and other services provided for senior citizens living alone in special plight without the ability to work, a source of income or a legal supporter and caregiver, as well as some senior citizens who cannot take care of themselves and whose families are unable to take care of them, under the leadership of the government and with the participation of all forces of society; Yuan and Jia [1] believed that senior care service is a social service provided for senior citizens and essentially should be a professional non-profit social service, and thus senior care service falls under the category of social welfare service for senior citizens[4]; Dong [6] put forward the idea of establishing a "shared, complementary and harmonious" social senior care service mechanism and continuing to give play to and strengthen the function of family-based senior care and energetically exploring policies on the establishment of family services; in December 2010, when Minister Li Liguo of the Ministry of Civil Affairs was giving an exclusive interview to a Xinhua News Agency reporter, he gave a more detailed explanation. He claimed that the senior care service and security system in China should be a basic system frame based on home-based senior care, that relies on community services and is supplemented with institutional care; home-based senior care services focus on protecting advanced senior citizens, senior citizens living alone, disabled senior citizens, low-income senior citizens and empty nesters, providing daily care, household services, rehabilitation nursing, medical nursing and other services with the help of professional senior care service organizations and introducing at-home services, emergency aid, day care, health care and rehabilitation, entertainment and other services according to local conditions, thus improving the capacity of community-based senior care services; institution-based senior care services should break the traditional model in which the government runs and manages the institutions directly and social forces are guided and encouraged to initiate apartments for senior citizens, welfare houses and senior care institutions.[5] The Ministry of Civil Affairs provided an explanation for the connotation of the senior care service system in *Soliciting Public Opinions on the Construction of the Social Senior Care Service System During the 12th Five-Year Plan*,[6] that is, the social senior care service system is a network consisting of facilities, organizations, personnel, technologies and other elements that adapts to the level of economic and social development, aims at meeting the demands of senior citizens for senior care services and improving the life quality of senior citizens and provides daily care,

---

[4]Yuan and Jia [1].

[5]A speech given by Minister Li Liguo of the Ministry of Civil Affairs on the National Senior Care Service Boost, website of the Ministry of Civil Affairs, http://ww.mca.gov.cn/article/, November 2010.

[6]Website of the Ministry of Civil Affairs, http://www.mca.gov.cn/article/zwgk/mzyw/201102/20110200133797.shtml.

rehabilitation nursing, mental solace, emergency rescue and social participation for all senior citizens, as well as supporting service standards, the operating mechanism and supervision system; the social senior care service system should be based at home, rely on the community and be supported by institutions, focusing on the demands of senior citizens, providing professional senior care services with appropriate technology, giving priority to ensuring the service demands of childless senior citizens, entitled groups, senior citizens without the ability to work, a source of income, a supporter or caregiver, households enjoying the five guarantees and low-income senior citizens who are of advanced age, who live alone, who are disabled and who have difficulties in senior care; the construction of social senior care services is a long-term strategic task to deal with an aging population and sustainable development which maintains the government's dominant position, encourages social participation, keeps perfecting the management system, enriches service content, and improves service standards to meet the people's growing demands for senior care services.[7]

3.1.1 Senior Care Services and Security

White Book of 2006 *China Aging Development*.

The author believes the connotation of senior care services and security can be explained in the following five aspects: ① the subject of the development of senior care services is stressing the governmental and social responsibility on the basis of giving play to traditional family-based senior care; ② objects of senior care services are all the senior citizens in society because senior care services are inclusive and shared; the government focuses on safeguarding the livelihood of senior citizens without the ability to work, a source of income, a supporter or caregiver in cities, households enjoying the five guarantees in the country and senior citizens in poverty who are of an advanced age, who live alone, who are disabled and so on, and combines free services and low-cost services with paid services according to the differing demands of senior citizens for senior care services; ③ contents of senior care services are as follows: social senior care services as a kind of public service are applied to meeting the basic demands of all senior citizens for senior care services and providing diversified services including daily care, household services, medical rehabilitation, long-term nursing, legal aid and psychological

[7]Department of Planning and Finance of the Ministry of Civil Affairs, *Soliciting Public Opinions on the Construction of the Social Senior Care Service System During the 12th Five-Year Plan*, February 11, 2011, http://www.mca.gov.cn/article/zwgk/mzyw/201102/20110200133797.shtml.

consultation; ④ senior care services mainly rely on professional social workers combined with volunteers; ⑤ the senior service system is based on home-based senior care services, relies on community-based senior care services, and is led by institution-based senior care and supported by technology.

Senior care service models are divided into home-based senior care, institution-based senior care and community-based senior care according to different living places and the senior care requirements of senior citizens. Differing from traditional family-based senior care, home-based senior care refers to senior citizens that live in their own homes, where the community organizes professionals to provide services regarding domestic affairs, designated patient nursing, health and psychological consultation and so on, or provides necessary equipment for senior citizens. In view of the traditional culture and economic development in China, home-based senior care is a model suitable for senior citizens who are able to take care of themselves and willing to stay at home based on the current situation of senior citizens in China, and also accords with traditional concepts. Institution-based senior care refers to apartments for senior citizens, welfare housing, senior care homes and other institutions built via direct investment from the government or social capital investment providing senior care services for a portion of senior citizens that need concentrated nursing. At present, institution-based senior care primarily provides concentrated support services for senior citizens without the ability to work, a source of income or a legal supporter and caregiver, senior citizens enjoying the five guarantees, low-income senior citizens, senior citizens of an advanced age and disabled senior citizens, as well as senior citizens who need or are willing to accept institution-based senior care. Judging from the number of senior citizens who accept such services, institution-based senior care plays a supplementary role, but judging from its function, it is the technical backbone to providing technical and personnel support for community-based senior care and home-based senior care, and influences these services via training, setting an example, out-of-institution services and other models. Community-based senior care refers to providing daycare services for senior citizens, food and beverage services, entertainment services, community health services, psychological consultation, education services for senior citizens and other services for senior citizens living in the same community as the service objects, with the community residents committee as the management main body and community service center or service station for senior citizens as carriers. The community service platform supports home-based senior care services and provides support for professional door-to-door services for home-based senior care services. Moreover, it is suitable for actual demands for the increasing miniaturization of the family structure in China, and communities can provide door-to-door services for home-based senior care in the neighborhood.

3.1.2 Significance of Building the Elderly Care Service and Security System

Confucius said in *the Theory on Great Harmony*: "People not only provide for their own parents and raise their own children, but also allow all the elderly citizens to live to the end of their days, all the people in their prime to find their useful positions and all the children to grow up. Aged and wifeless men, aged and husbandless women, fatherless children, aged and seedless people, disabled people and seriously ill people should be supported by the society." The security system for elderly citizens includes three pillars, namely economic security, service security and spiritual solace. Comprehensively promoting the social elderly care services and security is an important system guarantee that meets growing care service demands of elderly citizens, solves increasingly acute elderly care service problems due to aging and a low birth rate in China, and effectively improves the living quality of the masses of elderly citizens; a humane choice that carries forward the fine Chinese tradition of respecting elderly citizens and respects the emotional and psychological demands of elderly citizens; an important measure that promotes harmonious families, communities and the intergenerational harmony as well as the construction of a harmonious socialist society and also an important way that speeds up the development of the elderly care service industry, expands employment channels and promotes the growth model transformation of a consumption-driven economy.

The significance of the establishment of elderly care services and security is as follows: ① The establishment of the elderly care service system is an objective demand for realizing the strategic target of "providing elderly citizens with a sense of security". The social security system should implement the value and idea of "fairness, impartiality and sharing" and allow different social classes and groups to enjoy the same social security rights and interests. Elderly citizens are a special group in the society. They leave their jobs because of regulations of the retirement system. Part of them lose the ability to work because of physical causes or cannot take care of themselves. They need effective helps with basic necessities of life. Elderly care services and security are exactly institutional arrangements that free the elderly citizens from the worries that they cannot take care of themselves and realize the target of "providing elderly citizens with a sense of security". ② Acceleration in the construction of elderly care services and security is of great significance for perfecting the pension insurance system in China. Pension insurance is the most important content of the social security institution system. Establishment and improvement of the welfare system for elderly citizens are targets put forward in the 6th Plenary Session of the 16th Central Committee of the CPC. The 3rd Plenary Session of the 17th Central Committee of the CPC further put forward developing social welfare with emphases on "supporting elderly citizens, helping the disabled, saving the orphans, giving financial help and relieving the people in disaster", adapting to the pace of China's economic development and social civilization, insisting on the synchronous development of economic security, service security and spiritual security and gradually establishing a socialized security system for elderly citizens that is suitable for demands for elderly care security in China. ③ Acceleration in the construction of the elderly care service and security system is able to expand the channels of employment in China. In response to the financial crisis, China put forward the development policy of stimulating domestic demands and expanding employment. After years of exploration, China put forward the elderly care service and security system that is based on home-based elderly care, relies on community services and is supplemented with institutional care, and actively proposed that social capital, civil society organizations, charities and other social forces should jointly participate in the elderly care service career, adapt to the policies put forward by China in order to expand domestic demands and improve and guarantee people's livelihood, thus making the domestic economy turn into the domestic-demand-driven one from the foreign-trade-dependent one, making economic development get rid of the negative impact of the financial crisis and taking a lead in obtaining stabilization and recovery in the world. China's elderly care service industry just started. It involves household services, daily nursing, medical treatment, psychological consultation and various professional fields and provides potential jobs. China is vast, so an aging population base is large. According to the calculation of the Department of Social Security, if each community of China is equipped with a professional of social security services, more than 7 million new jobs will be created in China.[8] Community social security professional services mainly focus on services for elderly citizens. As a result, elderly care services and security have huge potential employment opportunities. At present, rapidly developing the service industry for elderly citizens is a continuous good channel to alleviate employment pressure, and promote economic development in China.

3.1.3 Urgency of Building the Elderly Care Service and Security System

With the rapid increase of China's elderly citizens, the proportion of the disabled and semi-disabled elderly citizens is increasing as well. Elderly citizens have become the largest vulnerable group in society and demands for elderly nursing are increasing. Data display that in 2005, China's urban household size is 3.13 persons per household, 0.33 person less than that of the fifth National Population Census in 2000[9]; data of the Sixth National Population Census display that China's average household size is 3.1 persons, down 31.8% from 4.61 persons at the beginning of

[8]Website of the Ministry of Civil Affairs, http://www.mca.gov.cn/article/zwgk.

[9]Main Data Bulletin of the 1% National Population Sample Survey in 2005 Carried Out by the National Bureau of Statistics of the People's Republic of China [2006-3-16]. http://www.stats.gov.cno.

the Reform and Opening up. The proportion of urban and rural empty-nest families is rising constantly and the proportion of urban empty-nest families has reached 49.7%[10]; in 2008, the elderly citizens aged 60 and above of Shanghai's registered population reached 3.0057 million, accounting for 21.61% of the registered population; Shanghai has become a city whose aging degree is the most serious in China and has been close to the highest level of the countries in the world, nearly twice as high as the national average. Aging has brought not only a huge challenge to Shanghai's economic development but also a series of social problems. For example, at present, death incidents of elderly citizens who live alone at home or who are lost constantly appear in newspapers. According to media reports, an elderly citizen living alone in Second Fengcheng Village in a district of Shanghai died at home recently and his daughter opened the door and found it several days later after the notification of the neighborhood committee; an old woman in her 80s was found lying frozen and dead in a fountain pool beside Exit 1 of Shanghai Circus World Station and before some citizens dialed the emergency telephone number, but it didn't work; bones of an old man in his 60s were found in Jinguang New District. How to prevent such a tragedy from happening again? How to allow empty nesters, elderly citizens living alone and elderly citizens with cognitive disorder (senile dementia) to have a sense of security and happiness? In the final analysis, we should rely on the strength of the whole society and the government, and society and families should jointly undertake the responsibility for care and rescue.[11] Professor Yao Yuan from Population Development Studies Center, Renmin University of China pointed out that in terms of elderly care, today's society is in urgent need of care resources. First is the care of daily life such as helping elderly citizens do some cleaning and other household services; second is that when elderly citizens are ill in bed, they need someone to take care of them all the time; third is elderly citizens' spiritual demands for psychological and emotional aspects, for example, elderly citizens need someone to take them for a walk and contact the society. Care and nursing responsibilities for elderly citizens move towards the society from families and are becoming a hot spot of attention to the whole society and the problem the government must address, so building elderly care security has the urgency and necessity.

1. Ways to cope with the challenge of an aging or old aging population

China is and will remain to be an aging society for a long time. Between 2001 and 2100, China's aging population will be divided into three stages, namely fast aging (2001–2020), accelerated aging (2021–2050) and stable severe aging (2051–2100). By 2020 when the goal of building a well-off society in an all-round way is achieved, the aging population will reach 248 million and the aging level will reach

[10]The Ministry of Civil Affairs, *Soliciting Public Opinions on the Construction of the Social Elderly Care Service System During the 12th Five-Year Plan*, http://ww.mca.gov.cn/article/zwgk/mzyw/201,102/20110200133797.shtml.

[11]Xinmin.Cn, January 22, 2010.

17.17%, among which the population of elderly citizens aged 80 and above will reach 30.67 million, accounting for 12.37% of the aging population.[12] China completed the aging process of developed countries which lasted for nearly 100 years within only 18 years, but China's per-capita GDP just exceeded 4000 US dollars. In the case of a general insufficiency of pension reserves for elderly citizens, aging or old aging will surely lead to a severe lack of family material support, daily care and spiritual solace. The young and middle-aged groups and elderly citizens are greatly different in ways of life and values, so the number of empty-nest families is constantly increasing. Data display that empty nesters in China account for about 50% of the total aging population and those in Shanghai, Beijing and other metropolis reach above 70%. Over the next few years, empty-nest families will be the main form of aged families. However, the socialization service level of home-based elderly care in China is relatively low, so there is a large gap between services such as dining, nursing and medical care and the demands of the elderly citizens, and it is hard to improve the quality of life for elderly citizens. Besides, the current number of left-behind elderly citizens in China is about 40 million, accounting for 37% of the rural aging population, so it is necessary to solve their food, clothing, shelter, medical care and interment on the premise that the pension insurance system is improved. In 2009, the number of China's elderly citizens aged over 60 reached 167 million. The elderly citizens aged over 80 and part of the elderly citizens aged between 70 and 80 are the main groups that demand elderly care services provided by the society and their number is more than 40 million.[13] The report of *Home-Based Elderly Care Services in Cities of China*[14] shows that 48.5% of elderly citizens in cities of China demand all kinds of real elderly care services. Data (see Tables 3.1 and 3.3) on elderly care special investigations in Sichuan organized by Ge Liying and others show that elderly citizens' demands for services are mainly daily care and medical care. With the increase of age, their demands for services will increase. For the aging and old aging data in China, see Tables 3.1 and 3.2 (Chart 3.1).

2. Elderly care ability of the families weakens

Great changes have taken place in the family structure in China as a result of family planning and other policies. Multigenerational families have been replaced with nuclear families of three. DINK families, single-person families, empty-nest families and various patterns appeared. According to the data of the 3rd National Population Census in 1982, China's household size was 4.43 persons; it was 3.96 persons based on the 4th National Population Census in 1990; it dropped to 3.44

[12]Office of the National Working Commission on Aging: *Research Report on the Development Trend Prediction of China's Aging Population*, *China Civil Affairs*, Issue 4 of 2006.

[13]See Footnote 5.

[14]Web portal of Office of the National Working Commission on Aging (http://www.cncaprc.gov.cn/).

Table 3.1 Changes in China's aging population from 2005 to 2009

Year	Number of elderly citizens aged 60 and above (10,000 persons)	Proportion of total population (%)
2005	14,509	11.1
2006	14,901	11.3
2007	15,340	11.6
2008	15,989	12
2009	16,714	12.5

Source The National Working Commission on Aging: *China Report of the Development on the Aging Cause (2009)*

Table 3.2 Changes in the dependency ratio of China's aging population from 2005 to 2009 (Unit: %)

Year	Old-age dependency ratio
2005	10.69
2006	10.93
2007	11.17
2008	11.33
2009	11.60

Source The National Working Commission on Aging: *China Report of the Development on the Aging Cause (2009)*

Table 3.3 The relationship between age of elderly citizens and basic service demands (Unit: %)

Age group	Daily care	Medical nursing	Chart and diversion
60–70	69	80	23
70–80	76	85	27
Over 80	84	100	67

Source Ge et al. [11]

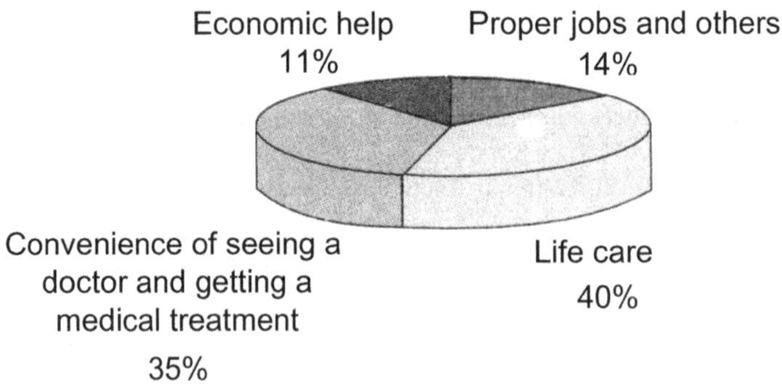

Chart 3.1 Problems elderly citizens most want to solve

persons based on the 5th National Population Census in 2000.[15] There is a close correlation between the family structure model and family security functions. The miniaturization of the household size, the diversification of family structure models, the aging of the family population structure and the disregard for the concept of marriage and family in today's China are shaking the Chinese traditional family-oriented security system in different degrees and urgently demand that we should establish a security system which takes the society as its main body and is suitable for the family structure model and its development trend. Director Sun Lujun of the Elderly Public Policy Research Center of the China Research Center on Aging said: "A large young and middle-aged rural population worked in cities and towns; young people no longer rely on the production skills and means of production passed down by the parents and the nuclear family is becoming more and more common. As a result, the elderly care function of rural families has been severely weakened." According to the survey data issued by the Gansu Province Aging Committee, among 2.19 million rural elderly citizens of the province, 23.8% were extremely worried that no one would take care of them, 23.65% complained that they had no one to talk with and 62.99% felt they had no economic guarantee (website of the Gansu Province Aging Committee). According to the survey data released by the Civil Affairs Department of Zhejiang Province in December 2009, households consisting of elderly citizens living alone and households consisting of only an elderly couple in cities and towns of Zhejiang Province account for as much as 67.6% and rural households consisting of only elderly citizens account for 50.34%; only 26.01% of the elderly citizens in cities and towns live with their children with the actual empty-nest rate reaching as much as 73.99, and 44.51% of the rural elderly citizens live with their children with the actual empty-nest rate reaching 55.49%; elderly citizens who lost self-care ability in everyday life and need to be taken care of account for 6% or so.[16] Data statistics show that China's average household size is reducing gradually and has dropped to 3.52 in 1996 from 5.9 before the founding of the People's Republic of China. The family structure of families with grandparents and a lot of grandchildren is now turned into inverted pyramid structures such as four-two-one families and four-two-two families. The only children in the families have to undertake the obligation to support at least four elderly citizens and the pressure is rather great. Meanwhile, as the society and economy are developing, job categories of people are changing all the time and the pace of work is accelerating, children are unable to guarantee the time and energy to take care of the elderly citizens, traditional intergenerational support and care functions of the families are weakened, the establishment of the elderly care and security system became an urgent demand of the broad masses of elderly citizens and their children and traditional family security functions are gradually weakening (see Table 3.4).

---

[15]The National Bureau of Statistics [2].

[16]Pan et al. [3].

Table 3.4 The age distribution of elderly citizens aged 60 and above in china and whether they can take care of themselves (Unit: person)

Age	People aged 60 and above	Can take care of themselves	Cannot take care of themselves	The proportion of people who cannot take care of themselves (%)
60–64	47,599	46,171	1428	3
65–69	40,062	38,132	1930	4.82
70–74	32,538	29,656	2881	8.85
75–79	19,159	16,436	2723	14.21
80–84	10,469	7809	2659	25.4
85–89	3727	2377	1350	36.22
90–94	1115	569	546	48.97
Over 95	234	105	128	54.7

Note The data in this table is the sample data of a sample survey of population changes in 2004. The sampling proportion is 0.966‰
Material source: *China Population Statistics Yearbook* (2004), China Statistics Press

Recently, the Social Survey Center of *China Youth News* launched an investigation into 1612 persons (among them the only children account for 40.1%) through http://www.minyi.net.cn/ and http://www.baike.com/ and the results show that 74.1% of the people said they are under high stress of work and life and powerless to take care of their parents; 68.4% of the people expressed that they have to bear the burden of providing for several elderly citizens; 50.1% of the people said they live in another place and are unable to take their parents to their place and take care of them; 42% of the people said pension insurance and medical insurance can't be used universally in different cities; 37.7% of the people expressed worries about old people's homes and social nursing institutions for the elderly citizens.[17]

3. Elderly citizens' demands for services increase

According to the data from the National Bureau of Statistics of China, in 1999, more than 10% of the population in China was aged 60 and above, and China became an aging society. In addition, it took a short time to transform the population age type from the adult type to the old age type. According to statistics, it took France 115 years to transform itself into an aging society, Switzerland 85 years, the US 60 years, Japan 25 years, while China only 18 years. Therefore, the aging speed in China is faster than general countries. Experts estimate that by 2050, China will have more than 400 million people aged above 60, accounting for about 26.53% of the total population.[18] With the sustainable development of social economy and the improvement of people's living standard in China, the average life span is increasing continuously, so nursing care for the elderly citizens has become

[17]STNN.CC, http://www.stnnxc, August 17, 2010.
[18]Qiu [4].

Table 3.5 Demand items of elderly citizens

Items	Demands of elderly citizens
Capital	Economic support that can ensure the basic living of elderly citizens
	Allowance that can improve the living standards of elderly citizens
Life care	Take care of elderly citizens in their house, such as preparing food and beverage, doing laundry, and cleaning rooms
	Get comprehensive life care in the institutions for elderly citizens
Medical services	Regularly come to check the cause of elderly citizen's disease, observe the disease and provide medical advice in a timely manner
	Observe and take care of the sick elderly citizens, and change fresh dressing for a wound, put on a drip for elderly citizens and inject them
	Provide door-to-door guidance and consultation about medication, nutrition and rehabilitation training for elderly citizens
	Regularly carry health examination for elderly citizens, provide health consultation and provide guidance about rehabilitation treatment for disabled elderly citizens
	Nursing services for disabled elderly citizens
	First aid services for elderly citizens
Culture and entertainment	Elderly citizens' demands for culture learning
	Cultivate elderly citizens' interest and carry out group entertainment activities
Mental care	Spiritual solace for elderly citizens living alone and empty nesters
	Psychological consultation for elderly citizens
	Deal with elderly citizens' family problems

a common demand. Most of the only children are unable to take care of bedridden or elderly parents in person. The traditional family care functions are weakening, while the demand for the model of social elderly care services is increasing[19] (see Table 3.5).

(1) Cope with the increasing demand for the nursing caused by the prolonged life span. Since the implementation of the family planning policy in China, "four-two-one families", "four-two-two families" and DINK families have weakened the function of family care. The contradiction between elderly citizens' lack of self-care ability in average life and children's failure to provide care for elderly citizens becomes prominent. Elderly nursing insurance can satisfy the demand for elderly services through the insurance company's risk sharing mechanism and the law of large numbers. The results of an insurance market survey done by the Shanghai Economists Association show that 45.33% of the respondents hope care insurance products would be developed.
(2) Due to changes in the family structure of China, such as the increasing empty-nest families, there is a growing demand for the elderly care. At present, they have become a ubiquitous family type in China. The data from Department

[19] Liu [5].

Table 3.6 Proportion of "Empty-nest Families" in China (Unit: %)

	National average proportion	60–69	70–79	Over 80
Elderly citizens living alone	7.1	4.8	9.7	12.4
Urban elderly citizens living alone	6.6	4.5	8.8	14.2
Elderly couple living alone	28	31.1	26.7	11.6
Urban elderly couple living alone	30.7	32.7	30.9	14.5
Empty-nest family	35.1	35.9	36.4	24
Urban empty-nest family	37.3	37.2	39.7	28.7

Source Compiled by China Research Center on Aging: *Data Analysis of Disposable Sampling Survey of the Urban and Rural Elderly Citizens in China*, Standards Press of China, 2003

of Population and Employment of the National Bureau of Statistics show that according to results of the fourth national census in 1990, the number of one-person households and a-couple ones accounted for 25% of that of the households with the elderly citizens. In 2002, this proportion reached 33.43%. In 2003, 14.03% of the households only had a couple and 12.35% of the households had single elderly citizen among the elderly citizens over 65 in China (see Table 3.6).

4. Aggravated burden of the elderly medical expenses

With the social and economic development and increasingly improved living conditions, people's life span has extended. The elderly citizens' spectrum of diseases has changed dramatically. The expenses for their medical care and diseases are also increasing and become a heavy burden of the family with elderly citizens. The living status survey report of the elderly citizens in Shanghai shows that in 2002, the annual average self-borne medical expenses of the elderly citizens in Shanghai were 1770 yuan, amounting to the pension income of about two months, of which 36.5% of elderly citizens had self-borne medical expenses of less than 500 yuan, 25.6% 500–1000 yuan, 19.4% 1000–2000 yuan and 18.5% more than 2000 yuan. The studies of Institute of Economics of the Ministry of Health in China show that from 1978 to 2002, the total health expenses in China averagely increased by 12.04% every year. After the current medical reforms in China, items, such as constant attendance expenses, nursing expenses and washing and attendance expenses aren't included in the basic medical insurance. The needed expenses for a family with an elderly citizen needing care will increase by 1000–2000 yuan. The long-term care expenses for elderly citizens who are of an advanced age and disabled are relatively heavy burdens for common families. At present, the measure on the basis of both social and commercial insurances in China can't avoid the very heavy burden of medical expenses of the elderly citizens.

3.2 Development and Practice of Elderly Care Service System in China

3.2.1 Construction Process of Elderly Care Service Security System in China

Since the founding of the People's Republic of China, the government has been carrying out the reform of pension insurance system and paying attention to the elderly group. In the new century, China has entered the aging society and the construction pace of elderly care service system is significantly accelerated. The construction process of elderly care service system has roughly experienced three periods, namely incubation period, exploration and development period and systematic construction period.[20]

1. Incubation period: 1951–1978

It was stipulated in *Labor Insurance Regulations* of 1951 that "birth, senility, illness, death, injury, disability and other difficulties which make the working masses the most painful are solved preliminarily". In this period, the Party and the government spared no effort to solve the national basic living security problem. The families were mainly responsible for the care of elderly citizens and only urban elderly citizens of no family and rural households enjoying the five guarantees could enjoy the elderly care services with the basic survival as the goal the elderly care institutions provided. In cities and towns, the basic livelihood of elderly citizens was mainly supported by organs, enterprises and institutions; in the countryside, there was land to the tillers through the land reform. Widowers, widows, orphans and childless couples could get more land appropriately. After the establishment of cooperatives and implementation of people's commune, the life of elderly citizens was guaranteed by the collective distribution of labor results and other measures. The original intention of the government to establish welfare institutions is to accept and find places for destitute and homeless people in cities and towns and reform them in economy, education and labor. At the end of 1953, there were 920 such institutions and about 100,000 elderly citizens of no family were accepted successively. Through the socialist transformation, production correctional institutions excluded those people with ability to work and changed the name into elderly care home, thus transforming the working content from transformation, education and relief to relief and education. In 1961, the Department of Interior specially rectified the welfare institutions, required them not to maltreat the adopted objects, reiterated the nature of social welfare and began to make clear the transformation to the direction of welfare services. In 1964, there were 733 welfare institutions and nearly 79,000 elderly citizens "without ability to work, source of income, supporter or caregiver" in cities and towns were adopted. And in the

[20]Dong [6].

countryside, the Department of Interior in China promoted the way of voluntary association and resettlement of elderly citizens of no family, the disabled and the children in Tanghe County, Henan Province in 1951 and initiated the model of elderly care home. The resettled people brought housing, land and property to the accepters and the accepters should be fully responsible for their birth, care, death and burial and might inherit their property. In December 1958, it was pointed out in *Resolution on Several Issues Concerning the People's Commune* which was passed at the 6th Plenary Session of the 8th Central Committee of the CPC that elderly care homes should run well to provide a better living place for those elderly citizens without child and support (households enjoying the five guarantees). At the end of 1958, there were a total of over 150,000 elderly care homes all over China and over 3 million people from the households enjoying the five guarantees were adopted. During the "Cultural Revolution", the development of elderly care institutions in cities and towns suffered serious setbacks.

2. Exploration and development period: 1978–2000

In this period, China began to carry out the socialist market economic system reform. Through the exploratory establishment of the social security system independent of enterprises, institutions and the collective, the basic life problem of citizens in old age was solved. With the welfare institution reform as a breakthrough, the socialization of social welfare was promoted. At the same time, with the rapid development of elderly care institutions and the rise of community services for elderly citizens, the elderly care services as the third industry began to be taken seriously and developed to a certain extent. After the "Cultural Revolution", China rebuilt the Ministry of Civil Affairs and began the system exploration on welfare nature and service direction. The Rectification Experience Exchange Meeting of National Urban Social Welfare Institutions which was held in November 1984 first put forward clearly that social welfare construction should be arranged from the society alone to the country, the collective and the individual, and the service objects in the institutions should extend to the elderly citizens in the whole society. In 1988, there were 17,000 elderly citizens in cities and towns and 9 million ones in the countryside. In 1998, the Ministry of Civil Affairs selected 13 cities as the socialization pilots of social welfare and promoted them as the third industry. During the same period, General Office of the State Council forwarded *Opinions on Accelerating the Socialization of Social Welfare* and further promoted the standard management of this work. Especially after 1990, the Ministry of Civil Affairs and other ministries and commissions successively issued *Interim Measures for the Management of Social Welfare Institutions*, *Basic Norms of Social Welfare Institutions for the Elderly Citizens*, *Code for Architectural Design of the Elderly Citizens*, *Interim Measures for the Management of Rural Elderly Care Homes* and a series of normative measures to strengthen the normative construction of elderly care institutions and integrate management with services. The service item developed from the single life security to the integration of residence, medical care, rehabilitation and entertainment and the elderly care service quality began to be taken seriously and improved.

3. Systematic construction period: from 2000 to now

The year of 2000 is a key year for the development of elderly care services. In this year, China generally entered the aging society. The country, the National Working Commission on Aging, the Ministry of Civil Affairs and other organs summed up and promoted the experience of home-based elderly care services. The home-based elderly care service better fixed the decreasing family care ability and other problems caused by small families and population mobility. The elderly care institutions were further improved in both quality and quantity. Therefore, in this period, the Party and the government attached great importance to the impact of aging on the economic and social development and started to construct the elderly care service system meeting the national conditions of China. The elderly care services began to develop in the systematic direction. Governments at all levels continued to promote socialization and accelerated the development pace of the elderly care institutions by increasing the financial investment and mobilizing the input of social forces, showing the diverse investment subjects, public service objects, market-oriented operation mechanism and the coexistence of various kinds of ownership, meeting the development pattern of socialist market economic system, better guaranteeing the support needs of the special objects, such as the disabled, the aged and the disabled elderly citizens of no family and satisfying the social masses' need for the welfare career to some extent. After we entered the new century, the number of beds in the elderly care institutions has increased at a rate of over 10% every year. As for the social elderly citizens outside the institutions, care and nursing are provided for them through the development of community services, thus developing towards home-based care services. From 2000, Shanghai, Ningbo and other places have begun to carry out the home-based elderly care service work successively, winning the recognition of various social groups. Through the accumulated experience, the Second National Aging Work Conference stressed that China should develop the construction of the elderly care service system with Chinese characteristics with the home-based elderly care as the basis, the community service as the support and the institution-based elderly care as the supplement in February 2006. In February 2008, Office of the National Working Commission on Aging issued *Opinions on Full Promotion of the Home-based Elderly Care Service Work* and made comprehensive deployment. In this period, the most important thing is that the development requirements of the elderly care service system were put forward for the first time. In December 2008, the Former Minister of the Ministry of Civil Affairs Li Xueju put forward the concept of establishing and perfecting the elderly care service system with home as the basis, community as the support and institution as the supplement at the national civil affairs working conference. And the elderly care service system construction has entered the development period of substantive advance.

In recent years, in the face of the present situation of rapidly entering the aging society, China has promulgated and implemented a series of policies to promote the construction of the elderly care service system. In March 2000, approved by the State Council, General Office of the State Council forwarded *Opinions on Accelerating the Socialization of Social Welfare* issued by the Ministry of Civil Affairs and other 10 departments, of which it was stipulated in Article 7 that when formulating the local health service development planning, all places should fully consider medical care, prevention, health care, rehabilitation and other health needs of the elderly citizens, the disabled, orphans and the disabled children, actively support the community health service institutions to carry out the elderly medical care, prevention, care, rehabilitation, health education and other work, encourage and support social forces to set up non-profit medical institutions with elderly citizens, the disabled and orphans as the service objects, put forward the guideline of promoting the social welfare socialization with elderly care and aid to the disabled as the key points and formulate many preferential policies in construction land, tax, public utilities fees, fee subsidies and other angles. In November 2005, the Ministry of Civil Affairs issued *Opinions on the Social Forces' Support of Establishing Social Welfare Institutions* and encouraged and supported enterprises, institutions, social organizations, individuals and other social forces to invest in the establishment of elderly care institutions. In February 2006, General Office of the State Council forwarded *Opinions on Accelerating the Development of Elderly Care Service* issued by Office of the National Working Commission on Aging, the Ministry of Civil Affairs and other departments, ordered to gradually establish and improve the service system with home-based elderly care as the basis, community service as the support and institution-based elderly care as the supplement in the principle of policy guidance, government support, social establishment and market promotion and put forward six key points on accelerating the elderly care service. In January 2008, Office of the National Working Commission on Aging and the Ministry of Civil Affairs introduced *Opinions on Full Promotion of the Home-based Elderly Care Service Work* and proposed to generally establish the home-based elderly care service network with various forms and extensive coverage in the nationwide urban communities. In the management of elderly care institutions, the Ministry of Civil Affairs formulated and promulgated *Interim Measures for the Management of Social Welfare Institutions* and made clear provisions in planning, establishment, daily operation and service of the social welfare institutions. On the basis of the reality, all places of China highlight the government's macro guidance and policy support and formulate and perfect many specific and very operable policies, regulations and standards, enabling China's social elderly care service system to embark on the rule-based track preliminarily. In November 2010, the Ministry of Civil Affairs held the National Social Elderly Care Service Promotion Conference in Wuxi, Jiangsu, implemented the spirit of the 5th Plenary Session of the 17th Central Committee of the CPC on "paying attention to playing the function of families and communities, giving priority to the development of social elderly care service and cultivating and growing aging service career and industry", put forward the requirement of "giving priority to the development of social elderly care service" and summed up and promoted the exploratory experience in elderly care service of Beijing, Shanghai, Chengdu and other cities. Governments of various places were required to list the development of elderly care service into the key contents of the "12th Five-Year Plan" and one of "the people's livelihood projects" of doing things concrete, so the construction of the elderly care service system achieves relatively rapid development.

After a comprehensive survey of the construction process of China's elderly care service system, we find that it's a process of evolving from the vacancy filling type to the common people benefiting type gradually. The system covers the whole group of elderly citizens in cities and towns from the rural "households enjoying the five guarantees" and urban elderly citizens "without ability to work, source of income, supporter or caregiver" and other special groups in the very beginning. With the increasing aging population, the elderly care service responsibility is transferred to the society from families. On the basis of the government's leading responsibility, more social forces participate in the service for the elderly citizens. The socialized elderly care service has become an irreversible trend. However, China's elderly care service system is still in the exploratory stage with inadequate professional talents serving the elderly citizens, slow infrastructure construction, imperfect management mechanism, intense contradiction between supply and demand, great urban and rural difference and other problems. During the development, these problems should be solved gradually, thus truly realizing the goal of "providing elderly citizens with a sense of security" put forward at the 17th National Congress of the CPC.

3.2.2 *Practice and Experience of the Elderly Care Service Security in Some Places of China*

In recent years, various places of China have actively promoted the socialized elderly care services and implemented the elderly care service system meeting the local actual conditions under the strong impetus of the Party and the government.

1. Beijing: construct the elderly care service system with characteristics of the capital[21]

As one of the Chinese cities with a higher degree of population aging, Beijing City has put forward the "9064" elderly care service development strategy in recent years. That is, by 2020, 90% of the elderly citizens can enjoy the home-based elderly care, 6% of them community-based elderly care and 4% institution-based elderly care. Adhering to the social reform as the driving force and the equalization of elderly care services the goal, Beijing City constantly improves the service development strategies for elderly citizens and makes great efforts to construct the

[21]Beijing Municipal Civil Affairs Information Website: http://www.bjmzj.gov.cn/.

elderly care service system with characteristics of the capital. The main measures are as follows:

(1) The government implements the elderly care subsidies and expands the coverage of elderly care services. To better meet the elderly care service requirements of the elderly citizens, Beijing City has formulated and introduced a number of favorable policies for the elderly citizens and constantly implemented the elderly care services bought by the government, greatly enhancing the benefit groups and security level of the elderly care services. In early 2007, Beijing City introduced *Measures for Strengthening the Preferential Treatment Work of the Elderly Citizens* and gave 11 preferential policies to the elderly citizens above 60 all over the city, including the policy for the elderly citizens above 65 who can take a bus or visit a park for free; the policy for those above 60 who can visit museums and cultural centers for free; the policy about the clear stipulation that community service centers, public stadiums, social law service agencies and large- and medium-sized medical institutions are required clearly to provide priority or preferential services for elderly citizens; the policy about practical settlement of the elderly care security problem of elderly citizens. From 2008, the elderly citizens above 60 and without social security in the whole city can get 200 yuan of welfare pension, solving the problem of providing elderly citizens with a sense of security of 700,000 elderly citizens without social security and marking the full coverage of the urban and rural elderly care security system in system. At the same time, the peasants participating in the rural elderly care insurance and meeting the conditions of the system can get additional basic pension of 280 yuan every month. In 2008, Beijing City issued *Notice on Deeply Developing the Pilot Work of Home-based Elderly Care Services* and developed the pilot work of the home-based elderly care services in 10 districts. To promote the home-based elderly care services, Beijing City introduced *Measures for the Elderly Care Service Subsidies of the Special Elderly Citizens in Beijing City*. In the pilot district, the elderly citizens with registered permanent residence in Beijing City, above 60 and meeting the conditions, such as advanced age, low income and pure elderly household can enjoy 50–250 yuan of the elderly care service subsidies. This policy can benefit 180,000 elderly citizens.
(2) Realize diversified investments and the development of the elderly care services. Under the guidance of the government and from construction capital, operating subsidies, land use and other aspects, actively encourage and guide social forces to participate in the establishment of the elderly care services and form the development pattern of diversified investment subjects. ① Strengthen the construction capital support in the elderly care service institutions funded by the society. The elderly care service institutions funded by the society with the approval of the Civil Affairs Department and meeting the funding conditions can enjoy one-time construction capital of 8000–16,000 yuan per bed. ② Improve the operation and subsidy policy of the elderly care service institutions funded by the society. Raise the capital subsidy standard, subsidize each old

citizen in accordance with the standard of 100–200 yuan per month and establish a scientific growth mechanism. In the whole city, 364 elderly care service institutions funded by the society were subsidized and the amount of subsidies reached 26.35 million yuan. ③ Improve the capital supporting policies of the governmental construction project. The municipal and district governments can bear it at the ratio of 1:1. As for those districts and counties with financial difficulties, the municipal government should give moderate inclination; construction and equipment purchase of the elderly care service institutions in streets, townships and towns and the municipal welfare lottery public welfare fund should be subsidized according to a certain ratio; the supporting elderly care service facilities and land acquisition and resettlement fees in the residential areas should be borne by the development company, the basic construction fees district and county governments and the municipal welfare lottery public welfare fund should be subsidized at a ratio of 30%. ④ Establish the effective land supply policy. In accordance with the land area of less than 40 m^2 per bed, the land for development of the elderly care service institutions should be reserved. Land of the fundamental and guaranteed elderly care service institutions funded by the government and general and comfortable ones with exemplary role should be allotted and that of other types of elderly care service institutions should be supplied according to the agreement.

(3) Improve the elderly care service facilities. In recent years, Beijing has continued to increase capital investment, pooled the coordinated development of the elderly care service facilities and strived to build the elderly care service system with perfect facilities, good operation and first-class service. ① Accelerate the development of urban and rural elderly care service institutions. To solve the problems, such as shabby welfare facilities, single function and low service level of the rural elderly citizens, Beijing City invested nearly 345 million yuan to transform 171 elderly care homes in townships and towns into rural social welfare centers with comprehensive service functions and newly increased 13,776 beds from 2001. From 2005 to 2007, the elderly care home project in streets was implemented. In 2008, the construction and subsidy long-term mechanism of the elderly care service institutions in streets, townships and towns was set up. ② Accelerate the pace of the construction of community-based elderly care service facilities. Under the unified deployment of the Ministry of Civil Affairs, the urban "Starlight Program" was carried out thoroughly and 2335 urban "Starlight Homes for Elderly Citizens" were built and expanded to the rural areas. On the basis of the construction of 1255 rural "Starlight Homes for Elderly Citizens", 2700 more ones will be subsidized in two years to realize full coverage of the urban and rural community elderly welfare service facilities. ③ Establish a development plan of newly increasing by at least 10,000 elderly care beds every year. To realize the goal of 4 elderly care beds per 100 elderly citizens by 2020, at least 10,000 elderly care beds should be newly increased every year. Focus on developing care and maintenance-type and ordinary-grade elderly care service institutions and meeting the elderly care service needs of the social masses. ④ Continuously

improve the management level of elderly care services. Speed up the construction of infrastructures and strengthen the standard management of elderly care services. Beijing City successively promulgated eight local standards, such as *Service Quality Standards for Elderly Care Services Institutions*, initially set up the elderly care service standard system, formulated a unified contractual model in view of the main issues related to elderly care services and promoted the healthy development of the elderly care service institutions.

2. Chengdu: promote the development of urban and rural integration and realize equalization of elderly welfare services[22]

Since 1992, Chengdu City has entered the aging society. It is a city with a relatively serious aged tendency of population. In the whole city, there are 1,816,600 elderly citizens over 60, accounting for 16.3% of the total population now. Now there are 218 various elderly care welfare institutions, of which there are 191 urban and rural state-run welfare ones and 27 private ones with a total of 28,000 elderly care beds. Now every 100 elderly citizens have about 1.6 elderly care beds. In recent years, Chengdu Municipal Party Committee and Municipal Government have attached great importance to the elderly care services, regarded it as an important measure to promote the urban and rural integration and build the harmonious Chengdu, made innovative ideas and scientific plans, integrated various resources, mobilized all the forces and created an elderly care service model with Chengdu characteristics, "namely governmental main role, social participation, rich forms and aid to the elderly in various aspects".

(1) The government is the main body in charge, and elderly citizens in urban and rural areas shall enjoy equal welfare services. In recent years, in order to cope with the challenge of aging of population, the Chengdu Municipal Government has listed the construction of urban and rural state-funded public welfare organizations in ten major "people's livelihood projects". According to the ideas of urban and rural integration, unified planning, unified standards and unified construction, city-level and county-level governments have invested a total of 700 million yuan in upgrading 4 city-level public welfare organizations and 93 rural elderly care homes, focusing on the transformation of rural elderly care homes in the city on the largest scale since the establishment of the People's Republic of China. The overall relocation of city welfare housing and city mental hospitals involving a total investment of more than 200 million yuan has been completed and put into use. The two hospitals cover an area of 100 mu and become the largest city-level state-funded welfare institutions in Western China. At present, the concentrated occupancy rate of urban and rural elderly citizens without the ability to work, a source of income, a supporter or caregiver in cities reached more than 80%, and the per-capita support standard has reached or exceeded the average life level of local elderly citizens. Meanwhile, constantly improve the community (village) home-based care

[22]Chengdu Municipal Civil Affairs Information Website: http://www.cdmzj.gov.cn/.

service infrastructure. Governments at all levels have invested more than 50 million yuan in completing and perfecting welfare service facilities for elderly citizens in the community (village). Governments buy services, so that more than 30,000 urban and rural empty nesters can be taken care of and loved, meeting most elderly citizens' demand for elderly care services.

(2) Encourage social participation, and actively explore new ways of elderly care services. With the continuous development of social economy, more and more private capital is expected to be put into "the elderly industry". While increasing financial investment, Chengdu City firmly grasps the chance of being listed as the National Pilot Zone for Overall Reform in urban and rural areas by the State Council, makes full use of various preferential policies to actively explore diversified investments in elderly care institutions. Private elderly care service institutions are first included in the overall planning of urban-rural integration development and construction in Sichuan Province. Allocate the amount of special land for construction. The private entrepreneurs jointly invest 20 million yuan in building the Xialingju Nursing Center with more than 1000 beds and an area of 150 mu next to the world famous Chengdu Panda Breeding Base. Various living facilities are high-grade and complete, and the greening rate reaches 70%. Elderly citizens live a comfortable life in the park-like yard every day. The Center is operated by a company and peasants. Every peasant household living around the Center can earn 3000 yuan each year additionally by working as nurses and becoming a shareholder. At present, the city has attracted private capital to build 27 high-grade, medium-grade and low-end elderly care service institutions with a total of more than 5000 beds. Therefore, the government has not only saved a lot of investments in hardware and costs of service personnel, but also attracted a large number of elderly citizens to live in it. It also cooperates with institutions of higher learning to establish "the Practice Base for Undergraduate Volunteers to Help Elderly Citizens" in China. The elderly care service institutions in urban and rural areas are open to undergraduates. The colleges and universities offer extracurricular teaching voluntarily, and undergraduates provide elderly citizens with "menu style" household services.

(3) Integrate effective resources to constantly improve community home-based elderly care services. Affected by the traditional concept and economic conditions, the majority of elderly citizens can only spend their later years in communities and homes. Chengdu City constantly improves community home-based elderly care services by integrating all kinds of resource advantages. ① Through organic integration with the integrated urban and rural assistance platform of Chengdu, Chengdu City invests more than 10 million yuan in building up an information website aiming at helping the elderly citizens in urban and rural areas of the city, inputs the basic information of elderly citizens without the ability to work, a source of income or a legal supporter and caregiver and elderly citizens living alone in the system, laying a solid foundation for the smooth implementation of providing the urban and rural elderly citizens with life relief, medical assistance, legal aid, home-based elderly care

services and other services and achieving the resources sharing of elderly care service information. ② Through the organic integration of the resources from telecommunications, health, public security and other departments, Chengdu City establishes PTT call system for elderly citizens to call for help in the community, so that elderly citizens can receive the fastest aid timely in an emergency situation. ③ Through the organic integration of the resources in resident units, Chengdu City makes use of idle land and factory buildings to build several hundreds of small, dispersed, convenient and applicable welfare service facilities for elderly citizens and provides elderly citizens with various services such as daily care, culture and entertainment, rehabilitation therapies, sports and fitness according to local conditions. ④ Provide professional services for elderly citizens through the organic integration of resources in professional organizations in the society. The district government of Wuhou District purchases services from professional domestic companies with more than 2 million yuan to provide the childless elderly citizens and poor empty nesters who can't take care of themselves every year. Strive to build "elderly care homes without enclosing walls". Chenghua District established the first Caixia HelpAge Service Agency. The government purchases services and private enterprises participate in it with the titles. It hires the laid-off, the unemployed (lose land) and minimal people enjoying the minimum living security to provide hourly paid services for elderly citizens without the ability to work, a source of income or a legal supporter and caregiver, poor and old empty nesters, implementing "one position and two kinds of care".

3. Guangzhou: establish a modest common people benefiting type of the elderly care service system[23]

As of the end of 2008, there were 1.0315 million elderly citizens aged 60 and above in Guangzhou City, accounting for 13.4% of the total population of the city. In recent years, in the face of the severe challenges of aging of population, the Municipal Government of Guangzhou has attached great importance to the development of welfare business for elderly citizens, continuously supported it with policies, accelerated the reform and innovation and actively guided the society to participate in it. The civil affairs departments at all levels in the city resolutely implement the decision and deployment of the Ministry of Civil Affairs and the Provincial Department of Civil Affairs and focus on building the elderly care service system that is based at home, relies on the community and is led by institutions, forming a modest common people benefiting type of the new elderly care service pattern with the policy funding in place, quality services and security, various development models, wide social participation and elderly citizens in the whole city enjoying it.

(1) The government vigorously promotes the construction of a modest common people benefiting type elderly care service system. The Municipal Government

[23]Website of the Bureau of Civil Affairs of Guangzhou Municipality: http://www.gzmz.gov.cn/.

of Guangzhou issued *On the Decision to Solve Several Problems Related to the Vital Interests of People* in 2007, promoting the conversion of social welfare business from relief type and vacancy filling type to common people benefiting type as a project in the public interest, which has become an important supporting policy to the development of elderly care service business. Guangzhou City has issued *Measures to Give Special Treatment to Elderly Citizens in Guangzhou City, Administrative Measures for Private Social Welfare Business Institutions in Guangzhou, Trial Measures of Pension Insurance for Elderly Citizens in Towns in Guangzhou City, Trial Measures of Social Pension Insurance in Rural Areas in Guangzhou City* and other policies and measures in succession and formulated *Proposal to Strengthen the Management and Service for Elderly Citizens in the Community, Notice of Further Promoting Community Home-Based Elderly Care Services in Guangzhou City* and other normative documents, laying a policy foundation for the care service systems. The Municipal Government of Guangzhou includes elderly care welfare business in *The 12th Five-Year Plan for Guangzhou* and clearly puts forward the goal of forming a social service system with Guangzhou characteristics in 2015. The Bureau of Civil Affairs organizes specially-assigned persons to carry out investigations into the homes for the aged and the families having aged members, visits elderly citizens having difficulties in taking care of themselves and elderly citizens who are of an advanced age to solve the problems in the development of elderly care services and strengthen the policy promotion.

(2) Promote the development of social elderly care services. In recent years, Guangzhou City has encouraged social forces to participate in the elderly care business, and the social elderly care pattern has basically formed. ① Actively guide social forces to set up elderly care institutions, diversifying the providers of elderly care services. Through policy guidance, financial support and work instruction, the government encourages social forces to run elderly care welfare institutions in the form of sole proprietorship, joint venture, cooperation, stockholding system, individual contract etc., providing multi-level services like health services, semi-nursing, nursing and other services for elderly citizens. A new pattern of diversified service providers has formed. As of the end of 2008, there were a total of 161 welfare institutions of different kinds for elderly citizens with 22,817 beds in the city. 76 of them were private welfare institutions with 13,702 beds, accounting for more than 60% of the total elderly care beds in the city. ② Actively guide social capital to fund elderly care services and consolidate the social elderly care facilities. Guangzhou City actively encourages all sectors of the society to take the initiative to donate money and objects to fund the construction of institutions for elderly citizens and activities for elderly citizens, promoting the development of elderly care services. In recent years, Home for the Aged, Guangzhou alone has received social donation of over 34 million yuan to improve the lives of the elderly citizens supported by the government. ③ Actively deepen the transformation reform of welfare institutions, marketizing the elderly care services system. In recent years, through the continuous deepening reform and the use of

market-oriented operation models, the stock of public elderly care institutions has been utilized. In Yuexiu District of Guangzhou City, the management model of "public construction and private running" is adopted. The completed street day care center is put out to tender and the street signs a contract with the hosted intermediary service organization. The District Bureau of Civil Affairs and the street are responsible for the supervision and guidance of daily management of the day care center. The privatization management brings good economic and social benefits. More than 90% of the city's district-level and county-level welfare housing also implement market-oriented operations. ④ Actively cultivate non-profit social community organizations to diversify the forms of elderly care services. Guangzhou City actively explores social management models, sets up and develops different forms of social community organizations for elderly citizens. Among them, the "Starlight Association for the Elderly" was set up in Yuexiu District in 2004. Instructed by the Office of the District Working Commission on Aging, more than 40 retired cadres and CPC members enthusiastic about social work patrol each starlight home for elderly citizens in the communities to ensure that welfare facilities and public welfare activity fields fully built are utilized, function well and give full play to their elderly care service function. Charities at all levels cooperate with medical departments to carry out "charity medical care in the community" activity in the city, providing door-to-door physical examinations, medical treatment and drugs for elderly citizens who have to stay at home. The elderly volunteer team and youth volunteers in each community have become a new bright spot of the city elderly care services, further integrating the elderly care service resources in Guangzhou City. Continue innovating the model of elderly care to build a platform for urban and rural elderly care services.

Guangzhou City takes new measures to promote elderly care in rural areas. While comprehensively promoting the elderly care services in cities and towns, Guangzhou City actively promotes the construction of "five guarantees villages" in rural areas and the reform of elderly care homes according to the elderly care needs of elderly citizens in rural areas, especially those who live alone and have difficulties in taking care of themselves. Meanwhile, send doctors and elderly care service personnel, and provide supporting facilities during the construction. Determine that the support standard of elderly citizens enjoying the five guarantees should be more than 30% of the minimum living security standard. The support funds shall be provided by the town-level governments and above, greatly improving the support level of elderly citizens enjoying the five guarantees in rural areas. At present, there are 33 "five guarantees villages" and 57 elderly care homes in the city. More than 500 elderly citizens enjoying the five guarantees live in "five guarantees villages" for relatively collective support and 1671 live in elderly care homes for collective support.

(3) Establish a specialized elderly care service network. Guangzhou City focuses on the construction of elderly care service facilities, improving the quality of

elderly care service personnel, accelerating the construction of the elderly care service network to constantly improve the level of elderly care services. ① Constantly improve the level of elderly care services. Guangzhou City takes advantage of the fact that it is adjacent to Hong Kong and Macau and actively introduces advanced concepts and methods of social work in Hong Kong to explore and improve professional elderly care services. Eight streets in Liwan District are selected as pilot streets in which management ideas and service models of social work are introduced. Services like home-based elderly care, services for extremely poor families and the disabled and "Safety Call" for elderly citizens are organized and carried out. In 2005, Home for the Aged, Guangzhou set up the Department of Social Work, set up positions for social workers and hired elderly social workers from Hong Kong Sheng Kung Hui to provide regular and professional supervision. The social work philosophy is applied to elderly care services to achieve more professional and occupational services. The training for the home-based elderly care nursing assistants is irregularly given in the city and district, and professionals from Hong Kong and other places are invited to give lectures so as to constantly improve the overall quality of caregivers and the level of elderly care services. ② Improve the elderly care service network. The linked home-based care service work network has been established. Leading groups of home-based elderly care services in the community and home-based elderly care service departments (stations, points) have been set up in the city, district, street and neighborhood committee to strengthen the leadership, coordination, supervision and evaluation of home-based elderly care service work; a community-based elderly care service network that contains portal services, day care centers, rehabilitation centers and starlight homes for elderly citizens has formed; the nonprofit community health care system has been improved. Health care service networks for elderly citizens with the community health service network as the main body and all levels of hospitals as the base have been established in 130 streets in the city. The city, district, street and neighborhood committee four-level legal rights service network for elderly citizens has been improved in the city with a total of 184 legal aid institutions established in urban and rural areas, providing convenient and efficient legal advice and legal aid services for elderly citizens. ③ Expand the function of elderly care services. Enterprises provide capital to establish, operate and manage the "Safety Call" service system, while the government supervises and purchases services of the "Safety Call" service system, providing elderly citizens with timely and effective rescue and assistance. Combine the "Safety Call" platform with 120, 119 and 110 service systems to construct four joint service networks including the community medical emergency rescue service network, the community nursing service network, the community household service network and the compulsory service network for elderly citizens. In addition, "96,909" hotline for the voice of elderly citizens and "12,348" hotline for legal consultation are also opened to dispel doubts and provide all kinds of advisory services for elderly citizens.

Beijing, Chengdu and Guangzhou have their own systems suitable to their social and economic development level while exploring and establishing elderly service systems. They have made valuable exploration in strengthening government responsibilities, carrying out community-based and home-based elderly care, supporting institution-based elderly care, establishing the elderly service information network and many aspects related to elderly services, and the experience is worth spreading. In particular, every place considers the elderly service sector as a special project and pays much attention to it, so gratifying progress has been made. However, these cities are located in the regions enjoying better economy and community infrastructure. Other cities and the vast rural areas in China still need to continue to explore and will first rid elderly citizens of the worries of life, gradually attaching equal importance to economic security as well as service and spiritual consolation security.

After several years of continuous exploration and development, Beijing and other places have accumulated some good experiences in the construction of the elderly care service system. The main experiences are as follows: ① The construction of the elderly care security and service system should be as important as economic security. Vigorously develop the elderly care service industry, especially the establishment of the home-based elderly care system composed of family-based elderly care and community-based elderly care, which is the main content of constructing harmonious communities. ② Determine the government's leading regulatory responsibility in the elderly care service and security system, and focus on strengthening the construction of the system and mechanism. Increase financial support and actively support the diversified and multi-channel social elderly care service and security mechanism at the same time. ③ Speed up the improvement of the infrastructure and system operation network system of community-based elderly care. Institution-based elderly care should be combined with the market and society, and industry service standards, industry access and regulatory systems shall be established. ④ Organize professional and high-quality elderly care service teams and improve service abilities and levels. Train outstanding social workers and improve the access system of the professional qualification examination of elderly care services and professional training system.

3.3 Problems in China's Elderly Care Services and Security System

In recent years, China has made some progress in the elderly care service and security system, but we understand clearly that China is still in the primary stage of socialism. Its basic national condition is that China has a large population base of elderly citizens. The aging population is increasing fast, and so is the empty-nest rate. The economic development in China cannot meet the multi-level service demands of the elderly citizens. There is a lack of professional service teams for elderly services. Elderly care facilities in communities and institutions are not perfect, and the imbalance between supply and demand is serious. The degree of marketization and socialization of elderly care is low. The elderly care service supervision system is not perfect. In a word, there are many problems in the construction of the elderly service and security system.

3.3.1 There Is a Lack of Professional Service Teams for Elderly Services

In the US, professional caregivers are usually trained persons like registered nurses, student nurses with occupational qualifications, personal care assistants, assistant nurses, domestic workers, qualified occupational therapists and social workers. These professionals offer services according to the advice from customers' doctors and are supervised by home-based service agencies. Elderly services cover household services, health consultation, psychological consultation and other professional fields. At present, because of the lack of all kinds of professionals, the overall level and quality of elderly services in China are unsatisfactory. Service practitioners lack professional accomplishment and a stable mechanism to guarantee their jobs.

1. Elderly services practitioners lack professional training

Vice Minister Dou Yupei of the Ministry of Civil Affairs pointed out in the 5th National Forum of Presidents of Home of the Aged held in December 2009 that the ratio of aging population and professional nurses is 3:1. 28.3 million elderly citizens cannot take care of themselves, so 10 million professionals are needed to take care of elderly citizens. However, there are only 220,000 workers in elderly care institutions in China, among which only over 20,000 people have professional qualification certificates for community work. Moreover, the overall quality of elderly care service teams is low and their professional level, professional proficiency and service awareness cannot effectively meet the needs of elderly citizens. Not long ago, the author used to be a volunteer in a public elderly care home in Chaoyang District, Beijing at weekends. The elderly care home has a total of 18 full-time caregivers, among which only three have learnt clinical nursing and medicine, only eight of them graduated from high school or higher schools and most haven't received professional training about elderly care before beginning work.

2. There is a lack of a stable incentive mechanism like social insurance for caregivers

The caregivers in the elderly care home the author interviewed said their fellow workers often quit and rarely work for 2 years continuously. The problem is that they are badly paid. Migrant workers are offered accommodation, but they just earn about 2000 yuan each month. They don't have any pension insurance, medical insurance or other welfare, so they have no job security. This is the main reason why the staff quit frequently. At present, most elderly service practitioners are laid-off workers born in the 1940s to the 1950s or housewives coming from rural areas to urban areas. Elderly care institutions lack an effective mechanism to improve the professional quality of practitioners and don't include social insurance into the compensation incentive mechanism for caregivers.

3. There is a lack of the management system for home-based caregivers

At present, our home-based caregivers are associated with the re-employment of the laid-off workers in the community. It provides social workers to take care of elderly citizens in the community, and the laid-off workers in the administrative region can have a job again. In addition, the workers living in the administrative region are more familiar with elderly citizens and the environment, so they are willing to accept elderly citizens. However, the service quality and effectiveness are poorly and ineffectively supervised. Home-based elderly care services and community services for elderly citizens are generally governed by related staff in the neighborhood committee, and there is a lack of professional service evaluation and supervision, which affects the further development of home-based elderly care services in the community. Foreign countries advocate of mutual care among elderly citizens. Healthy elderly citizens take care of the disabled elderly citizens. The service time is savings. They can enjoy the elderly nursing service of the same time when necessary in the future. The number, incentive mechanism and professional ability of caregivers in China remain to be improved.

3.3.2 *Infrastructure of Elderly Care Services Cannot Meet the Diversified Needs of Elderly Services*

Sound infrastructure of elderly services is essential to the realization of social elderly care services and security. However, in China, the design and layout of indoor facilities for home-based elderly care and infrastructure in the community and institutions are backward, and the imbalance between supply and demand is serious.

1. There is a lack of design and planning of facilities conducive to elderly citizens' life at home

Home-based elderly care refers to the situation in which elderly citizens live in their houses can take care of themselves, or be taken care of by their spouses and children. It also refers to the fact that professional social workers in the community provide the elderly citizens with door-to-door household services, treatment for patients, regular health tests, psychological consolation and other services. Elderly citizens are unable to move freely, so specialized design and reform of the facilities in their houses shall be carried out to facilitate the life of elderly citizens so as to reduce the risk of stumble, bump and the like. Singapore and other countries have indoor facilities designed specifically for elderly citizens, but China lacks it.

2. Elderly care infrastructure in the community is not perfect

Community is the carrier and platform of home-based elderly care. Now the number of day care centers for elderly citizens or elderly care stations in the community increases greatly in each place in China, providing elderly citizens with catering, relaxation, collective entertainment, elderly learning and other services. Meanwhile, organize the caregivers from neighborhood committees in the community (in Qingdao, Tianjin and other places) to provide door-to-door service for elderly citizens. However, due to the rapid aging speed in China, the backward infrastructure in the community cannot meet the service needs of elderly citizens. The facilities in some home-based elderly care stations in the community are old and small. There is a lack of capital and specialized management. Due to the huge difference in economic conditions, the infrastructure in the elderly care service stations and the elderly care service stations in community in townships and towns in rural areas in China are of varying quality. However, on the whole, the elderly care service infrastructure in rural areas is worse.

3. There is a serious lack of beds in elderly care institutions and the imbalance between supply and demand is serious

Vice Minister Dou Yupei of the Ministry of Civil Affairs pointed out in the National Forum of Presidents of Home of the Aged held in December 2009 that at present, there is a serious lack of elderly care institutions in China, and they cannot meet the needs of elderly citizens. Internationally, every 1000 elderly citizens share 50 elderly care beds on average, so all elderly citizens in China need a total of 8 million beds. However, China has only more than 2.5 million beds with a gap of 5.5 million beds. In per capita beds in elderly care institutions are less than 11,[24] which is much lower than the average level of the developed countries. On the one hand, there is a shortage of beds in public elderly care institutions, so elderly citizens have to wait for a long time before checking in; on the other hand, because of the high cost of private elderly care institutions, most of elderly citizens can't afford to live in, resulting in spare beds. There is much serious imbalance between supply and demand.

[24]See Footnote 14.

3.3.3 There Is a Lack of Effective Capital Guarantee for Elderly Care Services

The elderly service team is the key, and the infrastructure is a prerequisite, while the guarantee for the service capital is the premise of the success of the system. China rapidly becomes an aging society when the level of economic development is backward, and the meager financial resources must be used to solve many problems and invested in many fields, so the elderly system which is based on home-based elderly care, relies on community-based elderly care and is led by institution-based elderly care needs capital to hire professional social worker teams which provide door-to-door services, to construct service stations for home-based elderly care and "Homes of Elderly Citizens" in the community, townships and towns and to improve services and operation of institution-based elderly care. Although the government continues to increase the investment in the social security sector and vigorously encourages social capital in elderly care services and the security industry, now very limited finance is invested in the social security sector in China. In 2008, less than 12% of total financial expenditure was invested in social security. Besides, there hasn't been a systematic determined proportion and a growth mechanism for the financial capital invested in social security yet, so it is changeable. As for social forces engaged in elderly care services, a series of policies encouraging social forces to be engaged in elderly care service institutions are introduced, but it is difficult to have specific implementation measures and system guarantee. When the social capital is invested in elderly care service facilities in elderly care institutions and communities, it is hard to actually implement preferential policies such as land and tax policies, making it rather difficult to operate private elderly care institutions. It is reported by *The Beijing News* on February 21, 2011 that five dining rooms for elderly citizens were opened in a community in Haidian District, Beijing, but four of them closed down six months later because of the rising prices of vegetables and the untimely subsequent subsidies from the government. The institutional capital investment is the premise of the development of elderly services.

3.3.4 The Elderly Care Service and Security System Fails to Meet Multi-layered Needs of Elderly Citizens

In recent years, various regions in China are continuously exploring the institutional models of elderly care services, initially forming the elderly care service system which is based on home-based elderly care, relies on community-based elderly care and is led by institution-based elderly care. Due to the influence of the traditional view of "rearing children to support parents in their old age", urban and rural elderly citizens of China at present are mainly taken care of by family members. The elderly care service condition survey organized by Ge Liying and others shows that 99% of elderly citizens among 1000 families surveyed in Jinniu District, Chengdu City live a retired life at home, 92% of elderly citizens hope that the community can offer various elderly care services, 54% hope that at least two or more services can be offered and 17% hope that more than four services can be offered. In the survey, the families of married children are composed of 3.1 elderly citizens (including parents and in-laws) on average. However, traditional elderly care homes and other social elderly care service agencies seem to be open for the childless elderly citizens. Therefore, beds in elderly care homes are vacant, affecting the further development of elderly care institutions. In addition, because of the lack of practical preferential policies like tax policies, there is a lack of appropriate standards and incentive mechanisms for private elderly care institutions.

As the bridge of households and the society, the community has unique advantages in elderly services. At present, among 120,000 urban welfare service facilities built in the community in the cities, there are more than 20,000 elderly care services facilities, more than 600 apartments for elderly citizens, more than 1300 sanitariums for retired cadres, more than 5000 day care centers (nursing homes), more than 100,000 activity centers for elderly citizens and more than 8000 colleges for elderly citizens.[25] However, because of the large base of China's aging population and the differences in economic conditions in all regions, there is a great gap between the elderly services provided by the society and the needs. There is a great gap between the services elderly citizens actually enjoy and the needs. For example, 57.8% of the elderly citizens need regular physical examinations, but only 7.6% of them receive such services; 45.1% of elderly citizens need emergency services, but only 4% of them receive such services; 42.5% of elderly citizens need family beds, but only 14.8% of them receive such services; 36.8% of elderly citizens need health guidance, but only 9.3% of them receive it; 18.9% of elderly citizens need nursing homes, but only 0.5% of them live in nursing homes; 13.9% of elderly citizens need dining rooms, but only 0.7% of them have them.[26] Moreover, the elderly care service forms in most communities are monotonous, and a few service items can provide health consultation, psychological consultation, cultural entertainment and others. There is a lack of professional caregivers, and they often quit. These factors restrict the development of the elderly care service business in the urban community. Family-based elderly care still dominates in the vast rural areas. There hasn't been real social elderly care service except for those enjoying the five guarantees. All in all, the social elderly care service system in China is based on home-based elderly care, relies on the community and is led by institutions, but it is still in its infancy stage, so it hasn't performed its function as a multi-level elderly care service and security system.

---

[25] *Report on the Development of Elderly Care Business in 2009*, website of the Ministry of Civil Affairs.

[26] An Yuxue: *Construct an Elderly Care Service System Suitable for an Aging Society, China Collective Economy*, Issue 1 of 2010.

3.3.5 The Compliance Assessment and Regulatory Mechanisms of Elderly Care Service Institutions Are not in Place

Interim Measures for the Administration of Social Welfare Institutions and *Basic Norms of Social Welfare Institutions for Elderly Citizens* (MZ008_2001), which are the access standards for the elderly care service industry in China, were implemented in March 2001, but the current implementation of the access standards is not optimistic. In China, there is a lack of employment conditions and recruitment procedures for the service personnel and the business scope of folk organizations in elderly care services institutions. There is a lack of evaluation of service standards and performance appraisal methods in elderly care institutions, making elderly care services non-standard and unplanned. The operating mechanism of social elderly care services is not smooth. The elderly care institutions without an operating license and with poor service quality and substandard facilities can't be rectified timely. Elderly care services associations and organizations at all levels haven't been cultivated and developed, so they are not self-disciplined in the industry and cannot play a positive role in contacting the government.

There is a lack of effective supervision and management for all kinds of elderly care service institutions. According to *Basic Norms of Social Welfare Institutions for Elderly Citizens*, the supervised items mainly include the safety of the designed building structure in social welfare institutions, inspection and epidemic prevention of food hygiene, the complete set of fire-fighting equipment, the dynamic management system of support personnel, the heating in the living areas in winter, the standardized configuration of electricity, the security of the courtyard layout, the standardization of security systems and so on. At present, the elderly care institutions in China are of varying quality, so they need standardizing.

3.3.6 The Elderly Care Services and Security in Rural Areas Are Weak

Due to the long-term dual economic structure in urban and rural areas, the elderly care service models in rural and urban areas are rather different objectively. Compared with cities, the elderly care insurance system for peasants involves more complex groups and more problems. The peasants' elderly care still depends on the traditional family-based elderly care and land security. The elderly care services and security in rural areas are extremely weak. The detailed aspects are as follows:

1. The aging population in rural areas grows faster than that in urban areas

The rural areas in China have entered an aging society, and the aging degree is higher than that of the cities and towns. The data in *China Statistical Yearbook* (2002) show that in 2001, there were 807.39 million people in rural areas in China, accounting for 63.91% of the total population of China. 7.35% of them are aged 65 and above, which is greater than 6.3% in the cities in the same period. The family planning policy has made the families smaller and the average household size decreased from 6 in the 1980s to 4 at the beginning of the 21st century.[27] At present, it is common for a couple in the rural area to raise four aged persons, and couples have to raise children too, so the burden of rural laboring population is getting heavier and heavier. The family function of raising elderly citizens is becoming weaker day by day.

2. That rural labor migrates to cities exacerbates the aging process in rural areas

Since the implementation of the separation of the urban and rural household registration system in 1958, the residents were divided into "urban residents" and "rural residents". However, with the development of social economy, the strict household registration system has been broken by a large number of migrant workers since the Reform and Opening Up, so rural surplus labor force work in the urban areas, changing the original restriction solely on land security. However, the risk for their own survival also increases. Once they get weak or sick, cannot continue to work or return to their hometowns when they are old, they have no source of income, so the social insurance system for elderly residents in rural areas and elderly care services and security are especially important.

3. The degree of peasants relying on land security weakens

Over the years, peasants "have lived on farming". Land is the last security line of a family. However, the survey shows that expectation and dependence degree of living on the land are declining gradually. Although peasants haven't paid any taxes in recent years, the farming cost is still raising due to the rise in the prices of seeds and fertilizer. The net land income of peasants is decreasing. From 1978 to 1998, the income ratio of national peasants in the first primary industry had decreased from 91.5 to 57.2%, the net agricultural income of which only accounted for 42.9% of the total income.[28] Peasants cannot live on the land only, not to mention elderly care based on the land.

4. The construction of the rural elderly care service system lags behind

The rural community actually means the village committee. Some villages in China propose establishing the pension security system which is based on family-based elderly care, centers on the community-based elderly care and is supplemented by self-security, but the rural economic conditions in China differ widely at present. The collective economic condition of the village committee decides whether it will establish the elderly care service center and provide funds for its villagers. The aging of population becomes serious in those rural areas with relatively developed collective economy, while peasants are completely taken care of by their children in

[27]The National Bureau of Statistics: *China Statistical Yearbook* (2002).

[28]Website of the Ministry of Agriculture of the People's Republic of China.

the relatively backward Western China basically without village-level collective economy. The community-based elderly care services have not really been scheduled in rural areas.

5. Failure to establish the elderly nursing insurance system

The elderly nursing insurance system has developed rapidly since it was set up in the US in the 1970s. Health Insurance Association of America (also HIAA) gives a definition of the nursing insurance: designed for consumers, it provides security in case of the huge potential nursing fees arising from their long-term nursing. The aging tendency in China at present is increasingly serious, the development of health and medical treatment lags behind, fees increase and the need of elderly nursing grows, so the elderly nursing insurance system is required to be established objectively.[29] At present, regardless of social insurance and commercial insurance, China's elderly long-term nursing insurance system in the primary exploratory stage should use the experience of the nursing insurance systems of Germany, Japan and other countries for reference. An insurance risk spreading mechanism should be introduced to relieve the risk of aging risk care as soon as possible.

It's necessary and urgent for China to establish the nursing insurance. ① The aging of population in China causes the growth of elderly nursing needs. According to the material, it's estimated that the elderly citizens aged over 60 will reach 300 million by 2027. The disease spectrum of elderly citizens changes sharply with aging, for example, cardiovascular and cerebrovascular diseases occur frequently. Elderly citizens badly and greatly need the long-term medical nursing or daily nursing. ② The family structure changes and the function of family-based nursing weakens in China. Due to the implementation of the family planning policy, single-child families increase in China. The original multi-generational families have been changed into the inverted pyramid family structure of four-two-one families. The third generation shall undertake the obligation to support several elderly citizens simultaneously. Meanwhile, the fast-paced development of economic society has increased the difficulty for the young people to take care of elderly citizens. So many elderly citizens and their families eagerly request establishing the nursing insurance system. ③ The burden of elderly citizens' medical fees is heavy and the current system of medical insurance is imperfect. The rapid increase of medical fees for elderly citizens has been beyond of the affordability of their families. Take Shanghai as an example. In the public elderly care homes, the monthly fee for all-day nursing of each elderly citizen is about 1200 yuan, and that of half-day nursing is about 980 yuan. If every member of four-two-one families in the future needs to support six elderly citizens, they will surely be unable to afford that. At present, special nursing fees and daily nursing items for elderly citizens in China's medical insurance have not been included in reimbursable items. Commercial insurance companies also only reimburse a part of fees for disease treatment and outpatient services, and the additional hospitalization medical

[29]Chi et al. [7].

insurance can only reimburse the fees for a short term (180 days at most). So it is a matter of great urgency to meet the aging care challenge by developing and improving nursing insurance system.

3.4 Development Practice and Reference of Foreign Elderly Care Services and Security

Elderly care services mean to meet the elderly citizens' needs of living care, mental and psychological solace and provide various services like economic aid, care for the disabled and spiritual solace. At present, developed countries have entered the aging society earlier with the increasing proportion of elderly support (see Table 3.7) and much useful practice and exploration in the aspect of the elderly care service system, providing valuable referential experience for the development of China's elderly care service and security system.

3.4.1 The Development Practice of Foreign Elderly Care Services and Security[30]

1. German elderly care service practice

Germany, the cradle of the modern social insurance system, brought *Nursing Insurance Law* into effect on January 1, 1995, and started carrying out the compulsory long-term nursing insurance that became the "fifth pillar" type of insurance in Germany. It also had a significant impact on the establishment of the nursing insurance system in Japan and other countries. Germany adopts the compulsory social insurance plan for long-term care, covering all the citizens. The long-term care in Germany is supported by insurance and public expenditures (paid by the Financial Department after review), while the long-term care insurance includes the evaluation of personal nursing needs and personal care subsidies (see Table 3.8). Care facilities shall be surely established to meet the needs in all states and cities.

The German nursing insurance system shall abide by the principle of "nursing insurance subject to medical insurance", and all applicants of medical insurance shall apply for nursing insurance. The features of the German nursing insurance system are as follows: ① the insurance tax shall be calculated according to the income of applicants. Each half of the tax rate shall be borne by the employer and the applicant respectively; ② nursing objects include all citizens except state officials, judges and professional soldiers; ③ the German long-term nursing insurance coverage includes two categories, namely family-based nursing and

[30]Liu [8].

Table 3.7 Changes in support proportions of some EU countries (Unit: %)

Year	France	Italy	Britain
1950	26.4	21.2	24.2
1960	28.4	23.9	21.1
1970	31.3	28.1	29.0
1985	27.7	33.2	31.5
1990	28.7	35.5	32.1
2000	32.0	40.8	31.3
2010	32.4	45.0	31.9
2020	41.3	50.2	37.4
2025	45.4	53.8	40.4

Source Population in the Economic Development (1950–2025) by the International Labor Organization

Table 3.8 Standards of personal care subsidies in Germany (Unit: Euro)

Monthly subsidy	Level 1	Level 2	Level 3
	More than 90 min a day	More than 180 min a day	More than 300 min a day
Family care	205	410	665
Door-to-door care	384	921	1432

Source [Germany] Michael [12]

Table 3.9 Elderly nursing insurance in the US

Underwriting way	Optional underwriting should take age, medical service condition and medical history into consideration
Scope of payment	All kinds of nursing services, such as medical nursing and daily nursing which last for over 12 months
Underwriting period	Optional: aged from 40 to 79; aged from 50 to 84; aged from 55 to 79
Premium payment	Pay annual average rate according to the age, and take payment period and waiting period of the insured into consideration
Payment way	Payment in cash, with different options, such as different maximum payment amounts, payment period and waiting period (elimination period); "managing nursing" with direct nursing service

Source Yang [13]

nursing based on nursing institutions. According to the degree of needs, it can be divided into general nursing, at least 90 min every week, services provided for three times in different time at least a day, and full-time services provided for 24 h. The German long-term nursing insurance premium of family-based nursing shall be paid in material wages in accordance with the standard, while the hospitalization nursing fees shall be paid in cash.[31]

The nursing education system in Germany is very perfect. With a history of over 160 years, nursing education is divided into three levels, namely the specialized secondary education, specialty training and undergraduate education. Nursing education focuses on the specialized secondary education and the specialty training is the required course for continuing education, which cultivate professional nurses for clinical elderly nursing. Specialties include intensive care, psychiatric nursing, operating room nursing, cancer nursing, community care, public health and so on. In 1992, undergraduate colleges in Germany started to recruit students of the nursing specialty. For example, the Agnes Karll Nursing Institution in Germany recruits and cultivates students into masters of nursing and doctors of nursing. The nursing qualification examination is very strict in Germany, focusing on practical operation skills. Its outline is uniformly specified by the National Examination Committee and the nursing certificate is recognized across EU.

2. Elderly care service practice in Britain

In 1929, elderly citizens aged 65 and above accounted for 7% of the total population in Britain. It had entered the aging society, 13 years earlier than the US and 41 years earlier than Japan respectively. Over the years, Britain has accumulated rich management experience of elderly care services to form a relatively perfect elderly care service and security system. In 1990, Britain put forward a bill named after community care, specifying that local governments shall perform community care in special cases with special fund. A social elderly care service system organized and managed by local governments was established in Britain, while the government was mainly in charge of macroeconomic regulations and giving necessary financial support. In December 1997, the British Labor government established a special committee specializing in studying short and long-term care of elderly citizens and providing a sustainable financing system for elderly citizens who needed long-term care to let them be taken care of in their own homes or other places. In addition, the Labor government also strengthened management to improve the quality of elderly care services. In February 1999, the British government carried out the new national care strategies. In the process of the elderly care service system, local governments bore greater responsibilities to release a five-year plan for elderly care services, establish a new elderly care evaluation system and equip with elderly care supervisors. Social workers should voluntarily provide household services and night care for elderly citizens, run "nursing homes" and hold "Good Neighbors" assessment activities. Local governments in Britain issued orders to build apartments for elderly citizens, while private persons or volunteers organized and set up elderly care institutions. So, elderly citizens could make choices according to their own conditions. British elderly citizens could enjoy free medical treatment. The government had specially set up hospitals for elderly

[31] Li et al. [9].

citizens and equipped them with professional supervision interviewers who should visit elderly citizens regularly according to the advice and guidance of doctors and provide advice about treatment and recovery.

The Community Care Bill of Britain includes: ① the service provider is the Social Service Department. ② The objects of community care were originally sick and weak elderly citizens, adult psychopaths, retarded and disabled persons and the like. As Britain enters the aging population society, the community care almost represents elderly care services and covers all elderly citizens who need long-term care. ③ Service contents and forms: the community care is mainly divided into Health Care and Social Care. Health Care belongs to National Health Services (NHS) and provides free medical and nursing care. Social Care is under the jurisdiction of the Social Service Department of local governments.[32] The Social Service Department undertakes the main task of care. Britain has established the Primary Health Care System, while family doctors act as the first contacts of social life to maintain the basic health care. Moreover, community and regional nurses, healthy home interviewers, community psychiatric nurses and nurses for retarded persons will provide professional care within the community to show a higher professional service level of the Health Care System in Britain.

At present, nearly 130,000 people in Britain are engaged in elderly care service social work. There are over 600 nursing homes across the country. Solitary elderly citizens aged over 70 must live in elderly care institutions. Those with good economic conditions can hire senior nursing workers to provide long-term professional services in their homes. By virtue of mature institutional conditions in European countries, British elderly citizens can be taken care of in other places and countries very easily. The British government specially set up hospitals for elderly citizens, and equipped them with professional healthy interviews for elderly citizens. They should visit elderly citizens regularly according to the advice and guidance of doctors and provide advice about treatment and recovery for them.[33]

---

[32]The Ministry of Civil Affairs, Leading Group Office of Elderly Care Service System Construction of Office of the National Working Commission on Aging: *Situation Compilation of Elderly Care Service in Foreign Countries, Hong Kong, Macao and Taiwan Regions*, November 2010.

[33]Long-term nursing insurance system in the US began in the 1970s and its underwriting subjects are commercial insurance companies. When the insured accept various personal nursing services in their homes or in the elderly care institutions, the insurance companies bear the relevant nursing fees. Nursing services of the US include nursing home, auxiliary nursing facilities and family health nursing. National Association of Insurance Commissioners formulated *Long-term Nursing Insurance Model Statute* which specifies that insurance companies should comply with the minimum standard and the rights of the applicant when formulating terms for long-term nursing insurance, so that nursing insurance can develop rapidly in the US. According to the statistics of Health Insurance Planning Association of the US, in 2002, 104 insurance companies sold over 900,000 long-term nursing insurance products with average increment of market shares of 18% from 1987 to 2002. Moreover, the long-term nursing insurance policies and premium of individuals and associations reached one billion dollars. In 2004, over four million people purchased long-term nursing insurance. The participation rate of elderly citizens is 10% (see Table 3.9).

3. Elderly care service practice in the US

The pension security system in the US is composed of the Basic Social Security Plan, the Employer Supplementary Elderly Care Plan (the Corporate Pension Insurance Plan) and the Private Savings Supplementary Elderly Care Plan, featuring multi-agent social operation, wide coverage and high level. The pension system of the US was established from states. With the improvement of industrialization process and living conditions, the aging population increasingly grew. In 1915, The Elderly Employees Pension Law was enacted and put into practice in Alaska in the US. In 1923, similar laws were passed in Montana, Nevada and Pennsylvania successively. By 1933, The Elderly Employees Pension Law was approved in 28 states in the US. In 1942, the US entered the aging society and its elderly care responsibilities were mainly borne by the government and society. In 1965, a series of laws about elderly care were enacted in the US, including The Old Americans Act, Volunteer Work Plans for Elderly Citizens, Nutrition Plans for Elderly Citizens, Multi-objective Center Plans for Elderly Citizens and the Community Service Employment Laws for Elderly Citizens, which fixed the contents of the elderly services by laws and built a social safety net for elderly citizens to enable them to gain fund security through elderly insurance annuity and safety allowance. For elderly citizens who didn't have insurance annuity and applied for safety allowance, the government should apply to the authorities for public assistance to help those who needed medical assistance after surveys of their financial conditions and income and the confirmation of a poor life. Relied on social force to specially design and build residences for elderly citizens and let them live in the residences at a low price and even for free. The government encouraged social force to build elderly care institutions run by non-government institutions. Religious charities and other non-governmental organizations made great contributions to elderly care services. Elderly citizens can choose elderly care institutions of different properties and levels according to their financial conditions and health conditions. No matter which form of institution they choose, the government will offer them technical support and preferential policies like the exemption of land tax and business tax. The US specially establishes welfare houses and elderly care homes, elderly recreation centers and elderly day care centers for elderly citizens. Elderly care homes offer board and lodging, recreation, medical treatment, psychological consultation and other services. At present, over 20,000 elderly care institutions of different levels, properties and service items have been established in the US, forming the multilevel social elderly care service system. Elderly health care plan is implemented. The US citizens aged 65 and above, retired or employed, are all entitled to get health care services provided by the government. Welfare houses, elderly care homes, elderly recreation centers and elderly day care centers are established for elderly citizens in the US. Elderly care homes offer board and lodging, recreation, medical services, the arrangement of death and other services. Some elderly citizens with low income can receive food stamp and special favorable services from the government. The US government sets up universities for the elderly in which elderly citizens can get educated for free generally. Various adult education centers and elderly universities are in duty bound to let elderly citizens aged over 65 get educated freely. Some social institutions also provide elderly citizens with mental and legal services.

4. Practice of elderly care service in Japan

As early as in 1963, the Japanese government issued the *Welfare Laws for Elderly Citizens* to advocate guaranteeing the overall life interests of elderly citizens. This "Charter for Elderly Citizens" has been revised for several times to this day and its main contents are: the government invests in the construction of the special elderly care homes and provides the weak elderly citizens with services, including those who suffer from dementia and are bedridden, emphasizing guaranteeing the welfare and social rights of elderly citizens. In 1982, the Japanese government issued the "*Health Care Laws for Elderly Citizens*" to comprehensively promote the health care facilities of elderly citizens, and this document stipulates that the parties concerning the medical insurance are responsible for the medical fees of elderly citizens who are aged over 70 and emphasizes that elderly citizens should receive home-based elderly care and home-based nursing. Therefore, the government invested in the training of about 100,000 family nursing workers to look after elderly citizens who met inconvenience in daily life. The government also invested in the construction and popularization of the nursing homes, providing elderly citizens in need with short-term short residence, nursing and treatment services, developing a batch of houses in the type of "three generations under one roof" with a low price and so on. In 1989, the Japanese government made the *Ten-year Promotion Strategy of Health Care and Welfare for Elderly Citizens*, namely famous "Gold Plan". According to this plan, the local governments should actively build and improve different facilities related to the elderly citizens. With this opportunity, all kinds of apartments, activity rooms, hospitals for elderly citizens began to appear. The elderly care institutions in Japan, including the elderly care homes and the short-term or daily elderly care institutions, paid great attention to psychological care, education training and other aspects of elderly citizens. To ensure the smooth operation of the elderly care homes and avoid maltreatment of elderly citizen, the Japanese government decided to introduce a more objective and impartial third party evaluation system based on the self-check elderly citizen service institutions, including the buildings, equipment, personnel allocation in hardware, and the quality of service, the evaluation of elderly citizens, the "Gold Plan" issued in 1989, the *Welfare Laws for Elderly Citizens* revised in 1990 and the *Nursing Care Insurance* issued in 2000 in software, with the social or personal power, providing elderly citizens in need of care in the community with corresponding life service to meet their need in life, medical treatment, nursing and other aspects and the community services for elderly citizens with strong system guarantee and legal support, promoting the development of elderly citizen welfare and helping elderly citizens with their later life by providing them with diversified and all-around community services. A large talent team for providing elderly citizens with professional service was established in Japan. Before working, they obtained corresponding qualification certificate after strict exams. A variety of forms of universities for elderly citizens were founded in Japan. Now, there are more than 400 universities for elderly citizens in Japan. The Japanese government encouraged non-government institutions to invest in elderly citizen service and gave them tax preference. In 1970, 7.1% of the population of Japan was aged 65 or above, so Japan entered aging society. Now, one fifth of the population of Japan is aged over 65. The aging situation in Japan is as serious as that in China. Both China and Japan have the Confucian cultural tradition, whose people praise highly the virtue of "filial piety", with family-based elderly care as the traditional elderly care pattern. The elderly care service and security system in Japan has strong reference significance to our China.

The elderly nursing insurance system in Japan is implemented by the government and operated in different municipalities. In December 1997, the *Nursing Insurance Laws* was formulated in Japan and the national long-term nursing plan was carried out in April 2004. The funds of the long-term nursing plan of Japan are raised from two aspects: 50% is from the government's fiscal expenditures, and the other 50% is from the insurance fees raised forcibly from the citizens of the right age. There are six payment levels of the insurance fees, namely from the first level to the fifth level and the support level. Pay in material object mainly and in cash secondarily (see Table 3.10).

Table 3.10 Elderly nursing insurance system in Japan

Joining objects	Those aged over 65; those who are aged from 40 to 64 and have joined the medical insurance system					
Individual burden	10% of the nursing fee					
Resource of funds	50% is from the public fee: 25% is from the country, 12.5% is from the prefectures and 12.5% is from the municipalities. 50% is from the insurance fee: elderly citizens aged over 65 bear 17% and those aged from 40 to 64 bear 33%					
Nursing objects	Those who are over aged 65 and in need of regular nursing or daily help; those who are aged from 40 to 64 and are sick in need of nursing					
Nursing procedures	1. Application; 2. Interview survey; 3. Opinions of attending doctor; 4. Check and decision; 5. Decide whether he or she needs the nursing and the nursing level; 6. Make the nursing service plans; 7. Provide the nursing					
Nursing contents	Visit nursing, daily nursing, lending the welfare equipment and so on					
Nursing payment	Provide the nursing services directly					
Nursing level	In need of support	Level 1	Level 2	Level 3	Level 4	Level 5
Monthly nursing fee	60,000 yen	170,000 yen	200,000 yen	260,000 yen	310,000 yen	350,000 yen

Source Yang [13]

5. Practice of elderly care in Sweden

Sweden is one of the countries of the longest expectation of life in the world. 17% of the population is over 65 and the proportion still rises rapidly. The government and the social organizations are very concerned about the aging challenge. About 93% elderly citizens live in their own homes in Sweden. Over 70% of those who live in elderly care institutions suffer from senile dementia.[34]

To fully prepare for the growing service needs of elderly citizen population, the Swedish government made a special development plan. The specific contents are as follows: ① provide elderly citizens who are seriously ill with better care and social service. The purpose is to provide the citizens with a sustainable, efficient and high-quality care system. Provide family-based nursing doctor and make a special plan to strengthen the care for the dementia patients. ② design and build special houses for elderly citizens, ensure the safety of the living facilities of elderly citizens, organize a special team to analyze and make plans for the living needs and facilities of elderly citizens to make the residence life of elderly citizens convenient. ③ make a specific plan for elderly citizens whose mother tongue isn't Swedish and organize a special committee to carry out this plan and provide guidance. ④ carry out personnel training in dementia elderly citizen nursing, organize a commission to specially study the qualification to join elderly care and work of the service system, and establish skills confirmation system and professional qualification and title confirmation system.

Care institution was established in Sweden in 2001, and it became a window and network to provide elderly citizens with services in Sweden. In 2005, 650 care professionals from Japan accepted training in Sweden. In 2006, 2000 care personnel were trained by the care institution of Sweden in Japan.

6. Practice of elderly care in Singapore

Singapore is one of the countries with rapid aging of population in Asia. In the society of Singapore, those who are aged over 60 are jokingly called "happy elderly citizens". The elderly support policy in Singapore is a systematic project. Four levels, namely individuals, families, communities and the country are listed into the establishment of the elderly care system in which individuals must be responsible for their own elderly care and make a plan for their retirement life in advance; the family is the basis of the elderly care, and the communities are responsible for the assistance and support for the elderly care of the families; the government bears the leading responsibility for the elderly care, providing support in legal system and policy, creating conditions, and providing individuals, families and communities with development support.

[34]The Ministry of Civil Affairs, Leading Group Office of Elderly Care Service System Construction of Office of the National *Working Commission on Aging: Situation Compilation of Elderly Care Service in Foreign Countries, Hong Kong, Macau and Taiwan Regions*, Issue 11 of 2010.

The elderly care institution construction is vigorously supported in Singapore. On the construction of the facilities, the government is the investment subject, providing 90% of the construction funds. The government provides the elderly care institutions with different subsidies on the operating costs of different services. The task of the elderly care institutions is to strive to improve the living quality and life quality of elderly citizens and achieve specialization and refinement. The service objects are elderly citizens who meet the standard requirements of the government and really need the nursing with poor heath and low income. The design of elderly care facilities in Singapore is much humanized. The layout design of all indoor function zones, ventilation, lighting, courtyard and the induction bed and the plastic floor, a variety of armrests in the toilet and other equipment show the humane care and respect for elderly citizens. There are not only daily care, entertainment and rehabilitation health care functions but also psychological treatment and hospice care functions, which enable elderly citizens to enjoy professional care services at home. The nursing personnel must take nursing qualification exams, obtain the certificate before working and participate in the annual inspection training exams.

3.4.2 Reference of Practice of Elderly Care in Other Countries to China

With large population base and an acceleration of a large-scale aging society, based on the experience of the elderly care service and security in other countries and the national conditions in China, the elderly care service and security model of socialization and diversification should be promoted and its coverage should be enlarged in China step by step so that the goal and system of national "providing elderly citizens with a sense of security" can be achieved finally. The developed countries entered the period of the aging society earlier, so they have accumulated rich experience on dealing with the aging challenge and establishing the elderly care service and security system, which is worth learning in the construction of the elderly care service security system in China. The several major aspects are as follows:

1. Establish an elderly care service and security system to meet multilevel needs

In the process of the construction of the elderly care service and security system in all developed countries, the government's responsibilities are highlighted and the government fulfills the responsibilities of establishing and improving the legal system and supervision and management and providing necessary financial support. The governments of the UK, Japan and other countries actively encourage and support the social power and capital in establishing the commercial elderly care institutions by tax deduction and exemption and other preferential policies, vigorously supporting the charitable organizations, the non-governmental organizations, the private capital investment and the donation from all walks of life in society to reduce the government's financial pressure and obtain the good development of the elderly care service business. A relatively perfect policy and legal system of the community-based elderly care services is established in Japan with diversified organizational forms, including the government, enterprises and the non-profit organizations, which contains rich service contents of different levels for elderly citizens. All these enable elderly citizens in Japan to spend their later life without sorrow and anxiety and enjoy the social returns after making due contributions to society. As a special social group, there are many differences in the age, physical condition, cultural level and other aspects of the elderly citizens, so they have different needs for the community-based services. Therefore, it's necessary to establish an elderly care service system to meet the multilevel needs of elderly citizens, providing elderly citizens of different physical conditions with different community health service and medical service items, including regular health examination, door-to-door medical services, family-based nursing services, institution nursing or rehabilitation training. Meanwhile, different service items should be designed for elderly citizens at different cultural levels. Besides, what is worthy of learning from Singapore and other countries is that, physical and mental help is provided with elderly citizens by the community-based elderly care services, which also meets their needs of rendering a service after retirement and provides them with manageable jobs so that elderly citizens who are capable can participate in community services. The elderly care service system in Singapore covers four levels, individuals, families, communities and government, with complementary functions, forming a harmonious elderly care service system, which meets multilevel elderly care service needs.

2. Abide by the system and rule of legislation going first

The legal policy support system of elderly care services fixes the service contents and standards, defining the government's responsibilities and restraining the rights, responsibilities and functions of the administrative agencies in community-based elderly care to make clear the responsibility subjects and improve the service rate. In addition, guide and support other service subjects except for the government in a right way by laws. As a country governed by laws, the laws, as the powerful backing, cannot be overlooked. Legislation going first is the basic rule of the development of the social security system in developed countries, and the elderly care service system is no exception. For example, in the US, in 1915, the *Senile Employee Pension Laws* was enacted and put into practice in Alaska; in 1923, similar legislation was passed in Montana, Nevada and Pennsylvania successively; by 1933, the *Senile Employee Pension Laws* was approved in 28 states in the US; in 1942, the US entered the aging society and its elderly care responsibilities were mainly borne by the government and society; in 1965, a series of laws about the elderly care were enacted in the US, including *The Old Americans Act*, *Volunteer Work Plans for Elderly Citizens*, *Nutrition Plans for Elderly Citizens*, *Multi-objective Center Plans for Elderly Citizens* and *Community Service Employment Laws for Elderly Citizens*, which fixed the contents of elderly services by laws and built a social safety net for elderly citizens to enable them to gain fund security through elderly insurance annuity and safety allowance. In 1963, the *Welfare Laws for Elderly Citizens* known as the "Charter for Elderly Citizens" was enacted in Japan, which was the beginning of promoting the social elderly care in Japan. It fixed the service contents of the community-based elderly care for the first time. The *Health Care Laws for Elderly Citizens* enacted later further fixed the contents of medical health care services for elderly citizens. The *Ten-Year Promotion Strategy of Health Care and Welfare for Elderly Citizens* carried out in 1989 specified the dominant role of the community in elderly care services again. Particularly, the *Nursing Care Insurance* issued in April 2000 stipulates that all the citizens who are aged over 40 should pay the nursing insurance fees for solving nursing problems after they get old, and the local governments serve as the underwriter. When citizens want social nursing, they should submit the application to the local government first. The administrative agencies confirm the degree of needed care through investigation and authorize corresponding insurance compensation according to the provisions. The legal system is mandatory and binding. The *Nursing Insurance Laws* enacted in Germany in 1995 made the nursing insurance develop formally. We should learn the system and rule of legislation going first in the social security system in developed countries definitely.

3. Socialized operating mechanism of multi-subjects

The elderly care service system in all Western developed countries is dominated by community-based elderly care mainly with multi-level and multi-form elderly care institutions as the beneficial supplements. For a majority of elderly citizens, living in the community they're familiar with, conforms to their psychological homesick needs and has contributed to the improvement of the quality of their life. It's convenient to carry out various services in communities, including daily care of elderly citizens, household services, medical health care, psychological consultation as well as entertainment activities. As to the nursing, the social market operation mechanism is fully performed. The long-term nursing insurance system in the US began in the 1970s and its underwriting subjects are the commercial insurance companies. When the insured accept various personal nursing services in their homes or in elderly care institutions, the insurance companies bear the relevant nursing fees. In the US, under the leadership of the government, the social capital and organizational strength are made the best use of to support the elderly care service business. The government supports the elderly care service business by policies and tax preference. The community service center and elderly care institutions are operated by social capital generally. In the UK, the local governments build the apartments for elderly citizens, encourage and support individuals or volunteers to organize and found elderly care institutions. Elderly citizens are free to make their option according to their own situation, which has important reference significance to encouraging and mobilizing the social capital to participate in the construction of elderly care institutions and exploring the social elderly care service mechanism and other aspects of China.

4. Fulfill the function of community-based elderly care services

The community-based elderly care services cover household services, daily nursing, health consultation, medical treatment, spiritual solace and other professional services. In developed countries, the professionals must participate in the relevant training and exams and obtain the post qualification, and their services should be supervised and examined according to the provisions of the community-based elderly care service system. For example, a powerful community elderly care service network is established in Sweden. Its service contents include door-to-door housework services, housing maintenance, short-term care, daily activities and community medical nursing. The government encourages commercially-running elderly nursing institutions and provides them with great tax preference; in Britain, the local governments build the apartments for elderly citizens. Individuals or volunteers organize and found elderly care institutions. Elderly citizens are free to make their option according to their own situation. In Britain, elderly citizens can enjoy free medical services in the community medical treatment center. The government specifically establishes hospitals for elderly citizens, which are equipped with special invigilation visitors who visit elderly citizens at their homes regularly and provide them with advice on the treatment and rehabilitation as well as other aspects according to the advice and guidance of the doctors. Moreover, there are perfect community-based elderly care service networks in all regions.

5. Professionalized elderly care services

According to the experience of Japan, high-quality and professional community-based elderly care service workers are the important guarantee of elderly care services of high quality. Different from those in Japan, the communities in China don't have a great many housewife workers. However, we have a great number of primary, secondary and university student workers and plenty of healthy elderly citizen workers. Their participation will bring new vitality and energy to community elderly care services and become the potential force of community elderly care services. China should learn from the experience of Japan, enhance the educational training of professional elderly care services, and set and complete relevant professional systems to ensure that professional personnel can be provided for communities continuously and the economic and social status of this group can be guaranteed. In addition, China should enhance the educational training of family members. They are in contact with elderly citizens most frequently, so popularizing the nursing knowledge and improving their nursing level are of great significance. Meanwhile, the training and education of volunteers should also be enhanced so that they all can possess corresponding relevant knowledge before service. Only in this way can they serve elderly citizens scientifically. Moreover, we should let them fully realize the importance of volunteer service and appeal to more people to participate in the troops of elderly services by encouragement and publicity.

6. Play the role of elderly nursing insurance system

The world's commercial nursing insurance system began in the US in the 1970s. Later, it was founded in France and the UK successively. As a form of social insurance, legal nursing insurance system first appeared in Israel in 1986. Later, elderly nursing insurance system was founded in Austria, Germany, Japan and other countries successively. Making a comprehensive survey of the elderly nursing insurance system in developed countries, we can see their experiences of system construction are as follows: first of all, a comprehensive survey was conducted to completely realize the demands and risk to implement nursing insurance and nursing development problems before nursing insurance was carried out. Second, nursing insurance reflected the patient-oriented holistic view, which provided nursing mostly in the form of family nursing service, reducing the high inpatient nursing fees. Meanwhile, staying in the familiar community and receiving home-based nursing are good for patients to recover. Third, a sound nursing service and security system was established to provide a comprehensive security for various nursing demands of elderly citizens. Moreover, nursing services integration and regional network were realized. Nursing service can be provided at any time and the service contents cover medical nursing, mental nursing, life nursing and others. Nursing insurance system is an effective system to deal with the aging risk and solve the problem of elderly care. China must learn from the experiences of developed countries, increasingly develop and play an important role in elderly security system.

3.5 System Goal and Idea of Elderly Care Service and Security System

With the social and economic development of China, the liability subject of elderly care services is moving towards the society from the family. The ways of elderly care turn into a socialized and multilevel elderly care service system which is based on home-based elderly care, relies on community-based elderly care and is supported by institution-based elderly care from the traditional way of "rearing children for old age". To realize the strategic target for "providing elderly citizens with a sense of security" which was put forward in the 17th National Congress of the CPC, we have to take Deng Xiaoping Theory and the important thought of "Three Represents" as guidance, thoroughly implement and practice the Scientific Outlook on Development, establish the people-oriented idea and aim at meeting the basic elderly care service demands. According to the principle of "the government's dominant position, social participation, market operation and civil supervision", realize the development goal of aging career of "providing elderly citizens with a sense of security, medical treatment, education, a sense of worth and happiness" and provide professional elderly care services for elderly citizens to solve the problems of disabled elderly citizens. During "the 12th Five-Year Plan", social elderly care service system was basically established in accordance with aging of population and in line with the economic and social development level, and meet the multilevel and diversified elderly care service demands of elderly citizens to let them spend their remaining years in happiness and share the economic and social development achievements.[35]

3.5.1 Construct the Elderly Care Service System Goal with Chinese Characteristics

It is an urgent task to construct the social elderly care service system which meets the needs of Chinese elderly citizens. This system should be in line with the economic and social development level of China to meet the elderly care service demands of elderly citizens and improve their living quality. From the supplementary and selective system arrangement to the inclusive system development, all elderly citizens should be provided with living care, health consultation, patient physical therapy, disability care, spiritual solace and other services. Moreover, facilities, organizations, personnel and technologies should also be provided for elderly citizens for social participation. Relevant service standards, operating mechanism and supervisory systems should be established and perfected. The strategic target of the social elderly care service system develops from the supplementary type to the inclusive type. The foothold is to free elderly citizens from care worry. The contents are to establish a system which is based on home-based elderly care, relies on community-based elderly care and is supported by institution-based elderly care, focusing on the actual demands of elderly citizens, providing them with elderly care services of appropriate technologies, giving priority to the service demands of childless elderly citizens, entitled groups, elderly citizens without the ability to work, a source of income or a legal supporter and caregiver, households enjoying the five guarantees and elderly citizens who are of an advanced age, who live alone, who are disabled and who have difficulties in elderly care, so as to establish the elderly care service and security system with Chinese characteristics. See China's Elderly Care Service System in Chart 3.2.

3.5.2 Construct the Idea Followed by Elderly Care Services

China's elderly care services and security should be people-oriented, follow the idea of "Fair Sharing, Liability Sharing and Universal Benefit" and practically maintain the basic right of elderly citizens to live and develop. Only by directing

---

[35]Department of Planning and Finance of the Ministry of Civil Affair: *Soliciting Public Opinions on the Construction of the Social Elderly Care Service System During the 12th Five-Year Plan*, February 11, 2011.

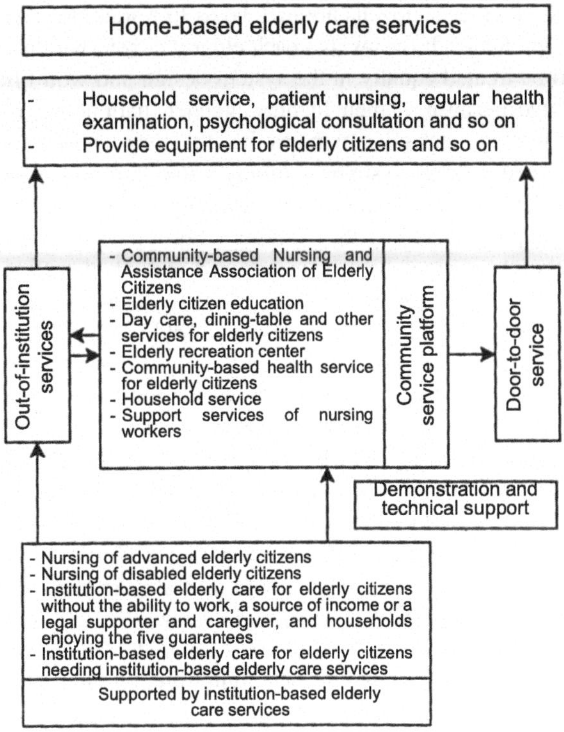

Chart 3.2 China's elderly care service system

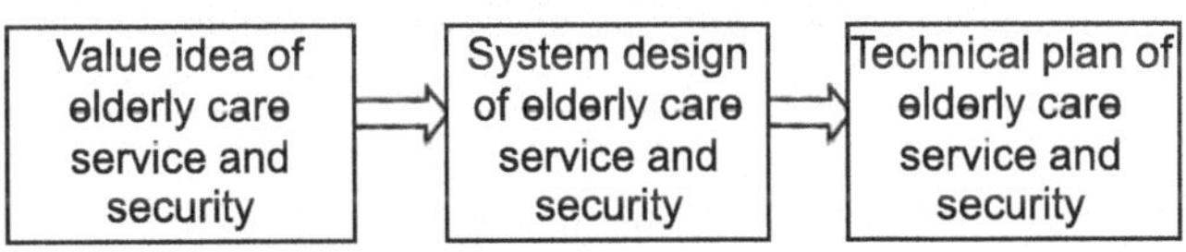

Chart 3.3 Elderly care service and security idea decision, system design and technical plan

China's elderly care service and security system with the value idea of "Fair Sharing, Liability Sharing and Universal Benefit" can we make the elderly care service and security system be in accordance with its essential laws and work effectively and comprehensively (see Chart 3.3).

1. Follow the idea of "Fair Sharing"

As stressed in the Report of the 17th National Congress of CPC, "all the effort and work of the CPC is for the benefit of people" and "all rights, care and benefit are for people" should be the basic idea that the government pays attention to people's livelihood and boosts the social harmony. The elderly care service and security system must follow the idea of "Fair Sharing" and ensure that the basic elderly care service demands of elderly citizens are met, so as to bridge the gap between urban and rural areas gradually. It means that each elderly citizen who needs elderly care services should be treated equally in the system design and won't be discriminated against or rejected because of their identity, gender, nationality, region and other differences. As market mechanism pursues the efficiency naturally, all the aims and end results of social security systems naturally reflect the pursuit of fair value and experience the development history from reducing the inequity to realizing and maintaining the equity, including that in different groups and between generations. Take whether the social equity can be maintained and whether the inequity can be reduced as the basic criterion to measure whether the social security system is effective. The equity value idea is the core idea of elderly care service and security system design. To realize the goal of "shared by all people" and "providing elderly citizens with a sense of security", the people-oriented must be established. For one thing, elderly care service and security system design must ensure that each elderly citizen has an equal and basic right to get access to elderly care. Whether elderly citizens live in urban areas or rural areas and no matter how their financial situation is, they should share the elderly care services equally and get service security of different demand levels. Status differences among different stratums or groups should be eliminated. The government should increase the investment of the basic facility construction of elderly care services in rural areas and remote towns. For another, as to security level, give priority to the goal of meeting the basic care demands of elderly citizens, improve their living quality step by step, bridge the inequity gap, and take much care of and give policies support to elderly citizens who are disable, who have low income in urban and rural areas, who enjoy the five guarantees in rural areas, who have no ability to work, no source of income, no supporter or caregiver in urban areas and other vulnerable groups to meet their basic elderly care demands, thus achieving the full coverage of elderly care service and security system in the whole urban and rural areas and exactly guaranteeing the life of all elderly citizens.

2. Follow the idea of "Liability Sharing and Universal Benefit"

Philosopher Rawls put in *A Theory of Justice* that justice is the first virtue of the social system. The elderly care service and security system follows the idea of "Liability Sharing and Universal Benefit", including: ① Follow the idea of "Liability Sharing", in other words, the most important basic standard in the development history of social security system is that the government plays leading role and the law reflects the will of all people and has the highest compelling force and authority. To establish the elderly care service and security system shared by all people, we have to regulate the right and obligation of the government, enterprises and individuals by laws to make elderly care services guaranteed and make those participating in elderly care services, service standard, organizational management and others institutionalized. In this case, people act abiding by laws with clear rights and liabilities. When the interests of related personnel are damaged during the operation of elderly care system, elderly citizens can protect their interests with laws. ② Follow the principle of "Universal Benefit". Pension insurance is the most important part of social security system. Elderly citizens devoted themselves to social and economic development in their lifetime and leave their posts due to retirement system or health conditions, so they rely on retirement pension or individual savings to support their retirement. Elderly citizens in rural areas can only rely on the traditional way of rearing children for old age to live through their retirement. The social security system frees elderly citizens from worries about survival and makes them share the results of economic development reasonably by redistribution mechanism. The secondary distribution should support elderly citizens and other social vulnerable groups partially to realize universal benefit between generations. ③ China put forward that the elderly care service system which is based on home-based elderly care, relies on community-based elderly care and is supported by institution-based elderly care should be established. On the condition that the government plays the leading role, social force and private capital should be brought in actively and the principles of Liability Sharing and Universal Benefit should be followed. To this end, we have to facilitate the cooperation of the participants, such as families, community elderly care service institutions, elderly care homes and other intermediary organizations for elderly services, so as to improve the operating efficiency of elderly care service and security system and guarantee the stabilized operation of the system.

3.6 Measures to Construct China's Elderly Care Service and Security System

3.6.1 The Government Plays the Leading Role

The history of elderly care service and security system construction of developed countries shows that the government is carrying out public policies to develop the main and regulatory responsibilities in the course of the realization of equalization of public service, performs duties in the law system implementation and the supervision and administration, and provides some necessary financial supports. Make policies to lead social forces to participation of elderly care services and provide preferential policies like revenue.

1. Construct the unified management organizations of elderly care service system

China should significantly learn from the experiences of the US, Singapore and other countries to establish nationwide unified elderly care service management institutions. From the competent department of ministries and commissions to the provincial and municipal and lower management organizations, management responsibility of all institutions should be definite, so that the organizational guarantee can be provided for the operation of elderly care service and security system. The competent department of elderly care service of China should be the

Ministry of Human Resources and Social Security. All subordinate functional departments should be responsible for elderly care service management of their administrative regions. Township and sub-district office are the primary-level organizations.

2. Develop and complete the relevant laws and regulations of elderly care service

In February 2008, Office of National Working Commission on Aging and other nine ministries and commissions jointly issued *Opinions on Comprehensively Boosting Home-based Elderly Care Services,* the first policy document to guide and support home-based elderly care service, and put forward a series of policy principles on project approvals, financial contribution, tax deduction and exemption, water and electricity preference and others for the development of elderly care service in urban and rural areas. However, the legal system to maintain rights and interests of elderly citizens is still so weak in China. China needs to further formulate policies and regulations on elderly care services, accelerate the formulation process of *Regulations on Administration of Elderly Care Institutions* and balance the development of elderly care system construction of urban and rural areas, government investment and private investment in elderly care institutions, for-profit and non-profit elderly care services in the range of legal regulations, so that *Basic Norms of Elderly Care Service Institutions*, national standards and industry access criterions can be formulated and published soon to standardize the elderly care service subsidy measures, support the social force in participating in the cause of elderly services based on laws, found elderly care institutions, and enhance the standardized management of elderly care institutions and community elderly care services, thus making the cause of elderly care services legalized and systematic.

3. Increase the support of elderly service policies and finance

In February 2009, the National Development and Reform Commission, the Ministry of Civil Affairs and Office of National Working Commission on Aging arranged a special fund for the development of experimental units of elderly care service construction in five provinces and municipalities, namely Jiangsu, Hubei, Heilongjiang, Chongqing and Gansu, officially started the basic elderly care service system construction, formulated *Plan for Social Elderly Care Service System Construction* and made overall arrangements for infrastructure construction of elderly care services during the 12th Five-Year Plan. The development course of social security systems in various countries in the world shows that policies and financial support of the government always play a very important role in institutional improvement, especially the elderly care service security system. ① The government should strengthen the planning work, formulate the mid- to long-term plan for elderly services and reasonably assign and make good use of social resources. When formulating the plan, according to the long-term plan and the reality, the government should take the current economic capability and needs of elderly citizens into consideration and avoid blind comparisons and overstepping ahead unrealistically. ② The government should significantly invest in the infrastructure of elderly care services and complete the community entertainment centers and health and psychological consultation service facilities for elderly citizens. When examining and approving the newly-built residential area, the Planning Department should involve houses for elderly care services in the residential area planning. The land involved in planning layout of socialized elderly care service institutions can't be used for other purposes or other projects. On the satisfaction of the housing condition plan for home-based elderly care of elderly citizens, the government should take their safety and suitability into consideration and show humanistic care to them; on service network, the government should carry out the professionalism and specialization of community workers and train existing available community workers. Subjects such as community work, nursing for elderly citizens and psychological consultation for elderly citizens are set up in colleges and universities to help gradually improve the overall quality of personnel for elderly services. ③ The government should mobilize all social forces to care for, support, participate in and invest in elderly service industry, give policy support and tax preference to private capital, give preferential policies to support the cost of the construction land of elderly care institutions, institution operating network and other infrastructure, and offer corresponding subsidy to elderly care institutions according to the number of beds. ④ The government should overall develop urban and rural areas. In particular, the government should pay much attention to the plan and development of the cause of elderly citizens in rural areas and gradually realize the goal of "providing elderly citizens with a sense of security" of the whole society. ⑤ The government should supervise elderly services, formulate unified evaluative criteria for the quality of elderly services and implement the industry access system aiming at the private capital for elderly care welfare services to guarantee the normalization and permanence of the operation of elderly care welfare institutions.

4. Implement government purchase of services

Government purchase of public services has been a basic system in Western countries. In the field of elderly care services, it is common that the government pays a certain amount to pension institutions and community pension service centers, and these organizations complete the elderly service work within the jurisdiction. The principle is "borne by the government, special commission, contract constraint, evaluation, and honoring the contract". Its core value is to introduce market competition mechanism into the field of public services to overcome the high cost, poor benefit, bureaucratization and connection network brought by the original monopoly supply or "hitchhike", and to introduce social funds into public service system, which ease the government's financial burden and realize optimized allocation of resources. Shanghai, Ningbo and other areas in China developed government purchase of elderly care services earlier and had a very good institutional innovation.

There are several suggestions as follows on introducing government purchase of service into the field of pension security services: ① considering the reality of difference in economic development level of our country, first of all, it should be developed in the local government having economic capacity and be implemented in the whole country gradually. According to foreign experience, the local government is usually regarded as the investor of elderly care services. As a result, the government's financial expenditure should be considered. Elderly citizens in Beijing, Shanghai and other big cities have relatively good economic conditions and a variety of elderly care service demands, so the government purchase of elderly care services is a good choice. ② Promulgate incentive measures to encourage private organizations, profit organizations and social communities to actively participate in elderly care services and become the major participants and providers of government purchase of service after obtaining pension service qualification through competition mechanism. ③ Establish a sound evaluation system of government purchase of elderly care service. Shanghai, Ningbo and other cities have relatively high economic development level in China, and their local governments have certain financial resources and initiatives to increase welfare investment. To make institutional innovation of government purchase of public service, actual conditions of other regions of the country should be considered. When evaluating public utilities of the government, governments in economically less developed areas must lack impetus to welfare investment and "government purchase of elderly care service". Therefore, we should sum up experience and make price and service standard of purchase of public service that match the economic development levels of different places, so as to improve service quality. In addition, implementation of policy should be operated by different levels. According to elderly citizens' economic and healthy conditions, different models of government purchase of services are adopted among different elderly citizen groups to help them to achieve elderly care service needs of different levels. The government should provide full-amount purchase service models to those elderly citizens who are of an advanced age and who cannot take care of themselves. Subsidy should be given in economically less developed areas, and funds, service tickets, purchase of service post and other forms can be provided directly. Or tax preference, loan with low interest and other measures should be taken in elderly care institutions to make elderly citizens enjoy more items of government purchase of elderly care services.

3.6.2 Realize Socialization of Elderly Care Service and Support Non-profit Organizations to Participate in Elderly Care Service

Socialization of elderly care service is to change the present situation that government "takes on all things", distribute the government's responsibilities and assimilate other social subjects to take part in elderly care service cause to realize "socialization of service objects, source of funds, management service models, service facilities and service teams".[36]

Non-profit organization (NPO) is independent organization that serves for public welfare not for power and profit, which is called also civil organization non-governmental organization or the third sector. NPO's participating in elderly care service can mobilize more social forces and accelerate the process of socialization of elderly care service cause. ① Play the role of non-profit organization to realize diversification of elderly care service and security. The welfare system of "corporatism pattern" in Western countries is relatively developed. We learn from their development experience, fully mobilize social forces, turning the providing way from single provider of government elderly care service to the providing way that combined type by multiple providers. ② Establish sound NPO-related laws to accelerate NPO's healthy growth. There are special laws about NPO in Germany, the UK, Singapore and other countries. China should give play to the functions of NPOs in the field of elderly care service and regulate them by legal form to confirm their social status, social functions and social responsibilities they should bear. Manage elderly care service professionally. Encourage and support the establishment of non-governmental elderly care service associations, strengthen internal cooperation and improve the overall quality of the industry. Meanwhile, study industrial development tendency to provide basis for the government to make scientific plans and policies, which will become the bridge and bond of the communication between the government and elderly care institutions as well as an important social force to develop socialized elderly care service.

NPO provides diverse social elderly care services, whose main contents include: ① support and life care services: first, there are places providing long-term and temporary elderly care (nursing) in streets and communities, such as elderly care homes, welfare housing, apartments for elderly citizens, day care centers for elderly citizens, canteens for elderly citizens, and so on; second, it sets up service accounts for elderly citizens' families and helps to manage life by door to door services; third, it subsidizes auxiliary equipment for elderly citizens' activities; fourth, it provides proper rehabilitation medical treatment knowledge education and consultation, making it better for families to understand elderly citizens' problems and needs; fifth, it adds elderly citizens' life service points in the social service industry to provide convenience for elderly citizens in daily life, such as stores and shops for elderly citizens and so on; ② medical care and rehabilitation services: communities should establish medical care service centers for elderly citizens, facilitate medical care of elderly citizens, establish health records for elderly citizens aged over 80 and provide nursing and treatment of common diseases for them; ③ education services: all kinds of schools for the aged and colleges should be established to provide a chance and convenience for elderly citizens to relearn and be reeducated; ④ social participation services: conditions should be provided for elderly citizens to continue to take part in social activities like strengthening the connection

[36]Han [10].

between elderly citizens and teenagers or the society, and allowing them to play a role in caring and educating the next generation; there should be service facilities suitable for elderly citizens in the construction of public places, bridges, roads and public facilities; elderly citizens should be organized to establish the Technical Service Department for elderly citizens, the Department of Scientific and Technological Consultation and Services and Associations for Elderly Citizens, and to voluntarily assist and take part in the work of the neighborhood committee in streets, etc.; ⑤ cultural, sports and entertainment services: all kinds of cultural, sports and entertainment facilities should be initiated and elderly citizens should be organized to establish all kinds of associations and seminars. Various cultural, sports and entertainment activities should be carried out, such as establishing activity rooms and centers for elderly citizens, calligraphy and painting associations for elderly citizens, the Tourism Service Department, drama associations, etc.; ⑥ other services: marriage service corporations for elderly citizens should be established to help elderly citizens to remarry and reestablish families and priority should be given to elderly citizens in medical care, riding, travelling and other aspects.

3.6.3 Strengthen the Construction of Elderly Care Service Professional Teams and Develop Intermediary Organizations of Elderly Care Services

In early 2000, the General Office of the State Council issued Guo Ban Fa [2000] No.1 Document, namely *About Accelerating the Overall Requirements of Socialization of Social Welfare*, and put forward the requirements of "diversification of the main investors, making service targets public, diversity of service models and professionalization of service teams" when China copes with aging. However, as mentioned above, the requirements of specialization of service teams in China are totally different in aspects of number, specialty setup, development planning, etc. Coping strategies are mainly as follows: ① the government should increase the training of professional talents of elderly care services. To provide elderly citizens with a sense of security, the government should try to improve the whole quality of elderly care service practitioners, position elderly care services as a professional post, set up majors like community nursing and elderly psychology and related courses in colleges and universities and vocational schools, include on-the-job training of elderly care service personnel into the work plan of social elderly care services, encourage rural labor forces and transferers, and laid-off workers in enterprises to work on elderly care services, and implement the qualification admittance system of the elderly nursing profession. Practitioners should take appointment with certificate and must receive professional training with fixed time at regular intervals every year; ② the government should set elderly care service posts as public welfare posts. Elderly care services and security are a system of public service nature that benefits numerous elderly citizens. Being engaged in community elderly care service nursing is as important as public posts like traffic assistants. At present, because of low salary and traditional prejudices against elderly citizens nursing, elderly care service practitioners are unstable and they are usually laid-off workers born in the 1940s to the 1950s and women from rural areas. If elderly nursing is set as public welfare posts and the government provides the salary of elderly care services, it can not only ensure that practitioners get salary on time and stabilize practitioner teams, but also increase elderly care service practitioners' salary, improve the salary level, correct prejudices against elderly nursing, and be beneficial to enlarging and stabilizing the elderly nursing service practitioner teams; ③ the government should take measures to stabilize the development of professional elderly care service teams. First, community services and elderly care institutions must accept practitioners who received professional training of elderly nursing services in proportion and includes this into the important indicator evaluation system of biding and annual inspection certificates, and bring in practitioners who major in sociology or related similar majors from colleges and universities and vocational schools to improve the present situation of practitioners' low overall quality and unreasonable structure; second, sign a labor contract with elderly care service practitioners, underwrite insurance for them, relieve their worries and make them really focus on their work as "city residents", laying the foundation for no more seasonal mobility and stability and normalization of elderly care practitioners.

To ease the present situation of shortage of elderly care service practitioners and unreasonable quality structure, the government should establish the volunteer service system and combine the voluntary spirit of "dedication, fraternal love, cooperation" with the elderly care services in China: ① The government should make great efforts to publicize volunteers' dedication and encourage more people to take part in the elderly care service. ② The government should reinforce management and professional training of the elderly care volunteers' team. Volunteers who take part in elderly care service should accept and know the professional knowledge in household service, health and medical counseling, psychological therapy, adjustment of marital relationship, legal aid and other professional fields and provide fine and professional voluntary services. ③ The government should establish sound rules and regulations about the elderly care service volunteers, standardize recruiting and training of volunteers, voluntary service standard and other management systems, establish networking and informatization of volunteers' management and realize sharing of resources in more communities. ④ The government should establish a long-term incentive mechanism of elderly care service volunteers, for example the community volunteers' "Time Bank" in Guangzhou. It can record and count the time of volunteers' taking part in the community voluntary activities. When they or their relatives need community services, they can enjoy community voluntary services by offsetting their recorded time equally.

The government needs to make great efforts to develop elderly care service intermediary organizations. Due to limited financial resources in China, the government can't afford to establish elderly care institutions and support the elderly care service completely. Drawing lessons from foreign countries, the government should push most of the elderly care institutions which were completely run by

China in the beginning to the market and realize the "state-founded and privately-operated" system besides keeping few model window institutions. The government should make great efforts to support development of the elderly industry, advocate, organize and offer necessary financial support to help and lead social associations, enterprises, units, individuals and foreign investments to donate and establish the elderly industries in various ways, thus establishing the elderly care service system taken part in by various subjects.

To support various subjects to take part in the elderly care service system, the government should make full use of the organizational role of intermediary organizations and elderly service societies and associations. ① The government should raise social capital by many ways and channels to establish or subsidize the elderly care service. ② The government should make use of the industrial management function of the social organizations such as the elderly care service association, draft industrial regulations and restrictions and reinforce industrial evaluation and inspection. ③ The government should make use of intermediary organizations to reinforce communication and cooperation among the people and hold the training in professional skills of the elderly care service and business communication activities to raise the service awareness and moral quality of practitioners in the elderly service industry constantly. ④ The government should make full use of the linking bond between elderly associations and the government and elderly care institutions and offer them advice on the elderly care service industry.

3.6.4 Accelerate Establishment of the Elderly Care Service Evaluation System

Due to the existing urban-rural dual structure in China, there is a relatively large difference of social and economic conditions in areas where elderly citizens live. During the process of establishing the elderly care service security system, the government should take multi-layer needs of the elderly care service into consideration to establish the evaluation system of elderly care service quality. The evaluation contents include the service quality of elderly care institutions, qualifications and standards of elderly care institutions the government subsidizes to serve elderly citizens, evaluation aiming at elderly citizens' various services and other aspects. ① Entrance evaluation of elderly care service institutions. Privately-operated elderly care institutions are allowed to have some profits and increase competitiveness of elderly care institutions to realize socialization and marketization orientations of the elderly care service in China; the government should invest in the elderly care service institutions with rural "households enjoying five guarantees" and elderly citizens "without ability to work, a source of income or a legal supporter or caregiver" in cities and towns as the service objects. These elderly care institutions shouldn't be profit-oriented. When evaluating them, the government should mainly stress their service quality and give full consideration to the evaluation comments on the service quality given to these institutions by the served elderly citizens; "state-founded and privately-operated" elderly care institutions are allowed to have meager profits. However, the government should stress the evaluation of its service quality, offer them preference on tax and subsidy and encourage privately-operated institutions to take part in the elderly care service. ② The government should insist on combining low-income elderly citizens' service subsidy qualification evaluation and the elderly care service demand evaluation. The service subsidy policy of low-income elderly citizens is a necessary measure to defend the justice of "providing the whole elderly citizens with a sense of security" and realize social harmonious development. However, in the process of evaluation, it should give full consideration of actual economic affordability of the elderly citizens' families and children, which not only guarantees the justice of the evaluation system, but also can meet each elderly citizen's basic needs of the elderly life service. Elderly citizen need evaluation gives high priority to low-income elderly citizens' income condition and determines the way of elderly service need they can enjoy, such as the necessities of home-based elderly care, community-based elderly care or institution-based elderly nursing, the fact about whether they need to offer door-to-door services, provide psychological or physical health counseling service, etc. The government should combine institutional evaluation and low-income elderly citizens service subsidy evaluation with the elderly citizens' need evaluation in practice to greatly improve the elderly care service quality and attract social capital power to take part in the elderly care service.

3.6.5 Improve Elderly Care Infrastructure and Establish Informationalized Service Platform

Elderly care service is an undertaking that benefits thousands of elderly citizens and guarantees their happy life, which needs supporting by the operation technology combining institutionalized hardware facilities and informationalized software. ① Establish elderly care service infrastructure. The government should strengthen investment in construction of elderly hardware facilities, establish community elderly care service stations and equip televisions, computers, medical instruments, elderly activity apparatuses, entertainment equipment, site facilities, etc. According to the number of elderly citizens under their jurisdictions, the government should equip relevant hardware facilities of social elderly nursing service and form institutionalized management and operation mechanism. Under its guidance, the government should plan the community elderly care service facilities as a whole and realize unified planning, design and implementation, matching operation and full use of social resources regarding it as the main contents of urban planning. Land and idle resources of the original enterprises, schools or factory building of institutions in the community should be integrated and used reasonably on the basis of regulation of the government, providing places of fitness activities, amusement, entertainment, medical consulting service and elderly day-nursing center for elderly citizens in the community, making the limited investment of the government be used optimally, increasing utilization efficiency of resources and enabling elderly citizens in the community to enjoy the nearest elderly care services. ② The government should strengthen informationalized construction of elderly services in the community, make full use of modern information and technology instruments, establish information comprehensive service network covering the whole community step by step by various forms, such as elderly care service hotline, computer network and "safety doorbell", make e-records of elderly citizens' physical health condition, elderly care service needs and community services they once accepted to realize the seamless connection between community services and elderly care service needs. It can also make use of information network to release weather report, health consulting and lecture information, administrative notice and matter in the community, health care, financial management, marriage, legacy planning, testament and other knowledge about elderly citizens. These are convenient to meet the needs of elderly citizens in the community about weather information, educational training, transport, travel, household services, entertainment, etc., thus realizing the low-cost and high-efficiency information network road of the community elderly care service. ③ The government should establish the information network of community service professionals, realize sharing of different resources in the community, ease the present situation of shortage of professionals in medical treatment, psychological and legal counseling and so on and especially increase professional service posts in health and psychological counseling gradually to meet elderly citizens' two major need themes of elderly care, namely "nursing" and "treatment" and realize the informationalized system of the elderly care service volunteers in the community, make full use of professional specialties of volunteers and improve the service quality.

3.6.6 Explore and Establish the Elderly Nursing Insurance System

We can learn from the successful experiences of Japan, Germany and other countries in implementation of the long-term nursing insurance and combine the present economic and social development level and aging situation in China to establish an elderly nursing system with the elderly long-term nursing social insurance as the main part and commercial social insurance as a supplement.

At first, the overall assumption of the elderly nursing insurance system is to comply with the "people-oriented" principle and meet the nursing system goal of "providing elderly citizens with a sense of security". In the system design, the government should fully balance the rights and interests of the country, units, elderly citizens, nursing elderly care institutions and other main bodies who attend the elderly nursing system, comply with and meet the purpose of nursing and care needs of disabled elderly citizens, summarize and study the experiences of the elderly nursing insurance systems in the US, Japan, Germany, Israel and other countries, study the universal law of the elderly nursing insurance system, analyze the special needs of elderly citizens in China, and combine the home-based elderly care, community-based elderly care, institution-based elderly care and other ways in the elderly care service security system of China. Meanwhile, in the system design, it not only needs to meet elderly citizens' needs in medical nursing, home-based nursing, nursing prevention, nursing environment, nursing quality and other aspects, but also provides proper care security for elderly citizens needing special nursing, especially those who get apoplectic hemiplegia, senile dementia, high paraplegia and other diseases. As for elderly citizens in remote mountainous areas and minority nationality regions, rural "households enjoying five guarantees", urban elderly citizens enjoying basic living allowances and others, the assuming methods of special nursing policies and nursing fees should be made to meet their nursing needs.

Secondly, the government should formulate laws and regulations related to the nursing insurance system and establish the system of making the social insurance nursing system as the main part and the commercial nursing insurance as a supplement to meet the Chinese national conditions, respectively formulating Elderly Nursing Insurance Law and Elderly Nursing Commercial Insurance Law as institutional bases to carry out the nursing insurance, and clearly defining the insured object, premium standard, claim amount and both parties' rights and obligations. Meanwhile, considering that China is in the critical economic and social transition period and the condition of getting old before getting rich, we find that there is a relatively large difference between urban and rural areas in the realization condition of the elderly care services. So the national unified elderly care and nursing insurance system can't be established. It can be only established at a successive and different level. Therefore, under the framework of the nursing insurance legal system, the government should make use of tax, financial subsidies and other policies to care for vulnerable groups who need elderly nursing services and realize justification and equality of the system.

Finally, the government should design the elderly nursing insurance system. In view of the relatively large difference in the social and economic development of regions, urban and rural areas, career and others in China, the government should take its design restrictions into full consideration and design the diverse nursing insurance systems in terms of different groups. So far, the elderly care insurance system in China hasn't been improved and realized the full coverage truly yet, so it can't carry out the obligatory nursing insurance system for the whole people like Japan and other countries. Summarizing factors in all aspects, we find that the pressure of elderly care services for urban population in China is higher than that of rural areas. Therefore, the government should first make experiments of the nursing insurance system and put it into effect from cities, popularize and improve it gradually and realize the system unification later.

1. Design different nursing insurance rates according to different economic development levels

China should establish the government-dominated social insurance nursing system and adopt the way of paying the premium by the government, enterprise and individual jointly. Considering the differences in region and payment capacity of China, the government should design reasonable and respective payment proportions with province as the unit. The provinces in East China with relatively good economic conditions can have the governmental part of premium borne by the county-level finance; those in Central China can have governmental part of premium borne by the prefecture-level city (city) finance. In case there is a poor condition with the prefecture-level city finance, it can be supplemented by provincial finance; as for provinces in remote mountainous areas of West China and minority nationality regions, the provincial finance bears the governmental part of nursing insurance. In case there is a poor condition with the provincial finance, it can be supplemented by national finance.

2. The confirmation of underwriting liability of elderly nursing

Nursing insurance product is different from medical insurance. Medical insurance provides security for the expenses of disease treatment of the insured, while nursing insurance provides security for nursing expenses that are caused by chronic diseases or deterioration of health conditions of the insured. The responsibility coverage of long-term nursing insurance should include treatment nursing and home nursing. According to the actual situation of China, there're more demands for home nursing. Home nursing generally refers to the fees caused by the insured as they lose their ability to take care of themselves and have to hire professional caregivers to provide nursing in their homes. The fees of home nursing mainly depend on its items and the self-care ability of the insured. Daily nursing items and service quality are stipulated in the nursing insurance contract.

3. Nursing insurance institutions based on the community

For the convenience of the elderly nursing services, nursing institutions should be based on the community where the elderly citizens are, make full use of the original community service staff and realize the rational allocation of resources. In this way, it can not only facilitate the needs of the community elderly nursing, but also improve the efficiency of the elderly services. Community service public welfare posts that are bought by the government can be used to solve the problem of community employment.

As a kind of useful supplement of the nursing insurance security, the nursing products developed by the commercial insurance company encourage insurance companies to carry out system innovation and product design innovation actively in terms of nursing insurance products. Medical services and insurance services are separated in the traditional Chinese medical insurance, which makes medical expenses difficult to be controlled and the moral hazards between the insured and medical institutions happen frequently. As for the nursing insurances developed by the commercial insurance company, the insurance company should involve in nursing institutions by shares or invest in them and authorize the agreed nursing institutions or service institutions to provide diverse nursing insurance services for the insured. As the nursing products of the commercial insurance company are more flexible than those of the social insurance, the commercial insurance company can design diverse nursing insurance products and realize product innovation in accordance with age, payment, economic ability and so on of the insured.

3.6.7 Strengthen the Rural Elderly Care Service Security

Because of the long existence of urban-rural dual structure in China, the effectiveness of the social security system and mechanism in rural areas is very weak. In the aspect of the elderly care services, the rural areas have mainly focused on family-based elderly care service at present. The social elderly care mechanism, including public elderly care homes and others, is very weak. China stipulates that the collective shall carry out the “Five Guarantees System,” namely food, clothes, housing, medicine and funeral guarantees as to elderly citizens of no family, ability to work, legal supporter and caregiver or stable source of income, but it hasn’t set up any system in terms of the rural elderly care services. In August 1996, *Law of the People’s Republic of China on Protection of the Rights and Interests of the Elderly*, a law on the protection of rights and interests of the elderly citizens was promulgated. In September 2009, China launched the pilot work of the new rural elderly care insurance, having milestone significance to the great mass of elderly citizens in rural areas. After their pension security was included into the national social security system, it has been of importance and urgency to build and explore the rural elderly care service system.

From the trend of social development, the weakening of the family-based elderly care function is irreversible. In extensive rural areas of China, the tradition of elderly care by family, land and children has still prevailed. The government shall take the lead to solve the elderly care service problem of the rural elderly citizens, give full play to the leading function of rural family-based elderly care, advocate establishing various elderly care services in rural areas with village committee as the carrier, institution-based elderly care as the supplementary form and so on and set up the elderly care service security system meeting the local actual condition in rural areas.

1. Build rural elderly care service system with voluntariness and suitability

In extensive rural areas of China, there is a great disparity of economic and social development level in rural areas, so it is impossible to implement the unified system or the one-size-fits-all elderly care service system. Under the guideline of government's policies and according to the local conditions, the elderly care service is formed at three levels, namely "full dependence on family care", "family care as the principal factor and self-care as the subsidiary factor" and "family care as the basis, community care as the core and self-care as the subsidiary factor" are chosen to meet the local conditions. Through the elderly services by children, public elderly care institutions run by the government, elderly nursing insurance bought by individuals and other forms, the elderly care service security of the rural elderly citizens can be achieved. In those rural areas with relatively developed rural collective economy and neighborhood committee-based communities with stable economic sources, the form "family care as the basis, community care as the core and self-care as the subsidiary factor" can be taken into consideration so as to reduce children's time and economic burden of taking care of the elderly citizens. In remote mountainous areas of West China and areas with relatively backward rural economy, people mainly carry out the family-based elderly care service and the government encourages and supports the elderly care institutions founded by social funds and give preferential support in tax, land and other aspects so as to develop the elderly services and the nursing system.

2. Strengthen the enforcement of relevant laws and guarantee the elderly care service rights and interests of the rural elderly citizens

Legalization and institutionalization are the fundamental guarantees of the elderly citizens to realize the goal of providing elderly citizens with a sense of security. *Law of the People's Republic of China on Protection of the Rights and Interests of the Elderly* which was promulgated in 1996 as well as *Marriage Law*, *Civil Law*, *Criminal Law* and others in China clearly stipulate the elderly citizens' maintenance and support right, but there is generally a problem of poor binding force in these laws. No clear definition of punishment on the people liable who fail to bear the elderly citizens' maintenance and support obligation has been established yet. Learning from the methods of Singapore, we can establish the rural elderly care insurance system in China gradually. In August 1994, Singapore promulgated a famous law *Maintenance of Parents Act*, which stipulates that anyone who refuses to support parents or take care of aged incapacitated parents will be fined tens of thousands of Singapore dollars or sentenced to one year's imprisonment. Nowadays, as many young couples working in the city leave their children in rural areas and ask their parents to take care of their children, they are too busy to care about their parents' life. This kind of phenomenon should be changed immediately through severer laws.

3. Develop the leading function of rural grass-roots organizations

To realize the harmonious development of the whole society, we must attach importance to the social and economic development in the vast rural areas of China. The rural areas of China are vast and there is a very great uneven development level in the regional development. Rural grass-roots organizations must play a leading role in the construction of various systems. In terms of the elderly care services in the rural areas, the neighborhood committee in rural areas should coordinate the forces in all aspects actively to know the basic situation of all elderly citizens in its jurisdiction, archive it and give special care to the poor elderly citizens who are incapacitated due to illness and not taken care of by their children. The village neighborhood committee should actively encourage and attract the elderly care institutions founded by social funds and give them preferential policies in tax, land, water, electricity and other aspects. The village committee should support the work in the rural elderly association, wedding and funeral councils and other social associations and organizations and give free rein to the authoritative coordination function of the traditional rural clan. The village committee with good economic conditions can appraise and elect "Families of Respect and Filial Piety" and award them with incentives. Rural grass-roots organizations should provide subsidies in a material way for rural family-based elderly care services, convenience home nursing and proper living subsidies low-income families which support the elderly citizens.

4. Develop the home-based elderly care service model in rural areas gradually

Home-based elderly care service in rural areas refers to that the rural neighborhood committee, as a kind of unit support establishes the home-based elderly care service institutions and trains and selects nursing staff to provide door-to-door services at a fixed point or time for the elderly citizens with service needs, such as household service, life care, health counseling and mental counseling. Service staff are local villagers selected through professional training. It will be best if they are rural women with free time, so they may take care of the elderly citizens in their village they are familiar with and the elderly citizens needing nursing may also accept them emotionally. At the same time, the nursing staff in the village should also report the health information of the nursed elderly citizens to the township elderly nursing management office in time. When some professional psychological and medical nursing services are needed, the professional nursing staff of the township or town elderly care home will complete them. As rural elderly citizens don't live as densely as those in urban areas, causing difficulty to door-to-door services, the nursing staff should base on the principle of "proximity service", improve the township or town elderly service facilities gradually, strengthen the home-based elderly service construction at the village level, arrange those incapacitated elderly citizens due to serious disability and advanced age suffering from apoplexia, hemiplegic paralysis and other diseases to the township elderly care service institutions. The objects of the home-based elderly care service at the village level are those elderly citizens who aren't so old and can live at home and receive door-to-door services at a fixed point or time. In view of unstable and unprofessional home-based elderly care service staff in rural areas and other problems, the township financial department should support to select and employ the staff with free time who can provide door-to-door services for the elderly citizens in their village in the way of purchase of the public welfare post. If necessary, the government should organize professional training on the elderly service and keep improving the service level and occupational qualities to provide better services for the elderly citizens.

5. Improve and enhance the service level of the township elderly care institutions

The rural elderly services still mainly depend on the institutional elderly care homes that are invested and founded by the government. Those elderly citizens of no family without the ability to work, a legal supporter or caregiver or a stable source of income and rural "households enjoying the five guarantees" should be gathered in elderly care institutions and given food, housing, medical care, rehabilitation, leisure, entertainment and other services. At present, because of the influence of traditional concepts, the occupancy rate of rural institution-based elderly care is low. Most of the elderly care institutions founded by social funds are still established in the urban areas. However, the township government still needs to improve and enhance the services in the institutional elderly care homes, improve the institutional facilities and equipment, stabilize the service staff team and raise professionalization, professional level and the quality of elderly services.

3.6.8 Publicize and Inherit the Traditional Culture of "Respecting the Elderly Citizens"

"Providing elderly citizens with a sense of security" is our basic goal and meets the excellent traditional culture of our Chinese nation. To create the socialized elderly care service system enjoyed by the whole people, we must cultivate "people-oriented" elderly care service security culture. ① Develop the good tradition of "respecting the elderly citizens" and advocate the system goal of "providing elderly citizens with a sense of security, medical treatment, education, new knowledge and techniques, a sense of worth and happiness" to cerate an atmosphere of respecting the elderly citizens and care for them in the whole society. ② Build the elderly care service security system comprehensively in the cultural concept of "family love and care", provide diverse elderly care services for the elderly citizens in different regions and with different incomes and health conditions, develop the leading function of the traditional family-based and family love-based elderly care, satisfy the needs of emotional attachment of the elderly citizens and combine community-based elderly care cultural and recreational services, hygiene and health care services, daily life care services, mental health counseling services and other various services on the basis of the professional elderly care service provided by the community. ③ Community-based elderly care makes up weak support and emotional alienation of the elderly citizens as their children fail to take care of them beside and enables them to enjoy their elderly life amid the living environment of neighbors and friends they are familiar with and around the strong family love and nostalgia.

References

1. Yuan, Ye, Xiaojiu Jia, et al. 2006. *Discussion on Theoretical Research and Operating Mode of Senior Care Service Socialization, Collection of Excellent Papers on Policy Theory Research into Civil Administration*, 1316.
2. The National Bureau of Statistics. 2000. *The Data of the National Population Census in 2000*, China: China Statistics Press.
3. Pan Haisheng et al. 2009. *Active Development of the Elderly Care Service Industry, Policy Outlook*, Issue 1, 41–43.
4. Qiu Peng. 2004. *Long-Term Nursing Insurance—The Future Development Point of China's Health Insurance Market, Shanghai Insurance*, Issue 3.
5. Liu Chang. 2010. The Construction of China's Elderly Care Service System—Based on the Perspective of System Theory. *Journal of Lanzhou Commercial College*, Issue 1.
6. Dong Hongya. 2010. *Development Process and Experience Enlightenment of Chinese Government Elderly Care Services, Population and Development*, The 5th issue.
7. Chi Qingyuan et al. 2006. *A Tentative Plan for Establishing Long-term Nursing Insurance in China, Contemporary Manager*, Issue 21.
8. Liu Wenfu. 2009. *Elderly Care Service Practice in Developed Countries and Its Enlightenment to Our Country, Legal System and Society*, Issue 3.
9. Li Jianfei et al. 2009. *Study on the Construction of the Elderly Nursing Insurance System in China, Insurance Studies*, Issue 11, 65.
10. Han Ainong. 2005. *Analysis on the Necessity of Non-Profit Organizations Playing a Role in the Field of Social Security, Inner Mongolia Science Technology and Economy*, Issue 20.
11. Ge Liying et al. 2009. *A Preliminary Study on Demand Characteristics and Industry Models of Home-Based Elderly Care Services, Economic Studies*, Issue 4.
12. Michael Koehlerhof. 2010. *Social Assistance in Family Care from Cities in Germany, Situation Compilation of Elderly Care Service in Foreign Countries, Hong Kong, Macau and Taiwan Regions*. China: China Social Press.
13. Yang Hongyan. 2004. *Elderly Nursing Insurance System in Developed Countries and Enlightenment, Foreign Medicine Health Economics*, Volume 21, Issue 1.

Chapter 4
Funding Guarantees for the Modern Pension Security System

Cuiping Jiang

4.1 Status and Function of the Pension Security Fund

4.1.1 Pension Security Fund

The pension security fund refers to the pension insurance premium collected from workers and employing entities or the fund concentrated from the direct financial expenditures of the government according to law that are then used to meet workers' basic living needs after they retire. Currently, China's pension security fund mainly consists of the social coordination fund and personal account fund of the basic pension insurance, the national social security fund as well as other supplementary pension insurance funds such as the enterprise annuity.

The social coordination fund is the most important part of the pension security fund. Based on the strong enforcement of state administrative departments, the pension fund is collected, managed and released in a unified way and shows the nature of the public pension fund. Before the pension system reforms of 1997, China's pension security fund was the social coordination fund. The social coordination fund is raised on a pay-as-you-go basis and paid according to a defined benefit model. At present, China's social coordination fund is paid by enterprises and equates to 20% of their total payroll.

The personal account fund is another part of the basic pension fund. It is raised on a fund-accumulation basis and paid according to a defined contribution model. As stipulated, the contribution rate of urban enterprise employees is 8% in China. Besides the personal accounts of urban enterprise employees, China is trying to set up a personal pension account for flexible employees and rural residents. What

C. Jiang (✉)
School of Criminology, People's Public Security University of China,
#1 Muxidinanli South Lane, Xicheng District, Beijing 100038
People's Republic of China
e-mail: binghan_39@163.com

Y. Wang (ed.), *The Development of Security and Whole Care System for the Aged in China*, https://doi.org/10.1007/978-981-10-6481-4_4

workers receive once they retire is decided by the accumulated fund in their personal account as well as the value maintenance and appreciation of the fund.

The enterprise annuity, also known as the enterprise's supplementary pension, is the accumulated fund provided by an enterprise for its employees in addition to the state-specified basic pension under the guidance of state policy. As an important supplement to the basic pension insurance system, the enterprise annuity is aimed directly at raising retirees' pension level.

A part of the national social security fund also goes to the pension security fund. The national social security fund was set up in August, 2000 as an important strategic reserve centralized by the government. It will be primarily used to satisfy social security demands during the future population aging peak. At present, only a few countries have set up a pension reserve fund, including Norway, France, New Zealand, etc. By the end of 2010, the total amount of all funds managed by the National Council for Social Security Fund reached RMB 856.69 billion.[1]

This chapter focuses on the study of the basic pension insurance fund.

4.1.2 Status and Function of the Pension Security Fund

The pension security fund serves as a material basis and the core of the modern pension security system. Its payment capacity is a guarantee and foundation of the pension security system that can be shared by everyone. Without the fund guarantee, the entire pension security system would make no sense at all.

4.1.2.1 Social Welfare Effects of the Pension Security Fund

From a macro perspective, the pension security fund can adjust the distribution of income to narrow the gap between the rich and the poor and maintain social stability. First, a portion of the national income can be transferred to the government budget via payments into the pension security fund. This kind of transfer payment will satisfy the basic living needs of some special members of society. Second, the self-insurance funds of enterprises and residents can be gathered by the pension security fund system through fund collection or tax payment and forms the social insurance fund. Therefore, the pension security fund can change the income distribution state of social members in their lifetime (before and in work, in periods of unemployment and in retirement) and narrow the income gap to some extent, to improve equity and efficiency of income distribution.

[1] *2010 Annual Report on the National Social Security Fund*, http://www.ssf.gov.cn/tzsj/201105/t20110519_3185.html.

4.1.2.2 Economic Effects of the Pension Security Fund

A portion of the pension security fund can be invested in the market so as to not only maintain or increase its value but also promote economic development. Economic effects of the pension security fund are embodied primarily in national savings, investment and economic growth.

Economic growth depends heavily on the increase in the workforce and capital as well as an improvement in productivity. Investment is a process of capital accumulation. Investment can increase the funds that are allotted to each person and promote the introduction of newer and more efficient production technology. In economically developed countries, pension assets account for more than 50% of the national GDP. A large portion of the pension fund is invested in the bond and stock market to increase the liquidity of the market economy and improve economic efficiency.

4.2 Major Raising and Payment Models for the Pension Security Fund

4.2.1 Major Models of the Pension Security System

Before the pension security system that emphasized the public functions of the state came into being in Germany, it was always seen as the responsibility of individuals and families to take care of the elderly. Therefore, the pension security fund was accumulated by individuals and families and distributed within the scope of the family. As early as 1601, the Parliament of England issued the *Poor Relief Act*, according to which the state should undertake the public responsibility for the welfare of the poor including the elderly in poverty. In spite of a small scope of benefits, this act had an inkling of the pension security system. The modern pension security system originated in Germany. It emphasized the common responsibilities and obligations of the state, the employer and the individual. The responsibilities shouldered by the above three parties varied in different countries, which brought different models of the pension security system.

4.2.1.1 Welfare State Models

(1) The Beveridge Model. This model originated from the *Beveridge Report* published in the UK in 1942. It proposed that the UK public should be offered the minimum living allowance and that there should be a single set of principles for the payment, allowance standard and management of social security. Based on this report, the UK government established the Department of Social Security and issued the *National Insurance Act* in 1946 to bring the national

pension system (in which pensioners paid a contribution and received allowances according to the unified standard) into the entire national insurance system, so that all British citizens could enjoy the nationwide high-welfare "cradle-to-grave" social security in various aspects including pension insurance. Later, Nordic countries such as Denmark, Norway, Finland and Sweden followed the UK's example and established this kind of high-welfare pension security model. After the 1960s, the UK carried out several reforms of this pension security system. By the end of the 20th century, a multi-pillar model had almost replaced this welfare state model. Nordic countries also made some small adjustments during this period. However, this nationwide and high-welfare concept and model is still used today and has become a good example of the welfare state. As all the four Nordic countries are located on the Scandinavian Peninsula, it is also known as the Scandinavian Model.

This pension security system primarily has the following features. First, the state undertakes the main responsibility for elderly care. This can be seen from the sources of the pension security fund. Before the pension system reform, about 50% of the pension security fund of Nordic countries came from financial allocations from the central and local government. This percentage reached 80% in Finland.[2] The second part of the pension security fund was paid by employers. The individual contribution accounted for a very small proportion. Second, every member of society can benefit. According to Danish law, all Danish citizens at or above the age of 67 can receive a pension from the government with no strings attached, whether they have paid any social insurance or not.[3] Third, unified standards are adopted. Before the reform, Sweden stipulated that all citizens would get 60% of their salaries as their pension security fund.

The Scandinavian Model has brought the concept of the welfare state into full play and become a good example of the welfare state. However, this high-welfare, high-tax and highly equalized pension security system also brings serious social problems. For example, it became hard to increase workers' enthusiasm for work; expenditures on pension security grew faster than the national GDP. After the 1990s, Nordic countries started one pension security system reform after the other. They focused on decreasing the proportion of government allocations in the pension security fund, raising the contribution rate of employers, reforming the tax system and relating pensions to retirement time.

(2) The Bismarck Model. This model originated from Germany's pension security system which had a profound influence on other continental European countries such as France, Austria and Italy. The pension insurance system was established by Bismarck in Germany at the end of the 19th century, making Germany a pioneer in the field of elderly care. According to Germany's original pension security system, different pension security plans were made for

---

[2]Ding [1].

[3]Xu [2].

different groups, namely white-collar workers, blue-collar workers and miners. Pensions of the three groups were managed by a special institution.[4]
The main feature of this model lies in a guarantee from the state, workers' pensions are also related to their jobs and contributions. Different groups are included in different pension security systems. The pension is connected with pre-retirement income. Compared with the Beveridge Model, the proportion of government allocations is lower and the contribution rate of employers is higher in this model, though the state still undertakes the main responsibility in the pension security system. Moreover, the criteria of pension distributions are differentiated. For example, France differentiates the pension distribution criteria of the public and private sectors. Generally, the pension security fund of states adopting the Bismarck Model have three sources, namely the national public pension, the ordinary occupational pension and the personal pension saving. The national public pension serves as a major source. The elderly participating in the pension plan can receive relatively high public pensions. The national public pension is jointly undertaken by the government, employers and individuals. The occupational pension and personal pension saving accounts for a very small proportion of the pension security fund. For example, in 2008, the private pension assets of Germany and Italy respectively accounted for only 5.2 and 4.1% of the national GDP.[5]By the end of the 20th century, continental European countries carried out a series of pension system reforms to cope with the aging population and the decrease in the rate of economic growth. Their reforms primarily focused on developing a multi-pillar pension security system, adjusting the payment years, the payment level and the pension level, encouraging the development of the private pension plan and implementing preferential tax policies towards the private pension plan.

4.2.1.2 Social Insurance Models

(1) Occupational Pension Model. This model mainly prevails in such countries as the US, Canada, Australia and Japan. After the pension system reform, the UK also adopted this model. One outstanding feature of this model lies in the fact that private pensions account for a very large proportion of the pension security fund. For example, the proportion of private and public pension assets in the national GDP was respectively 67.6 and 17.9% in the US in 2008. These two percentages became 62.9 and 8.5% in Canada, 82.3 and 5.9% in Australia and 73 and 3.8% in the UK. Because of the later national pension plan of Japan, the proportion of its public pension assets in its national GDP increased to 23.2%, while the proportion of its private pension assets in its national GDP was

[4]Jiang [3].

[5]OECD Global Pension Statistics.

25.2%.[6] However, the Japanese people distrust their national pension system which has many uncertainties in itself. They refuse to join the national pension plan or pay for it, which has become a general problem in Japan's pension system.[7]

In the Occupational Pension Model, the pension security fund is contributed mainly by individuals and employers. On the one hand, the pension insurance tax or premium is paid by individuals and employers. On the other hand, the government gives certain subsidies. Only those social members who have paid the pension insurance tax or premium can receive pension after retirement. For example, the pension security system of the US is supported mainly by its social security plan, occupational pension plan and individual retirement account. The social security plan is compulsory and has wide coverage, but the level of social security is rather low. 85% of the social security fund comes from the social security tax. As stipulated, any worker who has reached the minimum retirement age of 62 (the statutory retirement age is 65), paid the social security tax for at least 10 years and accumulated 40 social security points has the right to apply for and receive the social security payment.[8] The occupational pension plan is the most important pension system. Specific to different occupational groups, it is divided into the public pension plan and the private pension plan. The former mainly serves employees of government organs and the latter serves employees of enterprises. The contribution rates of employers and employees are both 6.2%.

(2) Compulsory savings mode. This mode is mainly adopted by such countries as Singapore, Chile and others. It means that the government forces members of the society to participate in the social pension insurance plan and the fully-funded system is adopted in pension security fund.

Singapore began to implement the Central Provident Fund in 1955, requiring the staff to pay 10% of their salary with employers and employees sharing it by halves. This ratio was raised to 13% in 1968 and then 50% in 1984, and it's 40% at present with employers paying 17% of it and employees the rest 23%. Besides, a personal account is established in accordance with this. Personal accounts of unretired staff are divided into ordinary accounts, medical accounts and special accounts, and only the fund of special accounts is used for pension. In 1987, in order to prevent too much of the accumulation fund from being used in other payment, the Singapore government regulated that a minimum deposit must be left in the accumulation fund account after one reached 55 so as to fully ensure that he could purchase the minimum lifetime annuity which was equal to 25% of the average social return after retiring at the age of 60. Chile started to implement the pension insurance system with the full accumulative system in 1980. Chile regulated that all individuals should pay at least 10% of the tax

[6]Liu [4].

[7]Feng [7].

[8]Chen [5].

revenues to establish a pension account in accordance with regulations. The fund of this account was invested in the capital market and one could receive the pension fund by month after reaching the legal retirement age or losing the working ability. Later, they made some adjustments to the personal account and established a supplementary account, encouraged individuals to raise the contribution rate voluntarily and compensated for the savings account, contractual savings plan and others.

The main distinction of the pension security systems of Singapore and Chile lies in different management institutions of pension security fund. The pension security fund of Singapore is uniformly managed by the Central Provident Fund Board established by the Central Government, while the pension security fund of Chile was managed by a pension fund management company which was a private institution and people could choose different pension fund management companies to establish their own accounts. Regardless of the different management methods, they were both a social pension insurance mode under the government enforcement in terms of the nature.

4.2.1.3 National Security Model

This pension security system is adopted by countries implementing socialist Command Economy. Originating from the Soviet Union, it was adopted by East European countries, the Democratic People's Republic of Korea, Mongolia and China before the implementation of the reform and opening-up policy. Similar to "the Beveridge Model" adopted by welfare states, national security model attaches more importance to justice with a strong equalitarianism tendency. In this model, pension security is the liability undertaken by the state and state-owned enterprises and the individuals need not pay any charges.

At present, China's social pension system is a combination of government social pooling and social insurance, but the former one is still dominant, so the development of occupational pension insurance and commercial pension insurance is quite limited. In view of the development of the pension security system mode and the pension security system reforms carried forward by all countries, from the primary family and private pension to the national public pension and then the multiple support pension mode commonly implemented by all countries at present, the state's function in the pension security system has been maximized; while in view of the reform of the pension security system, the state's function in pension security is constantly weakening and the modes of pension security are gradually getting diversified.

4.2.2 Fund Raising System of Pension Security Fund

In terms of specific implementation, different fund raising and payment modes were formed due to the different fund raising and payment methods of pension security

Table 4.1 Fund raising and payment systems of pension security fund

	DB	DC
Pay-as-you-go system	Defined benefit pay-as-you-go system applies to most of the countries that adopt the public pension fund system. Such countries include Germany, France, Netherlands, Finland and others. This method has also been adopted in the "social pooling account" in China's pension fund system	It's a nominal account system under the pay-as-you-go system. That is to say, a nominal personal account is established for account keeping without being used for any actual fund deposit. Countries adopting this system include Sweden, Latvia and others
Fund accumulation system	Defined benefit fund accumulation system applies to the enterprise annuity system of some developed countries including the US, Canada and others	Defined contribution fund accumulation system applies to most countries adopting the fund accumulation system. Such countries include the US, Australia, Chile, Singapore and others

fund. Theoretically, the pension security fund has three fund raising modes including the pay-as-you-go system, the full fund accumulation system and the partial fund accumulation system, and the corresponding payment modes are DB (short for Defined Benefit) and DC (short for Defined Contribution) (see Table 4.1).

4.2.2.1 Pay-as-You-Go System

The pay-as-you-go system refers to the institutional arrangement of paying the pension fund of a generation of retirees with the payment or tax payment of working people in the same period. The pension security fund of the same period is fully granted with some capital left basically. In view of revenue and expenditure, the pay-as-you-go system is a kind of short-term balance that generally remains for one year. In view of the revenue distribution of the entire society, the pay-as-you-go system is an adjusting method of wealth distribution between two generations of people. The pay-as-you-go system has quite a strong mutual aid function among society members, and the revenue of a retiree has nothing to do with or was in a weak correlation with his own payment level or the enterprise's benefit. Meanwhile, with the revenue and distribution realized simultaneously, there isn't any pension security fund balance in the pay-as-you-go system, so it's unnecessary to put the pension security fund into the capital market for preservation and appreciation, so it's subject to quite a slight impact of inflation. Generally speaking, a country that centers on the national public pension security implements the pay-as-you-go system, and such countries include Germany, France and others.

The Defined Benefit payment system is generally adopted in the pay-as-you-go system. If the elderly population increases in a society, the demand for pension security fund will also increase, which will certainly increase the payment pressure of the younger generation. Many people think that the pay-as-you-go system will result in a serious payment crisis in the face of population aging. Under such a pressure, many countries tend to change the pay-as-you-go system to the fund accumulation system for the national pension security fund. Take the pension security system reform conducted by Chile in the 1980s for an example. They completely changed the pay-as-you-go system into the fund accumulation system, which was regarded as a classic example of pension system reform.

China implemented the pay-as-you-go system before the pension security system reform in 1997. And after the reform, the pay-as-you-go system was implemented in the social pooling account, while the fund accumulation system was implemented in the personal account. The pay-as-you-go system was still dominant in terms of institutional arrangement.

4.2.2.2 Full Fund Accumulation System

The full fund accumulation system means that the personal and employer payments of the current period are fully kept in the personal account, and one can receive the pension fund in accordance with the payment record and the return on investment of the personal account fund with a slight but appropriate amount of allowance from the government after retirement. The retirement pension received upon retiring is depended on two aspects: One is the personal payment level; and the other is the return on investment of fund. In terms of revenue and expenditure, the full fund accumulation system maintains a long-term balance between one's expenditure during the working period and the income after retirement. In terms of income distribution, the full fund accumulation system is a type of distribution within the same generation of people. Under the full fund accumulation system, since the pension security revenue of the current period is to be consumed decades later, it's definitely faced with the problem of fund preservation and appreciation, the pension security fund must be used for investment, so it's also faced with some investment risks. Generally speaking, most countries with social insurance being dominant will probably adopt the full fund accumulation system.

Since the revenue and expenditure of the full fund accumulation system appear in the life cycle of a person, the population aging of the entire society has quite a small influence on the person. Therefore, people generally believe that the full fund accumulation system reacts to the population aging crisis better than the pay-as-you-go system. The full fund accumulation method is adopted for the "personal account" of China's pension insurance.

4.2.2.3 Partial Fund Accumulation System

The partial fund accumulation system is the combination of the pay-as-you-go system and the full fund accumulation system. It generally refers to preserving and accumulating part of the fund received on the basis of the high contribution rate of the pay-as-you-go system. For example, the current pension security fund of China can be regarded as a partial fund accumulation system.

The partial fund accumulation system can not only ensure the reallocation of the revenue of different generations under the pay-as-you-go system but also bring the fund accumulation system's function of stimulating economy into play. However, under the partial fund accumulation system, how to integrate the pay-as-you-go system and the fund accumulation system is a difficult problem that we must face.

4.2.3 The Payment System of Pension Security Fund

4.2.3.1 Defined Benefit

Defined Benefit refers to determining the pension fund benefit of every insured person by a certain equation under the pension security system. Take the regulation of pension fund's replacement rate to the salary for an example. A certain amount of cash payment or one-off payment is provided regularly based on the person's age and the scale of the past contribution he made. In view of practical use, countries that adopt a pension fund system with the pay-as-you-go system generally adopt the Defined Benefit payment system.

4.2.3.2 Defined Contribution

Defined Contribution generally applies to the fund accumulation system. That is to say, a personal account is established for every insured person with the payment amount recorded; it's a pension fund plan of calculating and granting the pension fund benefit in accordance with the pension fund earnings when the personal account owner retires. The amount of personal benefit is mainly depended on the accumulated fund and the earnings of the investment operation of the fund invested in the account. Some countries are also trying to implement Defined Contribution under the pay-as-you-go system. For example, Sweden is implementing the nominal account system.

Obviously, there is a wide variety of combinations of the raising system and payment system of the pension security fund. In the same country or region, different raising and payment systems are adopted for different pension fund plans implemented in the same period. For example, the pay-as-you-go system has been adopted in the "social pooling" account, while the Defined Contribution fund accumulation system has been adopted in the "personal account" in China's basic pension insurance. Different methods may be adopted for the same pension plan in different historical periods. For example, the social security plan which is the basic pension system of the US has experienced the reform from the full fund accumulation system to the pay-as-you-go system and then the partial accumulation system.

In short, no single system or mode is perfect and unchanged, a system that goes relatively well with population development and economic level should be considered in designing the pension security system.

4.3 Status and Adjustment of Revenues and Expenditures for China's Pension Security Fund

4.3.1 Revenues and Expenditures of China's Pension Security Fund

The proportion of a pension security fund in fiscal expenditure and GDP is an important figure used to judge the pension security system of a country or region. Compared to the current economic condition and growth rate of fiscal revenue in China, revenues and expenditures of China's pension security fund develop slowly, falling far behind developed countries.

By the end of 2009, as to basic pension insurance for China's urban residents, the revenue reached RMB 1149.08 billion, the expenditure reached RMB 889.44 billion and the balance was RMB 1.25 trillion (see Table 4.2). The revenue and expenditure accounted for 3.4 and 2.6% of the GDP that year respectively. In 2009, the expenditures for employment and social security accounted for 10% of the national fiscal expenditure that year.[9] From Table 4.3 we can tell that the proportions of revenue of China's basic pension insurance in GDP had been increasing year by year, yet with slight growth rates. In 1989, the total revenue of China's basic pension insurance accounted for 0.9% of the GDP. In 2009, this proportion reached about 3.4%. Compared to the growth rate of the overall economy and the yearly growth rate of GDP, which is about 10%, the proportions of revenues and expenditures of China's pension insurance in GDPs have only slightly increased.

Compared to other countries, the proportions are rather low. Table 4.4 shows proportions of the total pension assets in GDPs in some developed or developing countries in 2008. Of the countries, the proportion reached 88.2% in Australia, 71.4% in Canada, 76.8% in the UK, 85.5% in the US, 76.8% in Finland, 48.3% in Japan, 5.1% in France, and 5.2% in Germany. Countries such as the UK and the US mainly adopt the fund accumulation system, which leads to a large amount of retained earnings. Countries such as Germany, France and Italy pay pensions on a pay-as-you-go basis, leading to a small amount of retained earnings balance. On the whole, the proportion of the total pension assets in OECD countries reached as high as 73.5%, while the figures in Argentina and Brazil were also above 10%. China's

[9]*China Statistical Yearbook* (2009).

Table 4.2 Revenues and expenditures of China's basic pension insurance (2000–2009) (unit: RMB 100 million)

Year	Revenue	Expenditure	Balance
2000	2278.5	2115.5	947.1
2001	2489.0	2321.3	1054.1
2002	3171.5	2842.9	1608.0
2003	3680.0	3122.1	2206.5
2004	4258.4	3502.1	2975.0
2005	5093.3	4040.3	4041.0
2006	6309.8	4896.7	5488.9
2007	7834.2	5964.9	7391.4
2008	9740.2	7389.6	9931.0
2009	11,490.8	8894.4	12,526.1

Table 4.3 Proportions of revenues and expenditures of China's basic pension insurance for urban residents in fiscal expenditures and GDPs over the years (2000–2009) (unit: %)

	Revenue of basic pension insurance for urban residents			Expenditure of basic pension insurance for urban residents			Annual GDP growth compared to last year
Year	Annual growth compared to last year	Proportion in GDP	Annual growth of proportion in GDP compared to last year	Annual growth compared to last year	Proportion in GDP	Annual growth of proportion in GDP compared to last year	
2000	–	2.3	–	–	2.1	–	
2001	9.2	2.3	0	9.7	2.1	0	8.3
2002	27.4	2.6	0.4	22.5	2.4	0.2	9.1
2003	16.0	2.7	0.1	9.8	2.3	−0.1	10.0
2004	15.7	2.7	0	12.2	2.2	−0.1	10.1
2005	19.6	2.8	0.1	15.4	2.2	0	9.9
2006	23.9	2.9	0.2	21.2	2.3	0.1	12.7
2007	24.2	2.9	0	21.8	2.2	0	14.2
2008	24.3	3.1	0.2	23.9	2.4	0.1	9.6
2009	18.0	3.4	0.3	20.4	2.6	0.3	9.2

Note The proportions of revenues and expenditures of the pension insurance stipulated in the table all come from the statistics of basic pension insurance for urban citizens
Source China statistical yearbook (2010). Statistical communiques of relevant years

total pension assets accounted for about 6% of the GDP,[10] far lower than the average proportion in developed countries.

[10]OECD Global Pension Statistics.

Table 4.4 Proportions of the total pension assets in GDPs of some countries in 2008 (unit: %)

OECD country	Proportion	OECD country	Proportion	OECD country	Proportion
Australia	88.2	Ireland	57.8	Spain	13.8
Austria	4.9	Israel	46.9	Sweden	34.6
Belgium	8.3	Italy	4.1	Switzerland	101.2
Canada	71.4	Japan	48.3	Turkey	2.3
Chile	67.2	South Korea	28.3	UK	76.8
Czech Republic	4.6	Luxembourg	2.2	US	85.5
Denmark	43.3	Mexico	7.7	Average Proportion in OECD Countries	73.5
Estonia	6.9	Netherlands	129.8	Other Countries and Regions	
Finland	76.8	New Zealand	19.0	Argentina	11.5
France	5.1	Norway	12.2	Brazil	17.1
Germany	5.2	Poland	14.1	India	5.4
Greece	0.02	Portugal	19.1	Indonesia	2.2
Hungary	13.1	Slovakia	4.7	Russia	1.5
Iceland	118.3	Slovenia	2.6	South Africa	58.4

Note The total assets stipulated here contain public and private pensions
Source OECD global pension statistics

According to expenditures of public pensions (see Table 4.5), in OECD countries in 2007, the average cash expenditure of the national public pension accounted for 17.0% of a government's fiscal expenditure and 7.6% of the GDP while the average non-cash expenditure accounted for 8.0% of the GDP. According to governments' expenditures for public pensions, while the expenditures were less than 10% in Iceland, Ireland, South Korea and Mexico, other countries' expenditures all reached more than 10%. In countries where public pensions are a main measure used to support the elderly, such as Austria, Germany, France, Greece, Italy, Japan, Poland and Spain, fiscal expenditures for public pensions all exceeded 20%, while the figure in Italy already approached 30%. In countries where the elderly care is based on private models, the figure reached 12% in the UK, 16.3% in the US, 10.6% in Canada. The proportions of cash expenditures in GDPs all exceeded 10% in Austria, Germany, France, Greece, Italy and other countries, while the figure reached 8.8% in Japan, 6.0% in the US, and 5.4% in the UK. The non-cash expenditures reached a high level in all OECD countries. In China in 2007, the proportion of expenditure of employment and social security in total fiscal expenditure reached 10.9%[11]; the expenditure of the basic pension fund for urban residents accounted for 2.2% of the GDP (see Table 4.5). The fiscal support for the public pension and the pension security system in China fell far behind the developed countries.

[11]*China Statistical Yearbook* (2008).

Table 4.5 Proportions of public pensions in OECD countries in 2007 (unit: %)

Region	Proportion of cash expenditure in GDP	Proportion of cash expenditure in government spending	Proportion of non-cash expenditure in GDP	Region	Proportion of cash expenditure in GDP	Proportion of cash expenditure in government spending	GDP proportion of non-cash expenditure in GDP
Australia	3.4	10.1	4.5	Japan	8.8	27.0	10.1
Austria	12.3	25.3	12.7	Korea	1.7	5.7	1.9
Belgium	8.9	18.3	9.0	Luxembourg	6.5	18.1	6.6
Canada	4.2	10.6	4.2	Mexico	1.4	7.2	1.4
Chile	5.2	–	5.2	Netherlands	4.7	10.4	5.5
Czech Republic	7.4	17.5	7.7	New Zealand	4.3	10.9	4.3
Denmark	5.6	10.9	7.3	Norway	4.7	11.4	6.5
Estonia	5.2	15.2	5.3	Poland	10.6	25.2	10.7
Finland	8.3	17.5	9.2	Portugal	10.8	–	10.8
France	12.5	23.9	12.8	Slovakia	5.8	17.0	6.2
Germany	10.7	24.5	10.7	Slovenia	9.6	22.7	9.7
Greece	11.9	26.3	12.0	Spain	8.0	20.5	8.5
Hungary	9.1	18.3	9.6	Sweden	7.2	14.1	9.5
Iceland	1.9	4.5	2.3	Switzerland	6.4	19.9	6.7
Ireland	3.6	9.7	3.9	Turkey	6.1	–	6.2
Israel	4.8	10.7	5.0	UK	5.4	12.0	5.9
Italy	14.1	29.4	14.1	US	6.0	16.3	6.0
Average in OECD countries	7.6	17.0	8.0				

Source OECD global pension statistics

4.3.2 *Pooling Pension Security Funds: Pension Contributors, Pooled Funds and Required Years of Contribution*

4.3.2.1 Main Contributors to Pension Security Funds

As to the pooling models for pension security funds, transition from the pay-as-you-go system to the partial accumulation system is under way in China. Pension contributors mainly consist of the government, enterprises and individuals.

As to the government's fiscal support, gradual transition from county-level pension pools to province-level, and even nation-level pension pools is under way in China. Different levels of pension pools are adopted in different regions. It is stimulated in the Social Insurance Law of the PRC (2011) that the basic pension insurance will gradually be pooled at nation level and other pension insurances will gradually be pooled at province level, of which the exact dates are to be determined. By the end of 2006, in 13 cities in China, such as Beijing, Shanghai and Tianjin, pensions had been pooled at province level, while county-level pension pools were still in effect in other provinces.[12] In provinces where pensions are pooled at province level, all of the pension funds are collected and paid to pensioners by provincial financial authorities, which will give certain fiscal subsidies to pension security funds to balance the differences in all regions within a province and ensure the pensions being paid to retirees on time and at fixed amount. Provincial financial authorities are counted responsible for social pooling of pension. Municipal and county financial authorities are counted responsible for pooling at municipal level and county level respectively. On account of different levels of pension pools, there isn't any explicit requirement for the amount of subsidies the financial authorities should offer to the corresponding accounts of pension pools. When there are sufficient government revenues, pension payment won't be a problem. When there are not, local governments will put the pressure on enterprises or individuals to pay the pensions. If the contribution rate for enterprises is raised or funds in individual pension accounts are withdrawn earlier, there will be plenty of empty individual accounts.

Enterprises are important contributors to basic pension insurance. As to Chinese enterprises' contribution to the pension insurance, almost 100% of state-owned enterprises have insured their employees while there're still only a small proportion of private enterprises insuring their employees. For example, in June 2010, a survey of how many private enterprises in Binhai County of Jiangsu Province insuring their employees showed that only 27.5% of the private enterprises insured their employees, while most of the private enterprise owners there explicitly refused to insure their employees.[13] In 2009, 1.37 million new jobs were created by rural

[12]Zhang [6].

[13]Feng [7].

enterprises in China, while 167,000 employees were insured, which accounted for 12.2% of the total employees. That is ascribed to private enterprises' pursuit of profit and efforts to control costs, employees' undervaluation of social pension insurance, frequent personnel changes in private enterprises that make it difficult to carry out corresponding regulations, and so on.

As to individual contribution to the pension system, the coverage rates in urban and rural areas in China have been increasing year by year. However, there is still a great disparity between urban and rural areas. Table 4.6 shows the rates of coverage of urban pension insurance for urban employees and rural pension insurance for rural workers between 2006 and 2009 in China. By the end of 2009, the coverage rate for China's urban employees was 75.7% while the figure for rural workers was only 15.6%. Between 2006 and 2009, the coverage rate for China's urban employees increased by nearly 9.4% while the figure only increased by 4.6% for rural workers. According to the coverage rates in different regions, coverage rates of pension insurances are higher in developed countries (regions) than in underdeveloped regions.

4.3.2.2 Contribution Rates of Pension Security Funds

According to *Decision of the State Council on Establishing a Unified Basic Old Age Insurance System for Enterprises Employees* issued by the State Council of the People's Republic of China in 1997, usually the contribution rates of pension insurances for enterprises should not exceed 20% of their total payroll (including the funds transferred into individual accounts), which should be reported to Ministry of Labor and Social Security and Ministry of Finance for approval. The contribution rate for individuals is 8% of their insured salary. Of the enterprises' contribution to pension insurance equal to 20% of the payroll, 17% will be transferred to the social pooling account, while the other 3% plus the part contributed by individuals equal to 8% of the insured salary will be deposited to individual accounts, which amounts to 11% of the payroll in total. Compared to other

Table 4.6 Coverage rates of pension insurances in China's urban and rural areas

Year	Urban area			Rural area		
	Number of insured people (unit: 10,000 people)	Number of employees (unit: 10,000 people)	Coverage rate (%)	Number of insured people (unit: 10,000 people)	Number of employees (unit: 10,000 people)	Coverage rate (%)
2006	18,766.3	28,310	66.3	5373.7	48,090	11.0
2007	20,136.9	29,350	68.6	5171.5	47,640	10.9
2008	21,891.1	30,210	72.5	5595.1	47,270	11.9
2009	23,549.9	31,120	75.7	7277.3	46,875	15.6

Source China statistical yearbook (2006–2010)

Table 4.7 Contribution rates in some countries and regions in 2009 (unit: %)

Country/region	Contribution rate for enterprises	Contribution rate for individuals	Total contribution rate
UK	12.8	11	23.8
Sweden	11.9	7.0	18.9
US	6.2	6.2	12.4
Germany	10.0	10.0	20.0
France	9.9	6.8	16.7
Finland	4.5	17.1	21.6
South Korea	5.5	5.5	11
Japan	7.7	7.7	15.4
Hong Kong, China	5	5	10
China	20	8	28
Average of EU countries	11.2	8.4	19.6
Average of OECE countries	24.6	7.9	22.5

countries, the contribute rate of pension insurance for China's enterprises is very high (see Table 4.7). The International Labour Organization designates the warning line for an enterprise's contribution to pension insurance as 25% of its total payroll, while developed European countries designate the warning line as 24% of an enterprise's total payroll. As to the actual affordable tax rates calculated by the World Bank in 2009, China's social insurance contribution ranked first among 181 countries and regions, approximately doubling the average figure of other three BRIC countries, tripling the average of the five Nordic countries, 2.8 times the average of G7 countries and 4.6 times the average of neighboring East Asian countries.[14]

During the process of implementation, contribution rates for enterprises in some regions are higher than 20%, which brings heavy burdens to the enterprises and increases the costs for business operation. Yang and Liu [8] gave analysis on the burden of pension insurance for state-owned enterprises in Zhejiang, Jiangxi, Guizhou and Liaoning provinces that represent the economic development of China. Based on the contribute rate for enterprises, which is 20%, many state-owned enterprises' contribution to the pension insurance every year exceed 30% of their annual profits or even higher. For example, in 2007, the contribution of state-controlled enterprises to the pension system accounted for 35.6% of their profits in Zhejiang Province, 33.7% in Jiangxi, 31.0% in Guizhou, and 56.3% in Liaoning. Those are only state-owned or state-controlled enterprises, most of which are capital intensive companies. If we analyze labor intensive enterprises or some

[14]Yang and Liu [8].

Table 4.8 The central government's subsidiary expenditures directly used to fund the insufficient pension

Year	Amount of subsidy (RMB 100 million)	Total fiscal support (RMB 100 million)	Proportion of the subsidy in fiscal expenditure (%)
2000	337	15,886.5	2.1
2001	349	18,902.6	1.8
2002	410	22,053.2	1.9
2003	400	24,649.5	1.6
2004	524	28,486.9	1.8
2005	576	33,930.3	1.7

Source Relevant data from China Statistical Yearbook. Part of the data are cited from Wang [36]

small- and medium-sized private businesses, we'll find that they have much heavier burdens of pension insurance.[15] As to individual contribution, the contribute rate of pension for China's enterprise employees is stable and relatively moderate. The adjustment of the rate for employees fitting into the flexible employment model is still under way, which reached 10% between 2001 and 2005, increased to 12% in 2006 and has never changed since then. According to the current rules for regulating the contribution rate, there will be optimization of the future contribution rate for employees fitting into the flexible employment model, which is to, compared to the figure of 2006, increase the rate by 2% every five years until it reaches the rate for enterprises.[16]

As to the pooling of pension funds in China, on one hand, enterprises contribute too much to the pension system for them to afford. On the other hand, there is inadequate government's fiscal support. Every year, the total expenditure for employment and social security in China accounts for about 10% of the state expenditure. Only less than 2% of the total expenditure is used to fund the pension (see Table 4.8), which is far less than 17%, the average proportion in OECD countries.

4.3.2.3 Required Years of Contribution

According to the "Decision of the State Council on Establishing a Unified Basic Old Age Insurance System for Enterprises Employees" issued by the State Council in 1997, for individuals who have contributed to the pension system for a total of 15 years, they will be paid with basic pensions on a monthly basis after retirement. For individuals who have contributed to the pension system for less than 15 years in total, they will not be paid with basic pensions and a lump sum of the deposit in their individual account will be paid to them. Based on the current average income level in China, the required years of contribution are a relatively short period. By

[15]Yang and Liu [8].

[16]Han [9].

far, the required years of contribution in countries all over the world are basically as follows: higher-income countries require less years of contribution while less-income countries require more years of contribution. Lower-middle-income countries require an average of about 17 years of contribution.[17] Meanwhile, China is also faced with population aging and an increase in average life span. According to the prediction in relevant data, by 2050, the average life expectancy of the population aged 60 or over in China will be 22.5 years,[18] which, considering the 15 years of contribution required, will lead to a decrease in pensions the elderly receive.

4.3.3 Pension Paid to Pensioners in China

4.3.3.1 Average Pension Paid to Pensioners

Urban pensioners in China receive relatively large amounts of pensions. However, compared to the increase of urban employees' average wage (see Table 4.9), Chinese retirees' pension increases relatively slow, which is favorable for alleviating risk related to payment caused by population aging.

Different kinds of units pay different pensions to their employees. In 1990, the average pensions for retirees of enterprises, public institutions and government agencies were RMB 134, 148 and 143 per month, the ratio of which was 1:1.10:1.06. However, from the establishment of the basic pension system for urban enterprise employees, pensions paid to retirees from enterprises and government agencies/public institutions started to differ greatly.

In 2005, the average pensions paid to retirees from enterprises, public institutions and government agencies were RMB 714, 1220 and 1469 per month respectively, the ration of which was 1:1.89:2.06. After 2005, the basic pensions paid to personnel from government agencies and enterprises were changed so that the difference was reduced. In 2008, the above-mentioned ratio became 1:1.48:1.63.[19]

4.3.3.2 Pension Replacement Rate

Pension replacement rate is also an important indicator to judge the level of pension. On the whole, the pension replacement rates in China are too high and vary greatly from industry to industry.

[17] Wang [10].

[18] Wang [10].

[19] Yang [11].

Table 4.9 Pensions paid to urban residents in China

Year	Average pension for retirees/honored retirees (RMB/month)	Annual growth compared to last year (%)	Average wage of urban employees (RMB/month)	Annual growth compared to last year (%)
2000	594.8	–	777.8	–
2001	643.1	8.1	902.8	16.1
2002	733.9	14.1	1031.1	14.2
2003	783.9	6.8	1164.1	12.9
2004	809.6	3.3	1326.7	14.0
2005	889.0	9.8	1516.7	14.3

Table 4.10 Pension replacement rates in some countries (unit: %)

OECD country	Replacement rate	OECD country	Replacement rate	OECD country	Replacement rate
Australia	52.6	Ireland	34.9	Spain	81.2
Austria	76.6	Israel	85.3	Sweden	53.8
Belgium	42.6	Italy	64.5	Switzerland	59.3
Canada	48.5	Japan	36.3	Turkey	69.5
Chile	48.4	South Korea	46.9	UK	37.0
Czech Republic	57.3	Luxembourg	90.3	US	42.3
Denmark	84.7	Mexico	46.3	Average in OECD Countries	60.6
Estonia	50.9	Netherlands	89.1	Other countries and regions	
Finland	57.8	New Zealand	47.8	Argentina	81.1
France	49.1	Norway	52.9	Brazil	85.9
Germany	42.0	Poland	59.0	India	72.4
Greece	95.7	Portugal	54.4	Russia	65.1
Hungary	75.8	Slovakia	57.5	South Africa	13.1
Iceland	109.1	Slovenia	62.4	Average in EU countries	62.9

The pension replacement rates in China have always been high, of which the average is over 80%. Meanwhile, the average replacement rate is about 60% (see Table 4.10) in developed countries, 60.6% in OECD countries and 62.9% in EU countries. Generally speaking, the higher the income before retirement is, the lower the replacement rate will be. For example, in 2009, China saw an average replacement rate of 82.5%, 97.7% for low-income earners, 77.9% for middle-income earners and 77.2% for high-income earners.[20]

[20]OECD Global Pension Statistics.

The pension replacement rates in China are also different. Between 1995 and 2000, the pension replacement rates for government agencies and public institutions were on the rise while the replacement rate for urban employees was in decline. In 1995, the pension replacement rates for retirees from enterprises, public institutions and government agencies were 72.1, 92.1 and 94.3% respectively. In 2000, the replacement rates for the above-mentioned three became 66.8, 100.2 and 94.4%. In 1995, these figures all decreased while the replacement rate of enterprise employees' pensions decreased the most. In 2005, the replacement rates for the three kinds of retirees were 47.8, 86.3 and 84.6%.[21] Following economic development, replacement rates will be gradually on the decline, which will ensure the balance of pension in the era of population aging.

Therefore, in the reform of the pension system, it is also an important issue as to how to decide a reasonable pension replacement rate and ensure a high replacement rate. Given set years of contribution, the accumulation of pensions in individual accounts mainly depends on the growth rate of actual wages and the earnings rate of the funds. When the economy develops extremely fast in a country and the average wage grows fast, the pension replacement rate individuals receive will be relatively low. Under that circumstance, to maintain the high level of retirees' pensions, it is necessary to increase the pension replacement rate for retirees by increasing the earnings rate of the funds or via enterprise annuities and commercial insurances.[22]

4.3.4 Problems of Implicit Pension Debt in China

4.3.4.1 Implicit Pension Debt and Causes Thereof

(1) Implicit pension debt. It mainly refers to the fact that in the process of the pay-as-you-go system transforming to the full fund accumulation system or the partial fund accumulation system, when retirees or people who have contributed to the pension system in accordance with the pay-as-you-go system receive their pensions under the new system, while there are no corresponding funds and the government has to pay the pensions it has promised under the pay-as-you-go system, the shortfall then becomes debt. This kind of debt has no explicit records, will not be listed in the current budget deficits, and is subject to no explicit repayment deadline or amount. Thus it is called "implicit debt." It is actually the cost of the pension reform, which is a universal phenomenon in the pension reforms in all other countries.

    Implicit pension debt results from the promise of the government, which designs the pension system, to individuals of society who join the pension plan.

[21]Yang [11].

[22]Zhang et al. [12].

The pay-as-you-go system is the government's promise to pay set pensions to pensioners. The fund accumulation system is the government's promise to ensure the future profits for gradually accumulated individual accounts. Under the pay-as-you-go system, the individuals contribute to the pension system to bear the pensions paid to current retirees and enjoy the government's promise to pay the future profits to gradually accumulated individual accounts. When the system changes and the promised profits cannot be achieved, they become implicit debt. Under the fund accumulation system, the government promises to compensate the workers for the difference between the balance of their individual accounts when they retire and the minimum pensions supposed to be paid, which forms implicit pension debt for individual accounts.[23] The concept of implicit pension debt becoming explicit refers to the fact that the fund flow formed by contribution (a way to transfer purchasing power) is partially or completely diverted from the previous way of financing between people to the way of transfer through different time periods. That is the reason why the previously explicit debt hidden behind the fund flow consisting of contribution of employees becomes explicit.[24]

(2) Sources and consequences of China's implicit pension debt. The academia generally believes that China's implicit pension debt is a result of the pension reform starting from 1997 and continuous population aging.

① Pension security system reform. China's legal pension security system was first established in the early 1950s, which pooled funds and allocated pensions in accordance with the pay-as-you-go model. From 1997, the pension security system combining the pay-as-you-go with fund accumulation models began to be implemented, which is a combination of social pooling and individual accounts. In the same year, the State Council's Document No. 26 of 1997 was issued, which divided employees into "previous employees," "intermediate employees" and "new employees," and defined different pooling and payment methods for these three groups. "Previous employees" mainly refer to employees who had retired before the promulgation of the State Council's Document No. 26 of 1997. "Intermediate employees" refer to those who had started to work before and retired after the promulgation of the Document No. 26 of 1997. "New employees" refer to those who started to work after the promulgation of the Document No. 26 of 1997. In 2005, as an adjustment and improvement of this regulation, the State Council's Document No. 38 of 2005 was issued, in which the "previous employees" were defined as employees who had retired before the promulgation of the Document No. 38 of 2005, "intermediate employees" were defined as those who had started to work before the promulgation of the Document No. 26 of 1997 and retired after the promulgation of the Document No. 38 of 2005, and "new employees"

[23]Li [13].

[24]Liang [14].

were still defined as employees who started to work after the promulgation of the Document No. 26 of 1997.

The new pension system pools funds based on the principle of "old rules for previous employees, new rules for new employees, transitional rules for intermediate employees." Pensions are still paid to "previous employees" according to the government's previous rules and the adjusted standards for pensions. Pensions of "new employees" consist of individual accounts and basic pensions. Pensions of "intermediate employees" consist of not only basic pension and individual accounts, but also transitional pension calculated based on their working years. It is stipulated in the State Council's Document No. 38 of 2005 that transitional pension should be paid to those who have contributed to the pension system for over 15 years. The pension of "previous employees" and the transitional pension of "intermediate employees" were promises under the pay-as-you-go system, with no funds to support them under the new system, so the implicit pension debt has formed. As to new employees, funds pooling and pension payment are both based on the new system, which means there has not been any implicit debt for this group so far.

② Population aging. Population aging is also an important cause for implicit debt. So far, China's pension security system is still a pay-as-you-go system. Under the pay-as-you-go system, following the increase in the aging population and average life span, the needs for pension will continually increase. In the previous analysis, it is predicted that the population aged 60 and over in China will reach 17.4% by 2020, 29.4% by 2040, and 33.6% by 2060. The increase in the proportion of the aging population means the drastic raise of pension security welfare, which will lead to heavy burden of payment for basic pension under the pay-as-you-go pooling system. This part of expenditure has not been explicitly seen so far, so it is also considered implicit pension debt.

In the reform of the system, implicit pension debt, as a cost of the system reform, should be borne by the government's fiscal authorities. However, the policies adopted in China are as follows. On the one hand, long-term increase in the contribution rate for enterprises is ensured. By far, the contribution rate for Chinese enterprises is 12%, far higher than the average figure of the world, which means the enterprises have to both accumulate pension for "new employees" and afford basic pension and transitional funds for "previous employees" and "intermediate employees." That gives the enterprises more burden. On the other hand, the social pooling account and individual accounts of "new employees" and "intermediate employees" are not separated so that the funds of individual accounts will be diverted into the social pooling account to make up for the insufficiency, which leads to large numbers of "empty" individual accounts and decline of the pooled funds in individual accounts.

4.3.4.2 Amount of China's Implicit Pension Debt

Because of so many factors that influence the implicit pension debt, the actual situation of the implicit pension debt is very complicated. The major factors that influence the implicit pension debt are as follows. The first one is the age structure. Usually, the severe the population aging is, the more the implicit debts are. Under China's pension security system, with more "previous employees" and "intermediate employees," there will be more implicit debts. The second is the coverage rate of the pension security. Usually, implicit pension debts are higher in countries where coverage rates of pension are high than in those with low coverage rates. The third one is the pensions paid to pensioners. The higher the pensions paid to pensioners are, the higher the implicit pension debts are. Moreover, it is also necessary to take into consideration economic growth, system reform and other factors. As to specific calculation, different calculation methods will lead to different implicit pension debts. Based on different systems chosen, the previous studies on China's implicit pension debt had the following results:

(1) Calculation before the pension reform in 1997. In 1996, the experts from the World Bank did a survey of China's pension insurance system and then issued the *China—Pension System Reform* report. According to the report, considering the end of 1994 as the reference point, based on the international experience that implicit pension debt is 20–30 times the pensions paid to pensioners, China's implicit debt in 1994 accounted for 46–49% of the GDP that year (pensions paid accounted for about 2.3% of the GDP in 1994), reaching RMB 2.15–3.23 trillion. Meanwhile, according to the revenue of the pension and the age structure of the population, based on the prices of 1994, the accurate implicit debt was approximately RMB 1917.6 billion, of which the implicit debt for the "previous employees" and "intermediate employees" were RMB 681.3 billion and RMB 1236.3 billion respectively.[25]

(2) Calculation based on the pension system of 1997. Most of these calculations are carried out under the system based on the State Council's Document No. 26 of 1997, with the years before and after 1997 as reference years of calculation. The several authoritative results are as follows: ① Wang Xiaojun takes 1996 as the ending year of the old system to calculate the implicit pension debt. He delivers a forecast of the trend of pension debt accumulation over the next 50 years by taking the classified death rates listed in the 1989–1990 Life Table for China's Urban Employees as the estimated death rates for accurate calculation, and supposing the interest rates are 3, 4, 5 and 6%, pensions are paid on a yearly basis, pensioners who have worked for 35 years will receive a

[25] Li [13].

pension equal to 80% of their salary before retirement and the annual growth rate of the average wage is 3%. According to the calculation, the implicit pension debt is RMB 3611.8 billion.[26] ② He Ping takes the beginning of 1996 as the reference year to calculate the implicit pension debt for enterprise employees and gets a result of RMB 5720.42 billion, of which the debt for "previous employees" of enterprises is RMB 899.74 billion and the debt for "intermediate employees" is RMB 4820.6 billion. He also takes the years between 1994 and 2050 as the time period for long-term accurate calculation to calculate the implicit pension debt, which is RMB 2875.334 billion.[27] ③ Song Xiaowu takes 1997, when a unified basic pension insurance system was implemented across China, as the reference year for evaluation. Based on different contribution rates (16%:20%:24%), according to five schemes and three kinds of returns on investment, he calculates the implicit pension debt by supposing the average growth rate of wages would be 5% within the following five years, 4% within the following 6–10 years and 3% in 10 years, the average growth rate of pension is 6% of the growth rate of wages, average replacement rate increases by 0.66% year to year, and average support ratio is 3.25:1. According to the calculation, the implicit pension debt is somewhere between RMB 1.8301 and 10.826 trillion.[28]

(3) Calculation based on the pension system of 2005. In the State Council's Document No. 38 of 2005 issued in 2005, there are changes to the pension system of 1997, mainly consisting of increase in enterprises' contribution rate and more accurate calculation of pensions paid. Forecasts of implicit pension debt under this system are as follows: ① Based on the concept of "debt falling due", by supposing the growth rate of nominal wages is 10% before 2010, 6% between 2011 and 2020, and 4% between 2021 and 2050, the average inflation rate is 3%, and the interest rate is 4%, Liu Changping and others calculate the total implicit debt between 2006 and 2026 and get a result of RMB 2.1 trillion.[29] ② Li Dan and others calculate the figure of 2008 by supposing the employees start to contribute to the pension system since 20 years old, the average retirement rates for male and female employees are 60 and 55 years old respectively, the maximum life span of employees is 90 years old, and the growth rate of average wages is 8%. According to the calculation, the total implicit pension debt of China is RMB 26.5609 trillion, of which the debts for "previous employees" and "intermediate employees" are RMB 2295.2 and

[26]Wang [15].

[27]He [16].

[28]Cao [17].

[29]Liu and Yin [18].

2365.7 billion respectively.[30] ③ Based on the data in the life tables of males and females in the *China Population Statistics Yearbook* of 2000, by supposing the interest rate is 4%, pensions are paid on a yearly basis, pensioners who have worked for 35 years will receive a pension equal to 80% of their wages before retirement, the annual growth rate of average wages is 3%, annual growth rate of individual wages for seniority is 1%, Li Li and others calculate China's implicit public pension debt of 2010 and get a result of RMB 33.4 trillion, which accounts for 101.8% of the GDP that year.[31] ④ Without considering the factors including premature retirement, Fang Haiyan calculated China's implicit pension debt in 1997 by adding the accurate present value of the debt for retirees to the pension debt for "intermediate employees", and then subtracting the cumulative balance from the above-mentioned sum. According to the calculation, the result was RMB 1799.8 billion.[32] ⑤ With some presumptions, such as the employees start to contribute to the pension system since the age of 20 and the average retirement age is 58 year, Tan Xiangyu calculated and determined that China's implicit pension debt by the beginning of 2002 was RMB 9392.1 billion, which accounted for 98% of the GDP that year. Of the total debt, the debt for "previous employees", "retired intermediate employees" and "working intermediate employees" were RMB 1588, 1074.9 and 6729.2 billion respectively.[33] ⑥ Liang Junlin and others do calculation based on the figures of 1998. They simulate and estimate the total implicit debt for "previous employees" and the trend of implicit debt becoming explicit in China after 1998 by supposing the annual growth rate of average wage is 4%, "previous employees'" pensions are adjusted based on 50% of the wages' growth rate, the wage replacement rate of the explicit debt for "previous employees" is 60%, the retirement age is 60 years old and the number of years to receive pensions are 20 years. According to the calculation, the implicit pension debt in 1998 was RMB 6293.6 billion, accounting for 74.57% of the GDP that year, of which the implicit debt for "previous employees" and "intermediate employees" were RMB 936.3 and 5357.3 billion respectively.[34]

Based on different calculation methods, target time periods and presumptions about economy, populations and pensions paid, there come different results of the calculation of implicit pension debts. On the whole, China's current implicit pension debt is somewhere between RMB 2 and 4 trillion.

[30]Li et al. [19].

[31]Fang [20].

[32]Li et al. [21].

[33]Tan [22].

[34]Liang et al. [23].

4.3.5 *Problems About Revenue and Expenditure of China's Pension Security Fund and Solutions to Them in Future*

4.3.5.1 Fiscal Expenditure for Pension Security Is Too Low and Fiscal Expenditures at Different Levels Differ

The central fiscal authorities spend only 2% of the total fiscal expenditure on pension security, while the average figure in developed countries is about 17%. By far, there has not been any unified standard for the subsidies the financial departments of different provinces and cities in China offer to the pension security funds of their own regions. The different levels of pooling lead to difficulties in mutual assistance in pension security funds and frequent insufficiency of pension security funds offered in some regions. It is necessary to continually increase the fiscal input in the pension security fund and gradually achieve social pooling at province level, and even nation level in the end.

4.3.5.2 Excessively High Contribution Rate for Enterprises Brings Them Heavy Burden

While the fiscal expenditure for pension security is excessively low, the enterprises' contribution rate for pension security fund is excessively high. The average contribution rates for enterprises are 14.6% in EU countries and 11.2% in OECD countries respectively. The contribution rate of 20% in China is clearly higher than that of other countries. This undoubtedly increases the cost for enterprises and is unfavorable for their development, leading to severe phenomena of late payment and arrearages. Because of high contribution rate, small- and medium-sized enterprises and private businesses in particular are unwilling to buy pension insurances for their employees or delay the contribution to pension insurance. Therefore, while increasing fiscal input in pension security fund, it is better to reduce the contribution rate for enterprises to a reasonable level and encourage more enterprise employees to join the pension system, expanding the coverage of basic pension security.

4.3.5.3 Low Coverage Rate of Basic Pension Insurance

The basic pension insurance for urban employees, the social pension insurance for urban residents and the social pension insurance for rural residents compose three current systems of basic pension insurances in China. Except for the basic pension insurance for urban employees, which has a high rate of coverage as it is implemented compulsorily, the social pension insurances for both urban and rural residents always have low coverage rates as they are subject to no compulsory implementation. For increasing the coverage rate of pension insurance, it is suggested that the social pension insurances for both urban and rural residents should be compulsorily implemented, so as to increase the coverage rate of pension insurance and the population paying premiums, which will increase the revenue for the pension security fund.

4.3.5.4 Excessively High Pension Replacement Rate

China's current average pension replacement rate of 82.5% is relatively high. Meanwhile, the replacement rates in public and private sectors are different from each other. For example, the pension replacement rate for government agencies and public institutions is higher than that for enterprise employees. This kind of difference is unfavorable for income adjustment, which is likely to provoke social conflicts. According to universal rules for pension replacement rates, the replacement rate that exceeds 60% will easily cause pressure to payment of pensions and jeopardize the sustainability of the pension security system. Usually, the replacement rate for high-income employees should be 20% lower than that for low-income employees to ensure the fairness of the system. However, the replacement rate of China's high-income group is only 10% lower than that of its low-income group, which should be adjusted to a reasonable extent so as to ensure the fairness and sustainability of the pension security system.

4.3.5.5 Development of Supplementary Pension Insurances Is Encouraged

At present, China's supplementary pension insurances mainly consist of enterprise annuities and commercial insurances, which are seen as the second and third pillars of China's pension security system. However, because of multiple reasons, these supplementary insurances have not been implemented in an appropriate way. It is mainly because the supplementary pension insurances are not explicitly defined and the tax system for enterprise annuities still needs improvement. Thus, it is necessary to improve all of the relevant regulations for the supplementary pension insurances as soon as possible, improve the multi-pillar pension security system, and gradually increase the proportions of the second and third pillars of the pension insurance in the pension security system, so as to alleviate the pressure of the sole pension security insurance and deal with the risk of population aging.

4.3.5.6 Appropriate Investment in Pension Security Fund to Ensure and Increase the Value of the Fund

At present, the part of China's pension security fund that can be used to invest in the capital market consists of the individual accounts in the basic pension system and the elderly security fund diverted into the social security fund by the fiscal authorities. Because investment in the capital market comes with huge risk, appropriate investment strategies and channels become particularly important. The National Council for Social Security Fund is the only investment organization for China's pension security fund, mainly in charge of the investing issues for public pension security fund. Following the continuous improvement of the second and third pillars of the pension security system, there will definitely be a larger amount of pension security fund. Thus, more professional operation organizations for the pension security fund and more perfect laws and regulations are needed.

4.4 Population Aging and Pension Security Fund

4.4.1 Trend and Characteristics of China's Population Aging

Population aging refers to the process in which the young population reduces relatively and the aged population increases relatively in terms of the age structure of the population in the process of population growth. The direct results of population aging include an increase of the aged population, the society's expenditure in the elderly and supporting pressure. According to the international general standard, if population of the elderly who are above age 60 accounts for 10%, or population of the elderly who are above age 65 accounts for 7% of the population of the country or region, it means the country or region has entered into an aging society. According to this standard, many developed countries around the world have stepped into an aging society in succession. However, compared with these countries, China's aging is showing such characteristics as a massive aging population, a rapid aging speed, "getting old before getting rich" and others. Therefore, pension security is facing greater pressures in China.

4.4.1.1 The Overall Trend of China's Aging

From a worldwide perspective and compared with developed countries, China has just entered the population aging society stage early in the 21st century, but it's currently in a rapid growth stage of aging. And it's predicted that China's aging population ratio will catch up with and exceed that of developed countries by 2030, and that the ratio will reach its peak around 2070, accounting for almost one third of China's population. Although developed countries began to enter the aging society in the middle of the 20th Century and the aging population reached its peak around 2050, their population structures have maintained in a relatively stable status. In terms of time, it only takes almost 70 years for China's population to start entering the aging state to reaching the peak, while it takes almost a century for developed

Table 4.11 Population aging trends of China and the world (unit: %)

Year	China		World		Developed countries		Developing countries		Undeveloped countries	
	65+	60+	65+	60+	65+	60+	65+	60+	65+	60+
1990	5.9	8.9	6.2	9.2	12.5	17.7	4.5	6.9	3.2	5.0
2000	7.0	10.2	6.9	10.0	14.3	19.5	5.1	7.6	3.2	5.1
2010	8.2	12.3	7.6	11.0	15.9	21.7	5.8	8.7	3.4	5.3
2020	12.0	17.4	9.4	13.5	19.0	25.4	7.5	11.2	3.7	5.8
2030	16.5	24.4	11.7	16.6	22.4	28.6	9.8	14.4	4.5	6.9
2040	23.3	29.4	14.3	19.2	24.5	30.6	12.5	17.2	5.5	8.5
2050	25.6	33.9	16.2	21.8	25.7	31.9	14.7	20.2	7.0	10.5
2060	29.5	36.6	18.3	23.7	26.2	31.5	17.1	22.5	8.8	12.7
2070	30.2	36.8	19.5	25.0	25.7	31.3	18.5	24.0	10.6	14.8

Source Population Division of the Department of Economic and Social Affairs of the United Nations Secretariat, world population prospects: the 2010 revision

countries to do it, which shows that China's aging speed is particularly rapid and has exceeded that of developed countries and far exceeded that of the world. Meanwhile, due to the implementation of One-child Policy with a decreasing youth population, relatively speaking, the ratio of aging population during the peak period is almost 5% higher than that of developed countries. With a large population base, China's elderly population above age 60 will be more than 400 million, which will cause huge supporting pressure for China's economy and society (see Table 4.11).

The pressure of supporting the elderly is directly represented by the Elderly Population Burden Coefficient, namely the old-age dependency ratio which is used for representing the elderly population that every 100 workers of the right age are responsible for. Elderly dependency ratio and child dependency ratio together form the supporting pressure of the entire society. Relatively speaking, the term of supporting children is certain. But with the extension of human's average life expectancy, the age limit of supporting the elderly is also extended correspondingly (see Table 4.12).

From Table 4.12, we can see that China's old-age dependency ratio will reach the same level as the developed countries in 2040 and it will continue to rise to go far beyond the average level of developed countries. Taking child dependency ratio into consideration, the population of China's workers of the right age will suffer from great supporting pressure.

4.4.1.2 China's Aging Speed Is Faster Than the Economic Growth Rate

In terms of the degree of economic growth, the average per capita GDP generally reached over USD$4000 and the average per capita GDP of most developed European countries and the US reached over USD$5000–10,000 in general while

Table 4.12 Population dependency ratios of China and the world

Year	China			World			Developed countries		
	Total dependency ratio	Child dependency ratio	Old-age dependency ratio	Total dependency ratio	Child dependency ratio	Old-age dependency ratio	Total dependency ratio	Child dependency ratio	Old-age dependency ratio
1990	51	42	9	64	54	10	49	31	19
2000	48	38	10	59	48	11	48	27	21
2010	38	27	11	52	41	12	48	24	24
2020	40	24	17	52	38	14	56	26	30
2030	45	21	24	53	35	18	63	27	36
2040	59	22	37	55	33	22	68	27	41
2050	64	22	42	58	32	26	73	29	45
2060	76	24	52	61	32	30	75	29	46
2070	80	25	54	63	31	32	74	29	45

Year	Developing countries			Undeveloped countries		
	Total dependency ratio	Child dependency ratio	Old-age dependency ratio	Total dependency ratio	Child dependency ratio	Old-age dependency ratio
1990	68	61	7	91	85	6
2000	62	53	8	85	79	6
2010	53	45	9	77	71	6
2020	51	40	11	69	63	6
2030	51	37	15	63	56	7
2040	53	34	19	59	50	8
2050	56	33	23	57	46	11
2060	59	32	27	56	42	14
2070	61	31	30	56	39	17

Note Old-age dependency ratio = number of people aged 60 and over/number of people aged 15–59; child dependency ratio = number of people aged 0–14/number of people aged 15–59; total dependency ratio = old-age dependency ratio + child dependency ratio

Source Population Division of the Department of Economic and Social Affairs of the United Nations Secretariat, world population prospects: the 2010 revision

entering the aging society.[35] China basically entered the aging society in 2000 with an average per capita GDP of USD$949,[36] less than USD$1000. China entered the aging society in advance before the economic growth reached the level of developed countries. In recent years, although China's economic growth rate has maintained at a relatively high level, it still can't relieve the pressure of the increasing ratio of elderly population.

4.4.1.3 The Difference of Regional Aging Is Great in China

On the one hand, China's population aging still shows the characteristic of great regional differences. Judging from the ratio of the elderly aged 65 and over, the ratio of the elderly aged 65 and over was 6.96 in 2000 nationwide, Shanghai's aging was 11.53 to the greatest degree, while Qinghai showed the lowest ratio of 4.33. The population aging shows distinct regional features from West China to East China. On the other hand, the aging of urban areas and rural areas tend to appear to different degrees and the aging degree of urban areas is lower than that of rural areas. In 2000, the number of the elderly aged over 65 in rural areas accounted for 7.4% of the total rural population, while the ratio was 6.3% in urban areas and was distinctly lower than that in rural areas. Due to the huge differences in fertility policy, medical health and others in different regions, the aging of urban areas and rural areas in different regions shows huge differences. For example, the ratios of the elderly aged over 65 in urban areas and rural areas in Beijing are respectively 8.36 and 8.35, being almost the same; while the two ratios in Hunan are respectively 6.21 and 9.15,[37] showing a distinct difference. All along, urban social pension insurance has been dominant in China's social pension insurance and rural pension insurance has been developing slowly, but the aging speed of rural areas is higher than that of urban areas, which has also brought many problems for China's pension security system.

4.4.2 *The Pension Payment Pressure Brought by Population Aging*

The direct influence that population aging has on the pension system is an increased elderly population and increased pension fund payment pressure. From 1990 to 2007, OECD countries' expenditure in pension fund increased by 14.5%.[38] Table 4.13 shows the ratios of the insured of pension insurance among retirees and

[35]Jiang [24].

[36]http://data.worldbank.org/indicator/NY.GDP.PCAP.CD?page=2.

[37]*China Population Statistics Yearbook* (2001).

[38]OECD Social Expenditures database.

Table 4.13 Increase of China's retirees and pension welfare over the years

Year	Retirees/employees	Pension fund expenditure (RMB XX hundred million)	Increase compared with the previous year (%)
1990	0.2	149.3	–
1991	0.2	173.1	15.9
1992	0.2	321.9	86.0
1993	0.2	470.6	46.2
1994	0.2	661.1	40.5
1995	0.3	847.6	28.2
1996	0.3	1031.9	21.7
1997	0.3	1251.3	21.3
1998	0.3	1511.6	20.8
1999	0.3	1924.9	27.3
2000	0.3	2115.5	9.9
2001	0.3	2321.3	9.7
2002	0.3	2842.9	22.5
2003	0.3	3122.1	9.8
2004	0.3	3502.1	12.2
2005	0.3	4040.3	15.4
2006	0.3	4896.7	21.2
2007	0.3	5964.9	21.8
2008	0.3	7389.6	23.9
2009	03	8894.4	20.4

Note All retirees and employees listed here are the insured of urban basic pension insurance
Source China statistics yearbook (2010)

staff on active duty over the years as well as the ratios of pension fund expenditures. Since 1996, the ratio of retirees against employees has been growing. Because the ratio has only referred to the insured employees of urban pension insurance, it has relatively decreased compared with the ratio of the elderly population and workers. If the pension insurance coverage is expanded, this ratio will continue to increase greatly in the coming decades. Within the 20 years, the expenditure of pension insurance will also increase greatly.

China's urban staff retirement fund system has basically been the continuation of the pay-as-you-go system during the Command Economy Period, which will certainly aggravate the pension fund burden of employees on the premise of ensuring the pension fund level of retirees. Table 4.14 predicted the increase of China's pension fund burden in the coming decades. We can see that if the current pension fund level and pension fund system remain the same, the contribution rate of employees will be greatly increased in the future and become three times as it is now, bringing heavy pension payment burden.

Table 4.14 Pension fund burden resulting from population aging in China[a]

Year	Pension fund contribution rate	Increased percentage points compared with the previous year
2000	6.21	–
2010	7.35	1.14
2020	10.00	2.65
2030	13.98	3.98
2040	16.21	2.23
2050	18.50	2.29

[a]Jiang [24]

4.4.3 Population Aging and Pension Security Fund System Mode

Confronted with the pension payment pressure resulting from population aging, we all expect a good corresponding pension fund raising and payment system so as to guarantee the welfare of the labors group and the elderly group in society. As for the responses of different pension fund systems to the aging trend, there have been lots of arguments in the academia. It's generally believed that the pay-as-you-go system will result in a payment crisis that the elderly's primary pension fund level will be reduced without increasing the pension fund contribution rate of employees, while this won't happen in the fund system.[39] was the first to associate pension insurance fund with population aging. He applied the General Equilibrium Theory to prove the operating mechanism of the pay-as-you-go public pension system in a saving iterative model, pointing out that in the pure saving (without production or investment) economy, the growth of pension fund is mainly depended upon population growth. If the pure saving economy is implemented exclusively with the pay-as-you-go system in pension insurance, the interest rate of pension saving will be equal to the population growth rate. Auerbach and created the A-K model, making a positive analysis of how the transition from the pay-as-you-go system to the fund accumulation system can influence the macro-economy. As it proved, the privatization reform would increase the capital deposit and output level in the long-term economy, thus stimulating labor supply and increasing welfare. Robert Holzmann and others proposed the multiple support mode in pension insurance reform to relieve population aging in 1994, and it was applied to the pension insurance system reform by many countries.[40]

However, many people had questions about this. In Baal's opinion, if the pay-as-you-go system was adopted, the population aging would result in the reduction of workers, which would result in the decrease of gross output, and after the pension fund was reduced, the payment crisis would occur; if the fund

[39]Cheng [25].

[40]Guo [26].

accumulation system was adopted, with the development of population aging, although the pension fund wouldn't decrease nominally, the gross output would also decrease with the decrease of workers, when the social surplus value was less than the nominal value of pension fund, the actual gain of the elderly would still be decreased, thus finally reaching the conclusion that there was no substantial difference between the fund accumulation system and the pay-as-you-go system in reacting to population aging.[41] Yuan and Ge [27] believed that the risk of population change brought about by population aging under the pay-as-you-go system would certainly influence the balanced budget of pension fund, meanwhile, there were also many disadvantages in the fund accumulation system, so China wasn't well prepared for transiting from the pay-as-you-go system to the full fund accumulation system at present. Through a quantitative analysis of the pay-as-you-go system and population aging, Cheng [28] made the conclusion that population aging didn't necessarily lead to a payment crisis in the pay-as-you-go system and that the key lied in the gap between the economic growth rate and the population aging speed. Through comparisons of the ability of the pay-as-you-go system, the fund accumulation system and the nominal account system in reacting to population aging, Lin [29] believed that from the perspective of payment, the ability of any pension fund mode to react to population aging was actually not depended on whether the system's fund raising mode was the pay-as-you-go system or the fund accumulation system, instead, it was determined by whether the system's payment method was Defined Contribution or Defined Benefit.[42]

The challenges that population aging brings to all types of pension fund modes include whether these systems can ensure a balanced budget of the pension fund and whether they can increase the elderly's benefits on the basis of a balanced budget. Next, we'll analyze the pay-as-you-go system, the full fund accumulation system and partial fund accumulation system.

4.4.3.1 Population Aging and the Pay-as-You-Go System

In the pay-as-you-go system, the payment from working people is used to pay the pension fund of retirees of the same period. In such a pension fund mode, collection and payment are realized in the same period and the collection is determined by payment without balance. Supposing that in a certain period, Q_o represents the number of pension fund receivers, I the average pension fund level, Q_y the number of workers insured by pension insurance, W the average income of workers and R_c the contribution rate, then $Q_oJ = Q_yWR_c$

[41]Cheng [25].

[42]Lin [29].

Then take the population and pension insurance participation rate into consideration. Supposing that P_t represents the gross population of the society, P_y the population of workers and P_o the elderly population, pension insurance participation rate represented by $R_y = Q_y/P_y$, and the population ratio of the elderly receiving pension fund should be $R_o = Q_o/P_o$, then

$$PoRoI = PyRyWRc$$

The pay-as-you-go system generally adopts the Defined Benefit, meaning that the pension fund level is stable. On the left of the equation, if I representing the pension fund level remains the same and P_o representing the elderly population increases, the number of pension fund receivers will increase and the number will increase constantly. On the right of the equation, if the fertility rate remains the same or decreases, P_y representing the population of workers will decrease constantly. To maintain a balanced budget, R_y representing pension insurance participation rate, W representing the average salary and R_c representing the contribution rate should be increased. If the insurance participation rate increases, the population ratio of the elderly receiving pension fund also increases and the effects of both rates can be offset. Then what's left to do is increasing the average income and contribution rate of workers. The increase of the average income of workers requires the increase of the entire social productivity. However, if the population of workers continues decreasing for a long time, the social work comprehensive productivity will decrease. The contribution rate is adjustable, but a high contribution rate can impact workers' life welfare, thus decreasing the entire social welfare. In the long run, the pay-as-you-go system indeed carries some risks of payment crisis. Getting down to the left of the equation, we can also lower the population ratio of the elderly receiving pension fund. And this can be realized by raising the retiring age or adjusting the level of pension fund so as to slow down the growth rate of pension fund level.[43]

4.4.3.2 Population Aging and Full Fund Accumulation System

Theoretically, the full fund accumulation system refers to transferring part of one's income during the working period to be used after retiring, from the perspective of the entire society, population aging has hardly any influence on the full fund accumulation system. However, due to the extension of current people's life expectancy and the risk of fund preservation and appreciation, the full fund accumulation system cannot solve all problems resulting from population aging, either.

[43]Lin [29].

Supposing T_w represents a person's length of service, W the average income, R_f the contribution rate each year, B the average earnings rate of fund, T_s the length of receiving pension fund after the retirement and I the average level of pension fund. Then

$$T_sI = T_wWR_fB$$
$$I = T_wWR_fB/T_s$$

In the full fund accumulation system, the contribution rate is definite but the pension fund level is indefinite. Once the contribution rate is determined, the pension fund level will be mainly depended on the ratio of length of service to length of retirement, the pension fund earnings rate and the income level during the working period. The extension of life expectancy will inevitably lead to the reduction in the ratio of the length of service to the length of retirement, while the earnings rate of pension fund is indefinite, so determining an appropriate contribution rate becomes a difficult problem. Meanwhile, workers' population decreases due to aging. In the long run, social gross output is decreasing, which will also relatively reduce the average income of workers. If the pension fund earnings rate cannot remain at a high level, the income of the elderly will remain at quite a low level.

4.4.3.3 Population Aging and Partial Fund Accumulation System

The partial fund accumulation system is a combination of the pay-as-you-go system and the full fund accumulation system. The income of a retiree has two sources, one being the pay-as-you-go raising method and the other the fund accumulation method. The partial fund accumulation system can not only avoid the payment crisis of the pay-as-you-go system, but also decrease the investment risk of fund accumulation. Theoretically, it's quite a reasonable method to react to population aging. However, it's very difficult in the actual operating process because it involves calculating the life expectancy appropriately, determining the contribution rate and benefits as well as the degree of distribution between the pay-as-you-go system and the fund accumulation system.

On the whole, whether the pay-as-you-go system, the full fund accumulation system or the partial fund accumulation system is faced with the challenges of population aging, no problem can be solved appropriately by simply relying on any single system. However, population aging is a process, so we might as well adjust the current pension fund system before this process arrives so as to cope with the futuristic crisis.

4.4.4 International Lessons of Pension Fund System Reform to React to Population Aging and the Enlightenment to China

4.4.4.1 International Lessons of Pension Fund System Reform to React to Population Aging

From a global perspective, with the intensification of population aging, all countries are bearing increasing burdens on pension fund expenditure, which also endangers the balanced budget of pension fund. In view of the lessons from OECD countries, the ratio of public pension welfare expenditure GDP increased by an average of 14.5% from 1990 to 2007. In order to cope with the pension fund payment pressure resulting from population aging, many countries are conducting reforms of the pension fund system in succession. On the whole, it mainly includes the following aspects:

(1) Extending the retirement age. The direct influence of extending the retirement age is reducing the social old-age dependency ratio and thus reducing the support ratio of the pension fund system. In addition, raising the retirement age also extends the contribution length and increases the supply of pension fund to some extent, thus guaranteeing the balanced budget of pension security fund. Table 4.15 shows the retirement age and life expectancy of retirees of some countries, and men's average retirement age was 63.1 and women's 61.7 in OECD countries in 2009. In view of the trend of all countries, raising the retirement age and granting the retirement system greater flexibility are an important part of the reform. For example, the US has extended its primary retirement age of 65 to the present 67 and implemented the Flexible Retirement System, under which people who were born after 1960 should retire at the age of 67, those born before 1937 should retire at 65 and those born between 1943 and 1954 should retire at 66.[44] Germany raised men's retirement age to 65 between 2000 and 2001 with relevant policy for women implemented in 2004, planning to raise women's retirement age to 67 before 2030. France restricted employers from laying off employees under age 65 but encouraged retirees to continue to work while receiving pension fund.[45] Australia changed the age limit at which women received the pension fund every 2 years and 6 months, raising the age limit from 60 in 1995 to the predicted 65 by 2013. The UK is planning to raise women's retirement age to 65, the same retirement age limit with men.[46]

[44]The retirement age of people born between 1938 and 1942 should be 2 months older than 65 for every one more year born after 1938. For example, a person born in 1938 should retire when he was 65 years and 2 months old, and one born in 1942 65 years and 10 months old; the retirement age of those born between 1955 and 1959 should be 2 months older than 66 for every one more year born afterwards.

[45]Yang [30].

[46]Duan [31].

Table 4.15 Retirement age and life expectancy of some countries and regions (2009)

Country or region	Statutory retirement age		Life expectancy	
	Men	Women	Men	Women
The US	66.0	66.0	17.1	20.1
Canada	65.0	65.0	18.3	21.4
Australia	65.0	62.0	18.6	24.5
New Zealand	65.0	65.0	18.1	20.9
Germany	65.0	65.0	17.0	20.7
France	60.5	60.5	21.8	26.5
Finland	65.0	65.0	16.8	21.0
Norway	67.0	67.0	16.2	19.3
Sweden	65.0	65.0	17.9	21.1
Italy	59.0	59.0	22.7	27.3
Spain	65.0	65.0	17.9	21.8
Switzerland	65.0	63.0	18.9	23.9
The UK	65.0	60.0	16.9	24.5
Chile	65.0	60.0	17.3	25.0
South Korea	60.0	60.0	20.2	25.2
Japan	65.0	65.0	18.8	24.1
Mainland China	60.0	55.0	12.1	20.6
Hong Kong, China	65.0	65.0	15.2	21.4

Source Word population prospect, 2010 revision; OECD pensions at a glance

It's a common practice for most developed countries to set the same retirement age limit for men and women. In line of China's current retirement ages and life expectancy, it' not so realistic to extend the retirement age in the short run. For example, men's life expectancy after retirement is 12.1 with a huge gap behind developed countries. China is considering extending women's retirement age gradually and finally implementing a system under which men and women share the same retirement age.

(2) Increase the pension fund revenue. Faced with the huge pension fund payment pressure of population aging, increasing the pension fund revenue is an important strategy. According to the source of pension fund, this can be realized by increasing financial investment, expanding the contribution scope, increasing the contribution rate and extending the contribution length. In developed countries, the financial investment in pension security has always been maintained at a high level, leaving only a small space for increasing

financial investment. This can easily result in huge financial burden with a wide coverage scope of pension security. Therefore, raising the contribution rate and increasing the contribution length have become important measures in the reform. For example, the US raised employees' contribution rate from 1 to 5.2% between 1949 and 1984.[47] Sweden increased the personal contribution rate to 9.25% in its reform in 1999, while individuals basically didn't need to pay before that. Canada increased the total contribution rate of employees and employers from 5.6% in 1996 to 9.9% in 2003.[48] Germany began to raise its pension fund contribution rate in 2007, raising it from 19.5 to 19.9%. This contribution rate would remain the same until 2012 and then be reduced to 19.4% before 2019.[49] Finland's contribution rate was raised from 18.6% in 1994 to 21.6% in 2009. In terms of increasing the contribution length, for example, France unified the contribution length for public and private departments to obtain full pension fund. By 2012, the contribution length would be raised from 37.5 years before 2008 to 41 years. After that, it would try hard to establish an exponential connection between the contribution length and life expectancy, and the contribution length was predicted to increase to 41.75 years. Spain also extended its length of employment on which the calculation of pension fund was based. The salary base for calculating pension fund in 1996 was the average salary of 8 years before retirement, and it was increased to 15 years between 1997 and 2001.

(3) Lower the pension fund replacement rate. Lowering the pension fund replacement rate means reducing the pension fund amount and reducing pension welfare. Carrying forward a general pension fund reduction policy is relatively sensitive to the entire society, so some countries mainly take the measure of lowering the pension fund replacement rate of high-paid people without reducing the salary level of low-paid people. Because relatively speaking, low-paid people tend to mainly depend on social security for a living. For example, Canada collected income tax from high-paid people, Australia regulated that the pension fund each person received every two weeks should not exceed AUS$546.80 and the pension fund received by a couple should not exceed AUS$456.80 every two weeks.[50] With an average pension fund replacement rate of 48.5%, the pension fund replacement rate was 76.6% for low-paid people and 29.6% for high-paid people in Canada. With an average pension fund replacement rate of 52.6%, the pension fund replacement rate was 73.3% for low-paid people and 38.6% for high-paid people in Australia. Judging from the data of OECD countries, in addition to a few countries that maintained a rather high replacement rate, most countries maintained an average replacement rate of about 60%, while the replacement rate of high-paid

[47]Meng et al. [32].

[48]Duan [31].

[49]Chen [33].

[50]Meng et al. [32].

people generally wouldn't go over 50%. In China, the pension fund average replacement rate is 82.5%, and the rate was 71.2% for high-paid people, maintaining at a rather high level. Generally speaking, the elderly's basic life can be guaranteed as long as the pension fund replacement rate is maintained around 40%, and a pension fund replacement rate of more than 60% can easily cause the unsustainability of pension fund. In view of this, much space has been left for adjusting the pension fund replacement rate in China.

(4) Construction of a multiple support pension fund system. At present, most countries adopt the pay-as-you-go system as the financing system for their national public pension fund. Theoretically, under the conditions of population aging, the pay-as-you-go system will bring heavy financial burdens, while this problem can be avoided in the fund accumulation system. Therefore, an important direction for the pension system reform is turning the pay-as-you-go system to the fund accumulation system. For example, the pay-as-you-go system was turned into the full fund accumulation system and privatization management was implemented in Chile in the 1980s. Although a debt crisis once occurred in the early stage of the reform, notable achievements were made in the reform which was therefore regarded as a successful experience in promoting pension system reform worldwide by the World Bank. More countries integrated the fund accumulation mode with the pay-as-you-go system, and the main measure taken was constructing a multiple support pension fund system. Besides implementing the first support, the fund accumulation system was implemented in other cases and the development of the second, the third and even more supports was encouraged. For example, Germany and France implemented a tax-free policy in order to encourage the private saving or individual contribution in the second and third supports. Sweden included 2.5% from the 18.5% contribution rate into the personal account by means of establishing "the nominal account system". "The nominal account system" didn't change the fund raising mode of the pay-as-you-go system in pension fund. Instead, it just changed its payment mode of the Defined Benefit to the Defined Contribution. As a whole, the reform and improvement of the multiple support pension system is a process in which the liabilities of the national pension security constantly turns to a private department. This is effective in relieving the financial burdens brought by population aging.

4.4.4.2 The Enlightenment of Pension Fund System Reform on China

The pension fund reforms conducted in response to population aging by all countries around the globe have the following enlightenments on the pension security system reform in China:

(1) Full consideration should be taken in the age structure of the population in the pension security system reform. Meanwhile, it should coordinate well with the economic growth level. Although China has entered the aging society, our

economic growth level still fails to support the pension security system with a wide coverage scope and high benefits. It's quite a realistic choice to establish a multi-layer pension security system on the basis of ensuring the basic coverage scope. In view of the characteristics of China's population aging, the degree of population aging in rural areas is much higher than that in cities and towns, but the coverage of pension insurance in rural areas is much smaller than that in cities and towns. By the end of 2009, the insurance rate of basic social insurance for employees in rural areas was 10.2%, while the insurance rate of basic social insurance for employees was 37.9% in urban areas in China. Therefore, what we mainly need to do currently is continuing to expand the coverage of social pension insurance in both rural areas and urban areas.

(2) Adjust the calculation parameters of pension fund, including extending the payment length, lowering the pension fund replacement rate and others. Many scholars propose to delay the retirement age to correspond to the aging crisis. Delaying the retirement age is really an important measure in the pension fund system reform taken by many developed countries. The current legal retirement ages of China are 60 for men and 55 for women. But judging from our country's current life expectancy, China's life expectancy is much lower than that of developed countries. In particular, in terms of the current retirement age of men, the life expectancy of men after retirement is only 12.1 years (see Table 4.15) in China, while the life expectancy of men after retirement in developed countries is between 16 and 20 years. Therefore, it's not so desirable to increase the retirement age of men. However, we can consider gradually increasing the retirement age of women to unify the retirement age of men and women. This is a trend of the reform of most countries. In addition, the replacement rate of pension fund can be lowered gradually. The current average pension fund replacement rate is 82.5% with a rate of 97.9% from low-income earners, 77.9% from median-income earners and 71.2% from high-income earners in China, being a very high level in the world. Therefore, we can gradually lower the replacement rate of basic pension insurance while constantly developing and supplementing pension insurance.

(3) Further improve the multiple-support pension security system. At present, China's pension security system has basically been established in accordance with the "three-support" mode of the World Bank. Although the basic pension insurance system combining social pooling and personal account has been implemented, various commercial pension insurances and enterprise supplementary pension insurances are also constantly developing. However, the basic pension insurance of rural areas and migrant people shows a low coverage rate, and in the basic pension insurance of urban employees, many problems are outstanding. For example, the "empty account" phenomenon is serious in personal accounts, the positioning of enterprise supplementary pension insurance is incorrect, the commercial insurance market is in disorder, the law enforcement mechanism isn't sound enough etc. So the "three support" system of China's pension security system isn't sound. Therefore, in the futuristic reforms, the problem of "turning empty accounts into consolidated ones" in

personal accounts should be solved and the development of various social supplementary pension insurances should be positively encouraged. For example, we can learn lessons from such countries as Germany, France and others and encourage private saving pension by means of tax exemption so as to establish a multi-layer and diversified pension security system in practice.

4.5 Investment and Operation of Pension Security Fund

4.5.1 The Necessity of Investing in Pension Security Fund

In addition to the pension fund raised through the strict pay-as-you-go system, other pension security fund is mainly represented as reserve capital and the accumulated income is used to pay for the consumption of workers after retirement. China's current pension security fund is mainly divided into three parts: The first part is social pooling fund which is mainly used for paying the pension charges in that year; the second part is account pension fund that raises fund by means of accumulation and is an important part of one's income after retirement; the third part is special pension security fund which is the pension fund from the national social security fund that is the central financial special fund to correspond to the insufficiency of basic pension security fund and the fund reserve for futuristic population aging.

Because social pooling fund is received and paid at the same period, the risk of fund depreciation generally doesn't exist. Personal account pension fund and pension security fund are not used for payment in the same period. If no effective operation or management is conducted for this part of fund, the fund will certainly depreciate to a certain degree with the growth of economy. Therefore, appropriate investment should be made to ensure the fund value preservation and appreciation. Theoretically, the commentary earnings rate of the accumulation pension fund system should be higher than the growth rate of national income or the growth rate of labor forces in order to ensure that the monetary entitlement possessed by the fund owner doesn't depreciate.

On the one hand, the pension fund which is used as pension reserve fund is a type of financial asset with a financial nature. This part of fund must be put into the capital market in order to realize the preservation and appreciation of its value. As a method of directly distributing economic resources, the capital market generally has such functions as lowering the exchange cost of capital resources, increasing the capital liquidity, adjusting the restrictive mechanism of the economic structure and others. The development and constant improvement of the capital market will bring a continuous and stable investment earnings rate for pension fund. On the other hand, putting plenty of pension fund into the market so that the amount of circulating fund available in the market will increase will also promote the economic growth to a certain extent. But due to the uncertainty of the market, there is certain risk in investment, so we must make effective investment and management of the pension fund and expand the investment channel so as to lower the investment risk and increase the investment benefits. The major factors influencing pension fund investment benefits include the sustainability and stability of the pension system, the supervision and management of pension fund, the macroscopic economic environment, the financial market environment and others.

4.5.2 *Channels and Provisions on Pension Security Fund*

4.5.2.1 Investment Channels of Pension Security Fund

In view of the pension fund investment of all countries around the world, bank deposit, stock, bond, loan and others are major investment channels, and the distribution ratio of different investment products have experienced different development stages: Before the 1980s, most countries invested the pension fund mainly in bank deposits and national debts. This happens because bank deposits and national debts can ensure the safety and stability of pension fund the most. Besides, before the middle 1970s, capitalist economy grew at a high speed and the ratio of pension insurance expenditure was not so high whether in GDP or the government's total financial expenditure. And as for pension fund management, the improvement of asset allocation caused relative scarcity of drive and pressure for the appreciation of pension fund. From the 1980s to the early 1990s, the ratio of stocks started to increase in pension fund asset investment and more and more pension fund was invested in enterprise stocks and enterprise bonds. In the 1960s, the stocks held in the pension fund accounted for less than 10% of the market value of London Stock Exchange and New York Stock Exchange. But in the early 1990s, this ratio already exceeded 30%. By the end of 1994, the stocks held by the American pension fund accounted for 25.9% of the market value of all stocks and 56.1% of the amount of holdings of financial institutions.[51] After the 1990s, the channels of pension fund investment became more diversified. Moreover, with economic globalization and capital globalization, the ratio of pension fund invested in the overseas stock market grew constantly.

Table 4.16 shows different investment channels of some OECD countries in national public pension fund and private pension fund in 2009. Judging from the investment of public pension fund, stocks and bonds were the main investment channels. The US, Belgium, Mexico, Poland, Portugal and Japan regarded stocks as the first investment channel of pension capital, the ratios of their stock investment in public pension investment are respectively 100, 100, 74.8, 81.4, 70.1 and 82.5%; Canada, Australia, Sweden, Norway and Italy regarded bonds as their major investment channels with the ratios being 43.8, 44.2, 50.2, 61.4 and 49.0%

[51]Yu [34].

Table 4.16 Pension fund investment portfolio of some OECD countries in 2009 (unit: %)

	Cash and deposit	Bond	Stock	Others
Public pension fund investment channels				
The US	0.0	0.0	100.0	0.0
Belgium	0.0	0.0	100.0	0.0
Spain	3.3	0.0	96.7	0.0
Mexico	11.0	0.0	74.8	14.3
Poland	8.7	9.8	81.4	0.0
Portugal	0.4	16.4	70.1	13.1
Japan	0.1	17.4	82.5	0.0
Canada	0.0	43.8	29.9	26.4
Australia	14.5	44.2	23.7	17.6
Sweden	0.0	50.2	35.6	14.2
Norway	1.8	61.4	33.9	2.9
Italy	6.4	49.0	11.1	33.5
Germany	3.3	40.8	6.1	49.9
Private pension fund investment channels				
The US	2.2	31.4	45.4	20.9
Belgium	8.5	42.3	32.8	16.4
Spain	18.5	59.2	12.1	10.2
Mexico	1.0	80.6	14.9	3.5
Poland	2.3	66.5	30.2	0.9
Portugal	5.8	56.2	22.2	15.7
Japan	6.4	47.7	13.7	32.2
Canada	3.9	35.2	33.9	27.0
Australia	16.0	12.8	54.4	16.8
Sweden	4.0	62.4	24.3	9.2
Private pension capital investment channels				
Norway	3.9	58.9	30.9	6.4
France	7.0	44.2	47.3	1.5

Source OECD global pension statistics

respectively; Germany's investment ratio was 40.8% in bonds, 6.1% in stocks and 49.9% in other fields. In terms of cash and deposits, the US, Belgium, Canada and Sweden hardly distributed any pension fund investment share in cash or deposits. In comparison, Australia's ratio in cash and deposits was also 14.5% only though it was relatively higher.

Compared with the investment distribution of public pension fund, although bonds and stocks remained the major investment channels of OECD, other investment channels were more diversified and the distribution ratios of all channels were also more balanced. In private pension fund investment, only the US, Australia and France regarded stocks as their first investment channel. In the US, the stock investment ratio accounted for 45.4% of the total investment, but this was a lot of decrease compared with the 100% investment ratio of public pension capital, while the bond investment ratio accounted for 31.4%; in Australia, the stock investment ratio was 54.4%, which was obviously more than one time higher than the national stock investment; the investment ratio was 47.3% in France. Bonds were regarded as the major investment channel of private pension fund in the rest countries with an average investment ratio of about 50%, Mexico's bond investment ratio was 80.6% which was the highest ratio; while Canada's ratio was 35.2% which was relatively lower. The private pension fund of each of all countries mentioned above gave cash and deposits a certain ratio of distribution. Among them, the distribution ratios of Spain and Australia were respectively 18.5 and 16%, being relatively higher, and Mexico showed the lowest ratio of 1%.

Judging from the investment channels and distribution methods of pension fund of OECD countries, stocks and bonds were the main investment channels, the investment channels of private pension capital were more diversified than that of public pension capital with a relatively more balanced distribution. From Table 4.16, we can also see that the investment channel varied with the pension fund system. Such countries as Germany, Italy, Norway, Sweden and others that mainly invested in public pension fund regarded bonds as the main investment channel; while such countries as the US and others regarded stocks as the major investment channel. Maybe the reason was that bonds had a relatively lower earnings rate and lower risk than stocks, and a country that mainly depended on public pension fund system was more inclined to adopting a rather conservative investment method.

4.5.2.2 Investment Rules of Pension Security Fund

In view of the investment management and methods of pension security fund of all countries around the world, two main investment rules have come into being: "Prudent Person Rules (PPR)" and "Quantitative Asset Restrictions (QAR)."

"PPR" were mainly aimed at investment management personnel and meant that in the investment management process of pension fund plan and pension fund, the investment management personnel should have a necessary degree of caution. By "a necessary degree of caution," we mean the degree to which a normal and careful person should be cautious in the process of engaging in asset transactions. This principle originated from the PPR which was deeply rooted in the Trust Law of Anglo-Saxon countries and is mainly represented by countries including the US, the UK, Australia and others. It's characterized by the fact that the supervisees of pension fund investment is centralized on the behaviors of investment management personnel and is specifically embodied by the supervision of asset managers, institution investors and the decision process; emphasis of the honesty liability of

Table 4.17 Investment rules and investment distribution ratios of some OECD countries in pension fund

	Stock right	Bond	Fund	Loan	Real estate	Bank deposit	Foreign investment
Australia	None	None	None	None	None	None	None
The UK	None	None	None	None	None	None	None
The US	None	None	None	None	None	None	None
Japan	None	None	None	None	None	None	None
Canada	None	None	None	None	15–25%	None	None
Italy	None	None	20–25%	None	None	None	None
Spain	None	None	30% (privately offered fund)	20%	20%	None	None
Austria	50%	None	None	10%	20%	None	50% (OECD)
Czech	–	–	–	Not allowed	10%	10%	None
Germany (public fund)	35% (listed); 10% (unlisted)	50%	35% (retail fund); 5% (privately offered fund)	50%	25%	50%	10–35%
Germany (private fund)	None	None	None	None	None	None	None
Greece	70%	Unlimited (government); 70% (enterprise)	5%	Not allowed	None	None	EU; EEA
Hungary	10% (unlisted)	Unlimited (government); 10% (enterprise)	50%; 5%	MPF: Not allowed 5–30%	MPF: 5%; VPF: 10%	None	None
Norway	35%	30%	30%	None	None	None	None
Denmark	70%	None	70% (retail fund); 10%	10%	None	None	None
Finland	50% (listed); 10% (unlisted)	None	10%	70% (mortgage)	40%	None	None

Note "None" means no limit has been set

Source Survey of investment regulations of pension funds, 2007

investment management personnel and the transparency of investment management, requirement that the pension fund investment management personnel should have the honesty liability as it's the most fundamental liability that the pension fund investment management personnel should perform; requirement that a diversified method should be adopted for the investment portfolio management with the purpose of investment diversification so as to realize diversified investment, aversion and control of investment risks as much as possible; permission that the investment management personnel have a free choice of investment on the premise of adhering to the investment diversification rule, the investment management personnel enjoys full asset distribution rights, encouraging normal and orderly competitions among investment management personnel, preventing investment management personnel from operating the market and avoiding the similarization of investment portfolio. There isn't an exact boundary to the "PPR," but in specific use, it will play an important part when the government (legislature, supervisor and judicial officers) determines the rules for the management of different pension fund asset. And the rules themselves will also be constantly improved with the occurrence of new investment management theories, practices and financial products. In some countries, the "PPR" has an important effect on the asset allocation of pension fund and has created a cultural atmosphere of conducting with caution among the fund custodian, the trustee and other relevant institutions.

The "QAR" is mainly aimed at the investment mode and channel and means that the pension fund sets a quantitative or ratio limit to specific types of asset, it generally has prohibitions or a maximum ratio limit to risky assets with low liquidity and high fluctuation. Meanwhile, it sets the lowest ratio limit to the investment of some low-risk assets such as government bonds and others. The characteristics are: Centralizes on limiting the investment rather than the portfolio; sets a quantitative limit to "risky assets"; sets limits to single financial instrument; doesn't allow a free choice in investment. Its purpose is striving to protect the pension fund from the impact of the businessman's bankruptcy and investment risks while diversifying investment risks. The "QAR" is generally applied by most developed countries growing at a low speed.

Table 4.17 shows the rules of some OECD countries in pension fund investment process. In terms of domestic investment, the UK, the US, Australia and Japan followed the "PPR" without setting any limit to the pension fund investment distribution, while most of the rest countries followed the "QAR". Germany followed the "PPR" in its private pension fund, but the strict "QAR" in public pension fund. In terms of overseas investment, most countries don't follow the "QAR". This might happen because with economic globalization and capital globalization, the cost of cross-national stock investment is continuously lowered and the investment risk can be controlled more easily. Meanwhile, the capital markets of emerging market countries are becoming mature continuously, attracting the pension fund of all countries to get involved and pursue higher returns.[52]

[52]Yu [34].

In terms of stock investment, a country that follows the “QAR” mainly sets different limits to listed companies and unlisted companies. Generally speaking, the ratio of investment in listed companies is larger than that in unlisted companies, for example, Finland sets an upper limit of 50% to the stock investment in listed companies but 10% to that in unlisted companies; the limits are mainly set to enterprise bonds among bonds while there is less or even no limit to government bonds; in terms of the limit to purchasing fund, the ratio of purchasing private fund is more strictly controlled; in terms of loan, Czech and Greece don’t allow using pension fund for loans, Hungary doesn’t allow to use all the compulsory pension fund for loans, Denmark, Spain and Hungary also set a rather low limit to loans; Czech regulates an upper limit of 10% to the pension fund used for bank deposits, Germany sets an upper limit of 50% to the pension fund used for bank deposits, the rest countries haven’t set such a limit.

China follows the “QAR” investment rule, regulating that the ratio of investment in social insurance fund bank deposits and national debts shall not be lower than 50% and among that, the ratio of investment in bank deposits shall not be lower than 10%; the investment ratio of enterprise bonds and financial bonds shall not be higher than 10%; the investment ratio of bond investment fund and stock investment shall not be higher than 40%.[53]

4.5.3 Current Status of Pension Fund Investment in China

In China, the pension fund that can be used for capital market investment includes personal accounts and pension fund accounts in social security fund.

(1) Personal accounts. Since 1997, China started implementing a pension fund system combining social pooling with personal accounts. But in the process of practical operation, due to the existence of system transfer cost, the personal account fund of “medium people” and “new people” tended to be appropriated to pay for the pension payment, resulting in the “empty account” of personal accounts. In 2000, State Council of the People’s Republic of China released “Pilot Plan for Improving the Social Insurance System,” requiring to make consolidated personal accounts, implement an accumulative system for all personal accounts and implement the fund accumulation system to satisfy the demand of personal pension. On July 8, 2001, a social security pilot project was carried forward to consolidate personal accounts in Liaoning Province. In May 2004, the reform pilot project was expanded to Heilongjiang with the main task of consolidating personal accounts. In 2006, the pilots were expanded to eight provinces and cities including Tianjin, Shanghai, Shanxi, Shandong, Henan, Hubei, Hunan and Xinjiang.

---

[53]Yu [34].

Table 4.18 Financial appropriation fund in the national social security fund (unit: RMB 100 million)

Year	Amount	Year	Amount
2000	200.00	2006	574.23
2001	595.26	2007	308.14
2002	415.76	2008	326.95
2003	49.08	2009	825.89
2004	278.54	2010	634.44
2005	228.71	Total	4437.00

Source Official website of National Council for Social Security Fund of the People's Republic of China

On the premise of consolidating personal accounts, the accumulated personal account fund of Liaoning Province was increased from RMB 1.32 billion before the pilot project in 2001 to RMB 20.417 billion in 2005 and then RMB 60.417 billion in 2009 afterwards. After the accumulation of personal account fund, in accordance with relevant national regulations, we submitted RMB 12 billion for tender as contractual deposits and entrusted financial institutions to purchase national debts that worth RMB 5.3 billion, the comprehensive earnings rate was 2.92% in 2005 and respectively 3.01, 2.98, 2.87, 2.96% from 2006 to 2009.[54]

If the personal account fund in basic pension fund is completely consolidated, personal account fund will be faced with the pressure of appreciation and preservation. In addition, personal account fund will grow at a speed of RMB 100 billion every year. By 2020, the personal account fund will be accumulated to RMB 10,000 billion. For the moment, the investment methods of personal account fund mainly include bank deposits and national debts. From 2001 to 2009, the average practical earnings rate of national debt was 1.92%, while the actual investment earnings rate of the bank's one-year deposit was 0.46%.[55] With the deduction of management cost, the earnings rate of personal account fund investment in bank deposits and national debts couldn't realize preservation and appreciation appropriately. Therefore, how to realize preservation and appreciation of personal account fund by other means is still a big difficult problem faced with us.

(2) Pension fund account in social security fund. China started to establish its social security fund in 2000. By the end of 2010, the asset total reached RMB 856.690 billion with RMB 443.7 billion of net financial appropriation (see Table 4.18) among that. In view of the investment channels of social security fund, from 2001 to 2004, bank deposits and bonds were the overriding investment methods. In 2001, the bank deposits reached 64.6%, the bonds 33.6% and the stocks 1.6% only with an investment earnings rate of 1.7%; and

[54]Zhou [35].

[55]Zhou [35].

Table 4.19 Total asset, asset distribution ratios and investment earnings rate of the national social security fund (unit: RMB 100 million, %)

Year	Total asset	Bond	Bank deposit	Stock	Entrusted investment	Others	Investment earnings rate	Inflation rate
2001	805.1	33.6	64.6	1.6	–	–	1.7	0.7
2002	1241.9	22.5	75.6	1.0	–	–	2.6	−0.8
2003	1325.0	29.6	45.3	–	24.1	–	3.6	1.2
2004	1711.4	18.2	38.3	6.8	35.8	–	2.6	3.9
2005	2117.9	–	48.2	6.5	34.5	–	4.2	1.8
2006	2827.7	–	33.7	20.7	37.4	–	29.0	1.5
2007	4396.9	–	23.3	14.8	47.1	–	43.2	4.8
2008	5623.7	–	10.9	4.5	–	–	6.8	5.9
2009	7766.2	–	13.5	7.6	–	–	−16.12	−0.7
2010	8566.9	–	2.1	5.1	–	–	4.2	3.3

Source According to the statistics of the annual report of the national social security fund in relevant years, the annual average earnings rate was 9.17%

in 2002, the bank deposits ratio was 75.6%, the bonds 22.5% and the stocks 1.0% with the average investment earnings rate of 2.6% (see Table 4.19). China started to add entrusted investment with an investment ratio of 24.1%, and the ratio of bank deposit decreased to 45.3% in 2003. From 2004 on, bank deposit and entrusted investment have become the main investment methods. The ratio of bank deposit was 38.3%, entrusted investment 35.8% and bond 18.2% only in 2004. In 2005, the ratio of bond investment was 8.1%, bank deposit 48.2% and entrusted investment 34.5%. In that year, social security fund investment was more diversified with the addition of index investment, trust investment and others with an annual investment earnings rate of 4.2%. From 2006 to 2010, the ratio of bank deposit decreased continuously. By 2010, the ratio of bank deposit was 2.1% only. The ratio of stock investment began to rise in 2004. The ratio of stock investment was 20.7% in 2006, and the investment earnings rate was 29% in the same year. In view of the investment channels and earnings of social security fund through the years, with the constant diversification of investment channels, the investment earnings rate of social security fund shows a relatively rising trend.

4.5.4 Problems in China's Pension Security Fund Investment and the Countermeasures

From the previous analysis, we find that the following problems exist in China's pension security fund investment:

(1) With a single investment channel in personal accounts, we can only invest in bank deposit and national debt. Although it's of relatively lower risks, it can't help to achieve the goal of value preservation and appreciation. The interest rate of national debt is relatively higher than that of bank deposit, but because China's national debt interest rate is influenced by the bank deposit interest rate rather greatly, with the fluctuation of bank interest rate and the charges for personal account management, it's difficult to achieve value preservation and appreciation of personal account fund with the current investment method, so we should consider diversified investments.
(2) In view of the investment situation of the national social security fund through the years, a diversified investment method helps to increase the earnings rate of fund. Among that, the earnings of stock investment is the most obvious but with the highest risk. The ratio of stock investment began to rise in 2004. The ratio of stock investment was 20.7% in 2006, and the investment earnings rate was 29% in the same year with an increase of 25% points compared with 2005. Due to the financial crisis in 2008, China's pension fund lost RMB 39.372 billion in its stock investment.[56] Compared with developed European countries and the US, China's stock market started late with lots of systematic deficiencies, so we can't make stock our main investment method as European countries and the US do.
(3) The ratio of investment in bank deposit and national debt is too high. It's regulated in China that the ratio of social security fund investment in bank deposit and national debt should not be lower than 50%, and among which the ratio of bank deposit should not be lower than 10%. While in the pension fund asset allocation of developed countries, the ratio of bank deposit should not exceed 10% and should remain from 3 to 7% in general. Learning from international experiences, in the futuristic investments, we should consider further lowering the investment ratio of bank deposit, transferring part of the investment to bonds and then transiting to stock investment gradually in order to realize fund preservation and appreciation.
(4) Deficiency of fixed investment. Real estate investment is a main method for developed countries to make pension fund fixed investment. Although real estate can produce rather high earnings, to invest pension fund in real estate isn't suitable for China within a short period due to a large proportion of bubble in China's real estate industry at present. Nevertheless, we can consider investing in the construction of infrastructure including bridges, roads and others. In this way, it helps to realize aversion of risks and contributes to the increase of social public welfare.

[56]Zhou [35].

References

1. Jianding, Ding. 2003. *A Comparison of Social Security Systems of Northern European Countries, Academic Forum of Nandu*, no. 1.
2. Yan, Xu. 2009. *Enlightenment from Denmark's Elderly Welfare Policy and Reform to China, Tianjin Social Insurance*, no. 6.
3. Yonghong, Jiang. 2001. *Industrialization and Pension Security System.* A doctoral diss., Jinan University.
4. Feng, Liu. 2008. *The Social Security System of Japan: A Study Focused on the National Pension, Foreign Theoretical Trends*, no. 1.
5. Xing, Chen. 2005. Reform and Innovation of the Pension Fund System of the US. *Economic Herald* (10).
6. Lei, Zhang. 2011. An Analysis of the Effect of the Management System for Pension Insurance on Increase of Coverage Rate. *Journal of Xihua University* (4).
7. Guanyun, Feng. 2010. *Analysis of Pension Insurance for Private Enterprises.* Master thesis, Northeast Normal University
8. Yuanli, Yang, and Liu Xiaomei. 2010. Analysis of the Burden for Enterprises to Contribute to the Pension Insurance System in China. *Journal of Changchun University* (11).
9. Wei, Han. 2010. *Analysis on Optimization of Contribution Rates of China's Pooled Pension, On Economic Problems*, no. 5.
10. Hongchun, Wang. 2010. Systematic Defects of Individual Accounts in the Basic Pension System and Corresponding Remedies. *Journal of Hebei University* (6).
11. Lixiong, Yang. 2010. *Study on Gaps of Pensions Paid to Different Groups in China, New Horizons*, no. 1.
12. Xiantong, Zhang, Hao Yong and Wang Shanshan. 2010. *Calculation of Replacement Rates for Basic Pension Insurance Under the New System, Statistics & Decision*, no. Issue 5.
13. Dan, Li. 2009. Study on Mechanism of Payment of China's Implicit Pension Debt. Doctoral diss., Fudan University.
14. Junlin, Liang, Cai Hui and Song Yanqi. 2010. *Study on China's Implicit Pension Debt Turning Explicit, Chinese Journal of Population Science*, no. 5.
15. Xiaojun, Wang. 2002. *Estimation and Forecast of China's Pension Debts, Social Security System*, no. 6.
16. Ping, He. 2001. *The Report on the Calculation of China's Pension Funds, Social Security System*, no. 3.
17. Zuyao, Cao. 2004. *Study on Pension Fund Pooling and Repayment of China's Implicit Pension Debt.* Master thesis, Fujian Normal University.
18. Changping, Liu, and Yin Baoming. 2011. Fiscal Balance and Sustainability Research of China's Basic Pension System—Basic Pension System for Urban Residents Based on the State Council's Document No. 38 of 2005. *The Theory and Practice of Finance and Economics* (1).
19. Dan, Li, Liu Zuanshi and Zhang Yaling. 2009. Estimation of China's Implicit Pension Debt. *Finance and Economics* (5).
20. Haiyan, Fang. 1998. Calculation of China's Implicit Public Pension Debt. *Statistical Research* (4).
21. Li, Li, Liang Mingxing and Li Chao. 2009. Calculation of China's Implicit Public Pension Debt of 2010. *Statistics & Decision* (21).
22. Xiangyu, Tan. 2003. Accurate Calculation and Assessment of the Implicit Debt in the Transformation of China's Pension System. *Statistics & Decision* (7).
23. Junlin, Liang, Cai Hui and Song Yanqi. 2010. Study on China's Implicit Pension Debt Turning Explicit. *Chinese Journal of Population Science* (5).
24. Xiangqun, Jiang. 2006. Quantitative Research of Population Aging's Influence on Retirement Fund Burden. *Population Research* (2).

25. Yonghong, Cheng. 2005. Quantitative Research of Population Aging's Influence on Retirement Fund Burden. *Economic Research Journal* (3).
26. Lei, Guo. 2010. *Research of Population Aging and Pension Insurance Fund Supply and Demand Balance—A Case Study of Beijing*. A master's diss., Beijing Jiaotong University.
27. Zhigang, Yuan, and Ge Jinfeng. 2003. Economics Analysis of Transition from the Pay-as-you-go System to the Fund System. *Social Security Research* (4).
28. Yonghong, Cheng. 2005. Quantitative Research of the Relationship between the Pay-as-you-go System and Population Aging. *Economic Research Journal* (3).
29. Bao, Lin. 2010. *An Analysis on Population Aging and Pension Fund Mode, Population and Development*, no. 6.
30. Juan, Yang. 2009. Commentaries on Pension Fund Reform of Western European Countries in the 21st century. *Journal of Public Management* (4).
31. Meizhi, Duan. 2008. *The Enlightenment of OECD Countries' Pension Fund System Reform on China, Social Science Forum*, no. 2.
32. Weijun, Meng, et al. 2010. *International Experience on Pension Fund System Reform, Population and Economy*, no. 5.
33. Feifei, Chen. 2006. *Population Aging and Germany Legal Pension System Reform, Deutschland-studien*, no. 4.
34. Boyang, Yu. 2007. *Study on Pension Fund Asset Allocation*. A master's diss., Shanghai Jiao Tong University.
35. Xun, Zhou. 2010. *Study on Pension Fund Investment and Operation of Chinese Fund System*. A doctoral diss., Liaoning University.
36. Lijun, Wang. 2008. Analysis of the Effect of the Fiscal Funding for Insufficient Pension Funds in China on Economic Growth. *Journal of Liaoning University* (1).

Chapter 5
Research on the Elderly Mental Security System

Jinqun Shi

5.1 Significance and Function of Mental Security Within Pension Security

(I) Elderly security is a comprehensive support and care system
Pension security is the integration of all kinds of social support for the survival and development of senior citizens. It is a complete and comprehensive support system. As a comprehensive support and care system, it takes the needs of the elderly as its logical starting point. Western scholars who study senior citizens summed up the needs of the elderly as the three "Ms". That is, Money, Medicare and Mental Health, or, in other words, economic, security, emotional, esteem and self actualization needs. Historically, the concept of the "five things an elderly person should be entitled to" was put forward in China. These were economic support, medicine, worth, learning and enjoyment. Although these two concepts are different, the content is the same. Both western scholars and domestic scholars generally believe that senior citizens need not only economic support, life care and other material help, but also have psychological needs. "Economic support, life care and spiritual solace" are the three basic components of elderly care. A complete and comprehensive pension security system is an organic entirety that includes economic security, service system security and mental security. Any elderly care that neglects and omits any part of the above content will not be complete. In particular, according to Maslow's theory of a hierarchy of needs, people all have five hidden needs on different levels: physiological, safety, love/belonging, esteem and self-actualization; however, the urgency of these needs vary during different periods. People's needs

J. Shi (✉)
Institute of Sociology, Chinese Academy of Social Sciences,
#5 Jianguomennei Street, Dongcheng District 100732, Beijing,
People's Republic of China
e-mail: 1323914368@qq.com

Y. Wang (ed.), *The Development of Security and Whole Care System for the Aged in China*, https://doi.org/10.1007/978-981-10-6481-4_5

gradually transform from external satisfaction to internal satisfaction. Once needs at the lower levels are basically met, the incentive effect of these needs will weaken and their dominance will not continue. Needs at the high levels will replace those at the lower levels and become the main factors dictating behavior. High level needs have greater values than low level ones. As the highest level need for senior citizens, physiological needs will become more prominent once basic material needs are met.

(II) Urgency for mental security in China's pension security
The subjectivity of mental needs determines the different mental needs for different subjects. Age, physical characteristics and the unique social experience of senior citizens, a specific age group, determine that they have different mental needs from other age groups and that their mental security is different from that of other groups. Broadly speaking, elderly mental security secures their needs for social interaction and emotional communication and can prevent all kinds of mental diseases in the elderly such as Alzheimer's disease and depression.

China's pension security, in a broad sense, primarily includes elderly income security, elderly medical security and elderly service security (including life care and spiritual solace). Reviewing the development of China's pension security system since the founding of new China, one can easily discover that compared with economic security and medical security, China's elderly service system, especially the elderly mental service system, has developed much slower. For a long time, if one mentioned aging and pension security, the first thing people would think of would be economic elderly care and physical security, and then consider the raising, distribution and management of funds. Few people considered the mental support and mental security of senior citizens. This is related to China's level of economic development. "Getting old before getting rich" leads to scarce resources for physical elderly care, and arouses people's concern and attention. The mental security of senior citizens primarily relies on an informal support system, especially the family. However, with the development of the economy and the improvement of the social security system, senior citizens' mental needs increase. The strength of this informal support system for senior citizens' mental support is being challenged.

As the aging of China's population intensifies, there are more and more senior citizens. Psychological experts point out that as senior citizens age physically, their psychological quality becomes gradually weaker. There is a lowering of the abilities of enterprise and risk tolerance, and factors such as fear, loneliness and emotional vulnerability can appear. Compared with diseases, these psychological symptoms are more exhausting. Without the comfort of relatives, friends and society, a series of psychological and physiological diseases is an inevitable outcome. They need to communicate with people, participate in society and obtain emotional support through communication. Moreover, they require their children's and family's yearning and mental sustenance. Relevant data shows that due to the degeneration of brain functions and the dramatic changes in life before and after retirement, 85%

of senior citizens suffer mental disorder to varying degrees. The development and progression of 30–40% of common diseases in senior citizens are related to their psychological status. Moreover, in terms of the pathogenic factors involved in the "three major killers" of senior citizens, namely, cardiovascular disease, cerebrovascular disease and malignant tumors, psychological factors play a bigger role than physical factors. At the same time, according to research findings, senior citizens who are comforted and taken care of by society and their families can live 10–15 years longer than those who fail to get along with their children and do not receive any care. According to psychological experts, senior citizens who are happy will enjoy a satisfactory biochemical metabolism and neuroendocrine regulation, meaning they can live a long life. Conversely, senior citizens who are prone to negative emotions will suffer weaker immunity and an increased chance of disease. If their mental needs are not met, they will be solitary, depressed and lonely, which will accelerate their physiological and psychological aging and even cause depression, senile dementia and other mental and psychological diseases.[1]

Whether importance is attached and safeguards put in place to the quality and dignity of senior citizens' lives is not merely related to the quality of life of this expanding population. According to statistics, in the late 20th century, the number of people aged 65 and above and 60 and above had reached 6.196 and 10.13% of the total population respectively. The National Working Committee on Aging predicts that by 2020 in China, the number of people aged 60 and above will account for 17.2% of the total population and will reach 437 million by 2050, accounting for about 31% of the total population by then.[2] Meanwhile, this is also related to a society's moral standards and humanitarian values. This is a major social problem that cannot be ignored. The spiritual solace of senior citizens, as an important part of elderly care security, is an important manifestation of the life quality of senior citizens as well as an important measurement index of social development. Change to the social structure, economic development and the increasing demands of senior citizens impose greater challenges on China's aging problem. Only by establishing a complete and comprehensive senior care security system can such challenges be coped with.

5.2 Current Situation and Analysis of Mental Security of Elderly Citizens in China

(I) Mental security of elderly citizens via informal support system

The informal support system for senior citizens, consisting of families, relatives, friends, neighbors and other primary groups, is able to provide a more

[1]Wu and Mu [1].

[2]The National Working Committee on Aging: *Research Report on Development Trend Prediction of China's Ageing Population*, February 23, 2006.

comprehensive, faster, prompter and permanent support for senior citizens. In addition, its primary characteristic is to provide emotional support and comfort for senior citizens, meet the senior citizens' needs for social interaction and emotional communication and prevent mental disease.

1. Spiritual solace for senior citizens offered by the family

Filial piety is a core feature of the traditional culture of elderly care in China. According to *The Classic of Filial Piety*, "filial piety is an unquestionable moral truth that all people should follow". "Filial piety" differs from "support". The core concept of "filial piety" is to respect senior citizens and consciously preserve, cultivate and develop humanity's inborn respect for parents, to help them to become adults, sages and saints. Specifically, "filial piety" is about allowing parents to feel relieved and joyful and free from sorrow and worry. The traditional elderly care culture of "filial piety" still has a profound influence on China's elderly care practices. The family plays a very important role in the spiritual solace of senior citizens. In China, people's definition of the family is complex. Due to the influence of traditional culture, when it comes to elderly care in the family in China, people tend to think of elderly care offered by children, especially by sons. Therefore, it is best at this point to focus the discussion on mental support offered by children. The role of the spouse, another important member of the family, in mental support also cannot be ignored. Therefore, this discussion on the family will include the spouse.

In terms of the spiritual solace of senior citizens, the family can meet their needs for emotional communication. A survey[3] analyzed the composition of the mental support network for senior citizens by asking whom elderly people would turn to. These are the results. In urban areas, according to the average proportion, the spouse accounted for 23.7%, sons 19.4%, daughters 15.8%, friends, colleagues and neighbors 14.0%, daughters-in-law 10.4%, other relatives 5.7%, cadres of the neighborhood committee 4.8%, sons-in-law 2.8% and others 3.3%; in rural areas, according to the average proportion, sons accounted for 22.9%, the spouse 19.9%, daughters 17.4%, daughters-in-law 13.5%, friends, colleagues and neighbors 11.9%, sons-in-law 8.6%, other relatives 4.4%, cadres of the neighborhood committee 0.9% and others 0.5%. From these statistical results, it is clear that the role of the family in terms of support in offering emotional communication for senior citizens is much more important than non-family members. In urban areas, the average proportion of those surveyed as offering emotional support were as follows: spouse, sons, daughters and daughters-in-law were 23.7, 19.4, 15.8 and 10.4% respectively, accounting for 69.3% of the total proportion; in rural areas, the average proportion of sons, the spouse, daughters and daughters-in-law were 22.9, 19.9, 17.4 and 13.5%, accounting for 73.7% of the total proportion. In terms of spiritual solace, senior citizens in rural areas rely more on family members than those in urban areas.

[3]Zhang [2].

Senior citizens' need for emotional communication is mainly met by support from family members. Such emotional needs do not lessen with economic development and gradually improved social security. According to psychologists' analysis, emotional communication between family members enables senior citizens to feel free from anxiety, achieve a satisfactory biochemical metabolism and neuroendocrine regulation and have a healthy and long life. On the contrary, the lack of such communication between family members can lead to negative emotions such as solitude, depression and loneliness, accelerate physiological and psychological aging, and even induce depressive disorders, cardiovascular and cerebrovascular diseases, senile dementia and other mental and psychological diseases.

The family's emotional factor, economic factor, time factor and demographic factor contribute to the realization of emotional communication between senior citizens and family members. First of all, the emotional factor is the primary decisive factor. This refers to children's subjective filial piety for senior citizens. Emotional communication supporters must subjectively and emotionally approve and realize the importance of this practice. Second, this must rely on a certain economic basis and requires time. Some children want to show filial piety for their parents, but they are busy making money every day; some filial children can support their parents financially, but they live far away from their parents and are busy at work and occupied by too many other things. Lacking strength and considering time and social costs, they cannot even accompany and emotionally communicate with their parents, let alone offer them emotional solace. Last but not least, a certain demographic factor is required. This refers to the number of children and the situation of the spouse. An increase of the number of children does not necessarily enhance emotional support for senior citizens. Worse is that sometimes this can lead to family conflict whereby all children shirk their responsibilities and are indifferent to their parents. Nevertheless, the "4-2-1" family structure due to the one-child policy will undoubtedly impose challenges on the family's role in offering spiritual solace in the future.

With the development of the economy and the change in social structure, the following four factors regarding the family offering spiritual solace are being challenged:

(1) Emotional factor. People increasingly desire an independent and free life with personality equality and the pursuit of self-realization. Under the influence of such values, there are more and more "empty-nest families". According to the third census data of 1982, the ratio of one-couple households was 4.7% in China. However, as a household with a single couple are not necessarily empty-nest families, and accounting for some young couple households that temporarily are without children and single couple households that choose not to have children (DINK), in fact, there were less empty-nest families than this. In other words, the ratio of empty-nest families in 1982 was lower than 4.7%. In 2008, the Institution of Sociology, Chinese Academy of Social Sciences investigated family structures and family relationships in five urban areas with

the strict definition of an empty-nest family being one that the parents were old and their grown-up children had left home. According to investigation results, in urban areas, empty-nest families accounted for 16.3% of total families. On November 7, 2010, the minister of the Ministry of Civil Affairs announced at the Promotion Conference of the Construction of the National Social Senior Care Service System that by the end of 2009, there were a total of 167 million senior citizens in China, with over 50% of elderly households in urban and rural areas being empty-nest families, and in some large and medium cities this figure reached 70%; there were about 40 million rural left-behind senior citizens, accounting for 37% of rural senior citizens. The actual figure may be greater, as there are some statistically invisible "empty-nest families". Some senior citizens live in their children's houses to take care of their grandchildren and do the housework for their children. After their grandchildren grow up, they will be left-behind again. In addition, we should also pay attention to "psychologically empty-nest" families. In such families, although senior citizens live with their children, due to the differences in lifestyle and values and the weak sense of filial piety, they are empty psychologically. From this, it is clear that in over the past 20 plus years, great changes have taken place in people's ideology. The tendency to pursue independence and freedom and live separately becomes more and more obvious. There may be some objective factors such as housing and mobility, but such growth probability can still reflect the changing ideas that people have to a certain extent. In rural areas, young people work away from home, so there are more empty-nest families. However, the traditional concept of filial piety whereby one does not leave their parents home and move far away is fading and changing the traditional living pattern in rural areas. Some senior citizens and young people have begun to live separately, thinking that they will enjoy more freedom without suffering wrong and avoid a lot of family conflicts. Under the influence of such values, people pay more attention to their own self-actualization and ignore senior citizens' mental needs.

Case I The 68-year-old Mr. Xu filed a lawsuit against his sons to court, claiming that his sons should take care and visit him without mentioning any material requirements. Originally Mr. Xu worked for a pharmaceutical company in Tianjin City with generous pensions, so he didn't want his sons to bear the maintenance costs. In addition, he was in good condition and able to take care of himself. However, he said, "I don't need alimony but my children's care and greetings. It's so hard for me to meet him as meeting a big star, which makes me heartbroken." With the judge's efforts, his children agreed to take care of and visit him. This case eventually was closed in mediation. Before this case, a similar case the court tried weren't solved through mediation. Finally, the implementation of mental support was written into the sentence and certain "economic constraint" means were applied to ensure that elderly citizens could get necessary care and mental solace.

Case II Plaintiff of this case is Ms. Zhao who is over 80 years old. She has four sons and a daughter. She has been living alone without any source of income but a variety of diseases since her husband passed away 30 years ago. In the trial, her children basically agreed to pay alimony but always refused to pay visit care to her with excuses of being busy or health reasons. In this case, the court pointed out, children should give parents necessarily economic and material security as well as mental solace. The court decided that her five children should jointly pay her support fees and medical expenses while visiting and caring about her in her house. Those who fail to visit her in time should pay caring costs on a monthly basis for her.

Case III In a field survey, we found that two elderly citizens' families know each other and both have many children. Family A's children are quite successful and in very good economic situation. They bought luxury houses and the best living supplies for their parents. In addition, they gave their parents a large amount of money to let them travel abroad. However, they seldom go home to visit their old parents. The senor citizens said, "Even they went home, they just surfed the Internet, watched TV and talked on phones, being busy in their own business without communicating with us." In contrast, Family B is less successful and the economical support they gave their parents is far less than that of Family A. However, children of Family B will visit their parent's house every weekend or holiday, happily recalling their family matters together. Family A envies Family B so much. Such phenomena are common in modern society. Pursuing self realization, independence, equality and freedom of life make people easily forget the basic spiritual needs of elderly citizens - the need for emotional exchange.

(2) Economic and time factors. The emotional factors are subjective, while the time and economic factors are objective. Some young people also want to care about their parents, but they have no time to care about their parent's mental needs due to busy business or fierce occupation competition for making a living. The development of the market economy leads to the increasingly open society and growing flow of talent among occupations and regions as well as growing pressure and increasingly fierce competition. Especially in big cities like Beijing, due to the competition pressure, expensive housing and traffic congestion, many young people can only live in the outskirts while working in the downtown. They get up at 5–6 o'clock every morning to go to work and go back home at 7–8 o'clock in the evening. The home becomes a real "hotel", so it is very difficult for them to have time or to be in mood to communicate with their parents. Sometimes it's even a "luxury" thing for them to make a greeting

call. In rural areas, a large number of young and middle-aged people have to leave home to earn money to afford their children's tuition, marriage expenses and cost of building houses. Most of them often go home once a year or once every two or three years to save transportation cost and earn more money. Without children beside, two or one old parent stay in the rural home alone. This kind of people is often called as "migrant workers." Some people from rural areas become urban citizens through studying, doing business and working. Currently, they have a very nice name on the Internet—"phoenix man" or "phoenix woman." Some of them are very successful and take their parents to live in the cities. However, they have neither desire nor time to care about their parents. Elderly citizens who are used to rural life can't get used to the urban life. They have to always stay in the high isolate buildings like cornered birds, feeling quite empty and lonely. Some people can only basically support their life in cities while failing to afford the expensive housing fees, so they can't create conditions to live with their parents objectively. Besides, rural elderly citizens are even "supported separately". This phenomenon is originally caused by the traditional family separation system. Children separate and support their parents for the sake of fair supporting. However, in recent years, with the massive outflow of rural population, whether children leave home to work or do business, many parents are supported separately to take care of their grandchildren and do housework so as to allow their children to have more time and energy to work. The phenomenon that one elderly parent will stay at home to look after the house and do farm work, while the other will go to the cities to help their children becomes more and more common. "Lonely empty-nest support", "separated support" and "trapped rural support" become the three major problems of the spiritual support of the elderly citizens. Such phenomena can't be removed due to some subjective factors and objective factors like the economy and time. The elderly mental security has increasingly become an important problem which needs to be addressed urgently.[4]

(3) Demographic factors. This refers to the number of children and spouse situation. Although some researches show that the increase of the number of children sometimes will increase family conflicts and result in the phenomenon of mutual prevarication of supporting elderly citizens, some quantitative researches show that the number of children has a significant relationship with the elderly family support, which confirms an old Chinese sentence that "one more child, one more person to rely on". A research on the mental health of the elderly citizens in urban areas and related factors[5] shows that: the male elderly citizens who have spouses and children, live in families of two or three generations and are more satisfied with their economic situations, are in better condition and have better mental health than the corresponding elderly citizens.

[4]Li and Zhang [3].

[5]Research Group on Countermeasure of Chinese Population Aging: *General Report on Countermeasure of Population Aging in China*, August 2008.

At the same time, they think that marriage, number of children and the economic situation are important factors affecting mental health of the elderly citizens. However, with the implementation of the family planning, such a children resource declines increasingly. In addition, some couple families (lifetime infertility families) and single families (lifetime unmarried families) form challenges for the demographic factors of the family mental security. Results of the investigation by the "five-city family structure and family relationship" research group of China Academy of Social Sciences in 2008 shows: the one-child family accounted for 60.7%; families with two children accounted for 15.1%; families with more than three children only accounted for 11.5%. It is worth noting that 12.7% of families had no children because they were so young that they hadn't had children yet, or they were infertile, or they were unwilling to have children. In rural areas, due to objective reasons like labor forces, the one-child policy wasn't carried out in more regions; however, affected by educational expenses, living expenses and other factors, people's fertility concept is gradually changing out of economic rationality, thus greatly reducing the number of children. Villages in developed areas even show a phenomenon of being willing to have a child. The spouse condition is also an important demographic factor. As the saying goes, "be couples when young and partners when old." Living together for long, couples understand each other and spouse plays an important role in the mental world of the elderly citizens, being the main listener of emotional communication and talk. In the west, spouses are the most important spiritual pillar for the elderly citizens. However, due to the aging of the elderly citizens, different male and female expected life span, growing rate of single families (lifetime unmarried families) and divorce and the above mentioned phenomena like "separated support", the spouse which is an important source for family emotional solace is also encountering challenges. According to statistics (see Footnote 5), in 2000, the population of elderly citizens in China over 80 years had increased from 8 million in 1990 to 12 million in 2000 and was expected to reach 27.8 million in 2020, being the fastest aging group among the elderly. Due to the different male and female life span, women generally live longer than men, so many elderly widowed women live alone.

2. Spiritual solace of relatives, friends, neighbors and other primary groups for the elderly citizens

(1) Relatives. Western sociologists like Gooddy and others (1996) included the relationship between parents and adult children in kinship which is defined as a relationship outside the couple relationship. Because of Chinese special cultural tradition, many elderly citizens include their children in the family members and children play a greatly different role in supporting their parents from other relatives. Therefore, relatives discussed here exclude children and primarily include: ① lineal relatives of more than two generations such as paternal grandparents or grandchildren; ② other collateral relatives including consanguinity and affinity, such as brothers and sisters, maternal grandparents, uncle

and aunt and cousin. Due to the age reason, the relative network of the elderly citizens is mainly relatives of peers or younger generations. Like blood children, relatives of younger generations (such as nephew, niece, granddaughter and grandson) also face many challenges in the emotional factors, economic and time factors and demographic factors. Brotherly affection becomes the most important spiritual solace source in the elderly relative network. Results of the investigation by the "five-city family structure and family relationship" research group of China Academy of Social Sciences in 2008 shows that brothers and sisters are the most close relatives for the five-city families, more than 70% of families maintain daily interaction with them. More than 10% of families in Guangzhou, about 1/4 of families in Hangzhou and Zhengzhou and about 1/3 of families in Lanzhou and Harbin say they wouldn't seek for support and help from their brothers and sisters in critical events and economic assistance although they have a close relationship and maintain close interaction with them. That is to say, such a close relative relationship is not totally used for developing and providing utilitarian relative needs and mutual help but for needs of emotional comfort. Brothers and sisters are an important resource for the emotional life of the elderly citizens. They have the common living experience and blood relationship, so the establishment of brotherly affection can enable the elderly and their brothers and sisters to help each other. When the elderly citizens are in need, their brothers and sisters always sincerely help them without asking for any repayment. Some scholars believe that,[6] the best brotherly support is emotional support which can meet the needs of emotional communication of the elderly citizens. Because of this brotherly support, some elderly citizens who are trapped in emotional dilemma gain momentum and emotional comfort and face the life with a healthy attitude. For elderly citizens whose children don't live with them, the brotherly affection is more precious. However, such brotherly emotional solace depends on the living distance, health condition, family's conditions and emotional connection and other factors of the elderly citizens.

(2) Friends. Friends live together based on common life experience or common interests, or develop friendship due to study, employment and other factors. Friends are the mutual choice in interpersonal communication. They are usually of the same age and at the same life stage with similar educational background and social status. Friends rarely play the role in caring about each other but play a very important role in meeting the elderly needs of emotional communication and social interaction. Friendship helps the elderly citizens to re-recognize their self value and communication with friends can help the elderly citizens clean their negative self evaluation and feel the environment recognition. The friendship between friends is very important for the elderly support. Elderly citizens desire for friendship. According to the one-time survey data,[7] elderly

[6]Zhou et al. [4].

[7]China Research Center on Aging [5].

citizens who love making fiends in urban and rural areas respectively account for 65 and 61.2%. It's inconvenient for the elderly citizens to say some words to their children or spouses, or they can't say the words to their children or spouses, and friends are the best listeners in this case. However, due to the same age and with the aging and decline of health status, elderly friends will visit each other less and less, which will affect their intimate relations with friends. In addition, as they are aging, the number of their friends decreases naturally. Therefore, friendship between the elderly people will be interrupted because of something unpredictable, and the age also erodes the emotional support between friends.

(3) Neighbors. Neighbors refer to those who live nearby and often contact with you. Neighbors can not be chosen. As long as one continues to live somewhere, the neighborhood will be relatively fixed. An old Chinese saying goes, "Neighbors are better than distant relatives." The study result shows that neighbor is an early warning system of the informal system because neighbors know first things happening to the elderly citizens. When the elderly citizens have emergency incidents, respondents are usually neighbors. For example, they will call an ambulance and notify relatives or relevant organizations of the elderly citizens. Sometimes, neighbors also provide some temporary help, such as helping to care for the elderly citizens or providing living assistance. In addition, visit, chat and care between neighbors are also an important part of the elderly emotional comfort. However, the mutual help between neighbors are commutative. If neighbors don't contact each other, it will be hard for them to help each other. Or the asymmetric needs will also lead to the collapse of the neighborhood support. In addition, the reciprocal relationship between neighbors often ends because of moving home. In urban areas, the past neighborhood is rarely seen now. Due to the heterogeneity, high mobility and busy life in communities, neighbors living nearby or in the opposite apartment don't know each other and even never see each other before, let alone having mutual exchanges and mutual assistance. Many rural elderly people who live in the cities with their children find that, neighbors which are like the sayings that "they are families although they are not actually," or "neighbors are better than distant relatives" have become "road acquaintances." Many neighbors look familiar, but they never talk or chat with each other. Neighborhood relation is isolated by the locks or anti-theft doors, which becomes the best portrayal of neighborhood of the urban life.

(II) Mental security of elderly citizens offered by formal support system

Formal support system includes the building of political system, economic system and social policy and public or private social welfare institutions and community groups, involving caregivers who have no filial responsibility, emotional bonds and social ties with elderly citizens. Generally, facilitators of formal support system provide spiritual support and security for elderly citizens under the hierarchical structure and in a predictable and organized manner. Divided from the angle of structure, formal support

> system includes enclosed institution support and various forms of community support. The subject of formal support system is social service agencies at various levels. In addition, formal supporters may also be some non-service organizations, such as religious and cultural groups. They are not obligated to provide social services, but they help relevant persons occasionally.

According to Maslow's hierarchy of needs, in addition to emotional needs, elderly citizens also need respect and self-realization. In their old age, their physiological function is on the decline. Moreover, they have to constantly suffer losses, such as the losses of roles, incomes, relationships, loved ones and ideals. According to "activity theory" proposed by Robert Harvey, an American scholar, elderly citizens should participate in the society actively. They should keep the way of life of the middle-aged people as long as possible to deny the existence of the old age and replace roles lost due to the death of their spouses, retirement and other factors with new ones. In other words, they should get rid of depression arising from the interruption of social roles through new participation and new roles. Many scholars also positively affirm "activity theory" based on their own research findings. They hold that only by participation can elderly citizens know their own self from a new way and maintain a positive and happy mood. Elderly citizens, being cared, can't limit themselves in the cramped living space. Instead, they should walk out of their homes to participate in the society and find positive and healthy spiritual support. The society should create such chances for elderly citizens. Social participation and social interaction are other important aspects of spiritual needs of elderly citizens. To meet such needs of elderly citizens, formal support system creates a cultural atmosphere of respecting and cherishing elderly citizens and creates a variety of conditions for elderly citizens to participate in the society and realize social contacts.

In 2006, on the basis of *The One-shot Survey of the Situation of Chinese Urban and Rural Elderly citizens* which was carried out for the first time in 2000, China Research Center on Aging organized 20 provinces, autonomous regions and municipalities directly under the central government nationwide to carry *out Tracking and Survey of the Situation of Chinese Urban and Rural Elderly citizens.* The results showed that among 1000 urban residents' committees (communities) under survey, 66% had activity spaces for elderly citizens, 77% once health knowledge lectures, 77% often organized cultural and recreational activities for elderly citizens and 67% organized volunteer services; among 1000 rural committees under survey, 30% had activity spaces for elderly citizens, 31% had activity rooms for elderly citizens, 24% once health knowledge lectures, 21% regularly organized cultural and recreational activities for elderly citizens and 22% had workers for elderly citizens. Compared with *The One-shot Sampling Survey of the Basic Situation of National Urban and Rural Elderly citizens* carried out in 2000, the consciousness of participating in social economic development and community activities of urban and rural elderly citizens was gradually enhanced: the proportion of urban elderly citizens to participate in various social public service activities increased from 11.9 to 45.1%; the proportion of rural elderly citizens to understand the condition of collective economy increased from 11.9 to 18.8%; the proportions of urban and rural elderly citizens to be willing to participate in community activities increased from 56.4 to 66.2% and from 54 to 61.4% over the same period respectively.[8] From the results of the survey in 2006, we can know that Chinese urban and rural elderly citizens have strong consciousness of social participation. However, the contrast between subjective desires and objective conditions of elderly citizens to participate in society remains large. For example, there is a great difference between urban elderly citizens and those in rural areas in community organization. One reason is that the development of social organizations of elderly citizens in rural areas still lags far behind that in urban areas.[9] Compared with elderly citizens' growing mental needs for social interaction, both the quantity and the content are not enough. According to the survey, 69.2% of urban elderly citizens and 75.6% of rural ones increasingly felt that they couldn't keep up with the social development; 51.4% of urban elderly citizens and 59.9% of rural ones considered themselves as social burdens. Urban elderly citizens felt they couldn't keep up with the social development because the proportion of social burdens was lower than that of rural areas, which was related to the fact that most of them had pensions and more opportunities to participate in society. Moreover, 38.7% of elderly citizens suffered mild depression, 18.1% moderate depression and 7.4% major depression. 34.5% of rural elderly citizens often felt lonely, 10.8% higher than urban ones accounting for 23.7%. As to the question that "more and more young people respected elderly citizens", 48.3% of urban elderly citizens and 60.1% of rural ones held a positive attitude, while 26.9% and 15.8% negative. 24.8% of urban elderly citizens and 24.1% of rural ones answered that it was hard to say.[10]

1. The position and role of the government in mental security of elderly citizens

Under the huge impact of the aging of the population, the government has limited ability to provide limited senior care security resources for social members, but it has executive ability and policy means that a number of non-support systems and other nongovernmental organizations or profit-making institutions lack. Senior mental security is an unshirkable responsibility of the government when implementing its functions. The roles of the government in senior mental security are as follows: ① establish and perfect relevant legal policies to provide positive policy support and reliable legal protection policies for social mental senior care; ② invest talents and funds in mental service system; ③ macroscopically supervise and manage senior mental service system to guarantee the standards and quality of the market and services; ④ create a cultural atmosphere in which "elderly citizens are respected and their mental needs are emphasized".

---

[8]Zhang and Guo [6].

[9]Zhang and Guo [7].

[10]Zhang and Guo [6].

(1) Establish laws and related policies. Legal system is the foundation of social management. Our country is one of the developing countries which pay early attention to population aging and the aging problem. Since the founding of the new China, we have respected and protected the legal rights and interests of elderly citizens and made full use of various means such as laws and morality to strengthen the security of rights and interests of elderly citizens. A series of laws, systems, policies and plans have been made. Basic laws like *Law of the People's Republic of China on Protection of the Rights and Interests of the Elderly, General Principles of the Civil Law of the People's Republic of China, Law of Succession of the People's Republic of China, Marriage Law of the People's Republic of China, Criminal Law of the People's Republic of China, Law of Public Security Administration Punishments of the People's Republic of China* make clear the rights of elderly citizens and legal responsibilities one should bear if he infringes the rights of elderly citizens. The country should fully respect and take care of elderly citizens in the social life.

In addition, China also issued some national aging policies successively, such as the enterprise pension insurance system, medical system, rural insurance, socialization management of enterprise retirees and others of the Ministry of Human Resources and Social Security; the urban and rural minimum living allowance, regulations of five guarantees, measures for the management of rural senior care homes, Starlight Plan, temporary assistance system and others of the Ministry of Civil Affairs; *Opinions of Strengthening the Elderly Health Work*, the New Rural Co-operative Medical System and others of the Ministry of Health; *The Notice on the Elderly Education Work* of the Ministry of Culture; the national fitness program of the State General Administration of Sports; *The Notice on Publishing the Industry Standard Design Code of the Elderly Buildings* issued by the Ministry of Housing and Urban-Rural Development; the reward and assistance system for rural family planning households of the National Health and Family Planning Commission; the elderly legal aid of the Ministry of Justice; *The Notice on Speeding up the Development of the Pension Service Industry, the Opinion of Strengthening the Elderly Preferential Treatment, the Opinion of Strengthening Grass-roots Aging Work, the Opinion of Promoting Residents' Senior Care Service Work* issued by the National Working Committee on Aging. At present, 30 provinces (autonomous regions and municipalities directly under the central government) in China have drafted and implemented special local rules and regulations on protecting the legal rights and interests of elderly citizens. Provinces (autonomous regions and municipalities directly under the central government) have drafted the policy of providing preferential treatment for the elderly citizens. For example, the government takes various measures to facilitate basic necessities of life of elderly citizens; provide free or preferential policies for elderly citizens in urban public transport, long-distance passenger transportation, railway, waterway and air passenger transportation; plan and construct parks, exhibition halls, museums, libraries and other cultural entertainment places, increase the number of construction projects for elderly citizens and open to elderly citizens freely or preferentially, so that elderly citizens can fully enjoy the social respect and love.[11]

In 2000, China issued Decision of the Central Committee of the CCP and the State Council on Strengthening the Aging Work, in which the guiding ideology, principles and goals of the aging work were proposed and the significance of the aging work was emphasized adequately. In 2001, The 10th Five-Year Plan Outline of Chinese Aging Development was issued, which outlined the grand blueprint of Chinese aging work in the 21st century. In 2005, Suggestions for the 11th Five-Year Plan of National Economy and Social Development of the Central Committee of the CPC was issued at the 5th Plenary Session of the 16th Central Committee of the CPC, which proposed to "seriously study and lay down policies and measures for the aging of the population".

On October 18, 2010, the 5th Plenary Session of the 17th Central Committee of the CPC passed *Suggestions for Making the 12th Five-Year Plan of the Central Committee of the CPC* and proposed to "realize the complete coverage of the new type of rural social endowment insurance system, perfect the implementation of the endowment insurance system for urban workers and residents and realize the overall planning of basic pensions", "cope with the ageing of the population, pay attention to the role of families and communities, give priority to the development of social senior care service and cultivate and expand senior care service cause and industry".

These laws, systems, policies and plans have played significant roles in dealing with the aging problem in China. However, it is easy to find that the government pays more attention to economic endowment and material security and considers the raising, distribution and management of funds and other material levels. In recent years, the elderly service system is taken seriously gradually, but their mental support and mental security are rarely concerned.

The Protection of the Elderly Rights and Interests of the People's Republic of China (hereinafter referred to as *the Protection of the Elderly Rights and Interests*) which was issued and implemented in 1996 is the first exclusive law to protect the special rights and interests of elderly citizens in the history of China. It has played a positive role in protecting the legitimate rights and interests of elderly citizens, developing the elderly cause and promoting and carrying forward the traditional virtues of respecting and cherishing elderly citizens. It expounds mental security. For example, according to Article 3, the government and the society should adopt measures to improve the social security system for elderly citizens and gradually change the living conditions, health conditions and the conditions of participating in social development for them, so that they will be looked after and treated properly, do something and feel happy; according to Article 9, governments at all levels should praise or reward organizations, families or individuals which maintain the rights and interests of elderly citizens, respect elderly citizens and make outstanding achievements in senior care; according to Article 11, the whole society should carry

[11]Zhang and Guo [8].

out publicity and education activities concerning the respect and cherishing for elderly citizens and set up the social fashion of respecting, cherishing and helping elderly citizens, and youth organizations, schools and kindergartens should provide moral education of respecting and cherishing elderly citizens and law education of maintain the lawful rights and interests of elderly citizens for teenagers and children and advocate compulsory services for elderly citizens; according to Article 32, the government and the society should take measures to carry out mass cultural, physical and recreational activities suitable for elderly citizens to enrich their spiritual and cultural life. However, *the Protection of the Elderly Rights and Interests* is just about principles of mental security. It has no detailed provisions on the legal responsibility for the failure to fulfill the obligation of mental support. For example, according to Article 11, the supporters should provide economic support, life care and spiritual solace for elderly citizens and meet their special needs. However, in actual operation, elderly citizens can't ask for mental support. The request for mental support is provided legally, but it is difficult to enforce it.

In trial practice, many elderly citizens win the lawsuit but lose the feelings. It may make the contradiction fiercer. The judgment supports their claims, but the ultimate goal of the litigation isn't achieved. Some litigations failed. Mr. Zheng, 70 years old, lived in his wife's house after getting remarried in 2000 and sold his house. Later, he lived in the rented house alone because he was on bad terms with his second wife's children. Two months ago, he submitted a case to the court and asked his son to visit him once a month. As to his request that his son should visit him once a month, the judge thought that children should provide spiritual solace for elderly citizens. According to Article 11 of *the Protection of the Elderly Rights and Interests*, it is essentially to promote social morality and encourage children to provide "mental support" for elderly citizens, but it is not a mandatory regulation. If the court ruled that the adult son should regularly visit Mr. Zheng, it couldn't promote the communication between them and was impractical. The judge of this case said in the interview, children were obligated to support elderly citizens materially, which was also one of legal norms of conduct in China. However, mental support was advocated without legal rigid rules, which showed the helplessness of the law. At last, the elderly trial division of the court of Jing'an District in Shanghai made a sentence to invalidate Mr. Zheng's request for "mental support".[12]

Other laws and policies also involve mental security, but it is often seen in laws and regulations of other departments. Complete and perfect laws and regulations haven't been laid down, and there is a lack of integrity. Moreover, there are no specific definitions between mandatory provisions and guiding provisions, leading to insufficient legal behavior constraints and penalties. For example, according to Article 11 of *the Protection of the Elderly Rights and Interests*, the supporters should provide economic support, life care and spiritual solace for elderly citizens and meet their special needs, which clearly defines three obligations, namely

[12]Hu [9].

economic support, life care and spiritual solace. Nevertheless, according to Article 21 of the newly amended *Marriage Law*, "children should support their parents", and "if children fail to fulfill the obligation to support their parents, parents who don't have the ability to work or have difficulties in life have the right to require their children to support them economically". However, it only confirms that children should support their parents materially and in life, and excludes spiritual support. According to Article 49 of *The Constitution,* "the abuse of elderly citizens is prohibited". According to Article 3 of *Marriage Law*, "maltreatment or desertion of any family member is prohibited". *Criminal Law* has provisions of the crime of abuse. Theoretically, they should also prohibit family members from abusing elderly citizens spiritually, but they are unclear and ambiguous, so in judicial practice, people habitually think that family members can't abuse each other (including elderly citizens) physically.[13] According to *Criminal Law* of China, in flagrant cases in which children refuse to support their parents, they will be accused of offence of abandonment. Only when the offence of abandonment is as flagrant as what is ruled in *Criminal Law* will relevant departments actively intervene; under slight circumstances, relevant departments won't actively intervene. Moreover, it often refers to material or physical abuse without involving mental contents. Responsibilities of enterprises and units are more relaxed. For example, according to Article 44 of *the Protection of the Elderly Rights and Interests*, as to departments or organizations which don't fulfill their duties of protecting the legal elderly rights and interests, their superior departments should give criticism and education and order to correct". In the countryside, elderly care, especially mental support, is almost something private of the family, and it is hard for the public power to step in. In our field survey in the countryside, village officials told us that they did want to control those who don't respect or support elderly citizens, but they didn't dare to do this, because if they stepped in, they made it difficult for the young of the family, put their finger in another's pie and embarrassed them. Although Article 6 of *the Protection of the Elderly Rights and Interests* stipulates that people committees, villagers' committee and the elderly organizations established legally should reflect the requirements of elderly citizens, protect the legitimate rights and interests of elderly citizens and serve elderly citizens, given the "acquaintance society" in the countryside, the law, in front of traditional habits, often appears to be powerless. How to implement the law in the countryside is a problem worthy of further research.

(2) The capital investment and talent input to the mental service system. Since the early 20th century, the rich countries and middle-income countries of the world have established the basic pension system in succession as the quasi public product. The basic pension security provided by the government to the residents has become the consensus of the governments. Practice has proved that the establishment of pension security system not only is conducive to the stability of the Western society, but also promotes the development of social

---

[13]Hu [9].

economy in Western countries. From the world scope, in the countries which took the lead in establishing basic pension security system, the economic and social development has entered a period of stability. People are no longer worried about life in old age and can concentrate on work; from the point of view of family and individual, because of clear pension security, they are free from worrying the life in old age and can better estimate the disposable income, which promotes the consumption, savings and investment, stimulating the economic growth.[14] Pension security is one of the three basic elements, so the government should undertake the main responsibility of mental security.

① Capital investment. The government's capital investment in mental security is mainly the investment in mental service public facilities and investment in mental service costs. The *Civil Affairs Development Statistical Report 2009* released by the Ministry of Civil Affairs on June 10, 2010 shows that as of the end of 2009, the population aged 65 and above reached 113.09 million, an increase of 3.22% over the precious year, accounting for 8.5% of the total population, an increase of 0.2% points over the previous year. The report shows that as of the end of 2009, there were a total of 19,909 legal aid centers of the elderly, an increase of 2075 than the precious year, grew by 11.6%; 136,000 elderly rights protection and coordination organizations, an increase of 53,000 than the precious year, grew by 64.9%; 59,543 schools for the elderly with 5.415 million students, grew by 49.6 and 7.4% respectively; 329,000 activity rooms of all types for the elderly. Promote the establishment of the old age allowance system. As of the end of 2009, there were 38,060 elderly welfare institutions of all kinds in the country, an increase of 908 over the previous year; 2.662 million beds, an increase of 9.1% over the previous year, 2.109 million adopted people of all kinds, an increase of 7.7% over the previous year. There were 175,000 community service centers of all kinds in the whole country, 53,000 neighborhood community service stations and 112,000 other community service facilities. There were 693,000 urban service network stations which bring convenience to people and 289,000 community volunteer service organizations.[15]

The government's investment in elderly services has increased year by year, but there is still a big gap between the demand and supply. The *Civil Affairs Development Statistics Report 2009* issued by the Ministry of Civil Affairs of the Elderly shows that as of the end of 2009, the population aged 65 and above was 113.09 million, the various types of elderly welfare institutions had a total of only 2.662 million beds. The number of beds was just over 2% of the total number of the elderly population. It is far behind the general ratio of 5–7% in the international community. In recent years, we vigorously develop communities to establish

[14]Pan [10].

[15]Concerned About the Development of Civil Affairs: *Civil Affairs Development Statistical Report 2009*, http://www.chinapop.gov.cn/xwzx/rkxw/201006/t20100611_205282.html, June 11, 2010.

service facilities and sites for the elderly. In the country, there are a total of 175,000 community service centers of various kinds; 53,000 neighborhood committee community service stations, 112,000 other community service facilities as well as 693,000 urban service network stations that benefit the citizens. "The Star Plan of Community Welfare Service for the Elderly" launched in 2001 planned to use the welfare fund of about 5 billion yuan raised by various issued welfare lottery at all levels and the 5 billion yuan invested by governments at all levels and all walks of life within three years to establish or expand a group of urban community old-age welfare service facilities for the entertainment, fitness, learning, health care and medical care of elderly citizens, activity places and rural homes for elderly citizens in 100,000 community neighborhood committees and rural townships. However, compared with the huge demand of elderly services, it is a drop in the bucket. The elderly in the vast rural areas and outlying and poverty-stricken areas are still in need of activity and entertainment facilities. In some urban new residential areas, public venues and facilities are generally inadequate. Some developers did not build community public welfare activity facilities in accordance with the requirements of the country or did not build activity supporting facilities for the elderly; or are keen to establish various for-profit clubs which are so expensive that the elderly dare not show any interest in them. Especially, compared with elderly citizens' growing mental needs, such facilities are inadequate. For both institutional services and community care services, most content is confined to convenience provided in life services for elderly citizens. It meets their survival and security needs, but ignores the mental needs of elderly citizens. It cannot provide the corresponding facilities and services in cultural entertainment, education, reemployment and other aspects. Although there are many facilities and activities for the recreation and fitness of the elderly in the community, some activities are formalistic, ignoring the different needs and deep-seated needs of elderly citizens. Most facilities and activities aim at younger old people and the healthy old people, ignoring the mental needs of the elderly who are basically unable to take care of themselves and completely unable to take care of themselves.

In addition to investment in facilities, there is investment in services. Some mental services provided by formal support system must be bought, and the mental services provided by informal support system also need the sacrifice in the economic interests of informal supporters in other aspects, so the government should give appropriate compensation. Most urban elderly citizens have pension. However, compared to the relatively high cost of living in the city, pension of most retired elderly citizens can only deal with the basic life. Once they are sick, they have to assume expensive medical costs and the cost of care. As a result, even if they are healthy, many of them dare not spend money in the wrong way. They save the rest money in case of emergency. To spend money on mental services is a luxury and unimaginable thing. Old age allowance systems have been set up in our country now, but the standard is not unified. For example, Jiangsu, Zhejiang and other provinces have not implemented the provincial unified dispatch. The allowance amount standard hasn't been unified in a province (it is 80 yuan in some counties, while it is 100 yuan in other counties). Besides this, those who receive the allowance in all places in a province are at different ages. Some cities issue allowance to people aged over 90 and some cities issue allowance to people aged over 80. For example, each person aged from 95 to 99 in Huizhou, Guangdong can receive 100 yuan per month; each person aged over 90 in Ji'nan, Shandong receive 60–100 yuan per month. The issuing standards are very low. Some provincial capitals only issue 10 yuan to people aged over 80 per month. Such a low standard simply cannot meet the life care needs and medical care needs of the elderly, not to mention the mental needs. Compared with life care and medical expenses, mental service expenditure is often the next best choice for people.[16]

② Talent input. Different from other services, mental service for elderly citizens should be comprehensive, scientific, service and professional. Mental service for elderly citizens involves knowledge of gerontology, education, sociology, psychology, medicine, law, elderly kinematics, geriatric social work and other aspects. Only by carefully studying and scientifically analyzing geriatric mental problems should be solved and formulating specific work methods and measures according to these systematic theories and scientific knowledge can we properly solve the problems. In addition, elderly citizens are a vulnerable group of the society, who face not only the decline in physical, but also the decline of the role, so they need social love and care. The elderly mental service industry is a cause that needs love and enthusiasm. People who engage in this cause must have a good sense of passion and responsibility. However, the pension service personnel in our country are very unstable. Due to the lack of motive mechanism to promote the sustainable development of community service for elderly citizens, the lack of standardized and effective management, also due to the fact that many people cannot adapt to serving elderly citizens, the elderly service personnel in the communities are unstable. For example, in some residential areas, it is often the community manager who temporarily organizes persons of neighborhood committee or residents to provide services to elderly citizens. Service personnel of some small daycare institutions leave on their own in less than a month. Pension service personnel quality is uneven. Some service managers and service personnel lack caring for the elderly and professional ethics. At present, few people engage in gerontological social work in China, and they are relatively old. Because the elderly community work isn't normative and does not reflect the professionalism of social work, the gerontological social workers' specialized degree is low. Most workers active in gerontological social work are some people who engage in humanitarianism and experience participation. They have not received professional education or knowledge training related to services for elderly citizens, which not only affects the quality of community service for elderly citizens, but also restricts the gerontological social work. In addition, the volunteer team is also rather inadequate, because they have low public participation awareness and

[16]Research Group on Countermeasure of Chinese Population Aging: *General Report on Countermeasure of Population Aging in China*, August 2008.

lack enthusiasm.[17] Although some community service centers take elderly citizens, empty nesters and sick old people as a focus for the elderly work in the community, these aids to the elderly are only the most basic material assistance and holiday greeting, which neglect the mental needs of the elderly. The lack of talent in the elderly work in community is an important reason.

(3) The macro planning, supervision and management of mental service for the elderly system ensure the standards and quality of the market and services. The welfare pluralism originated from the 1980s is the reflection on the traditional welfare model as the response to the crisis of Western welfare states, and an alternative to solve the crisis of the welfare states. Welfare pluralism argues that the government is no longer the only provider of social welfare, and social welfare can be shared by the public sectors, profit organizations and non-profit organizations, families and individuals. The government turns into the regulator and buyer of welfare services, the arbiter of goods management and the role to urge other sectors to provide the services. The two most important aspects are participation and decentralization. Welfare pluralism has become the theoretical mainstream of Western social policy, multiple departments replacing the state to provide welfare. When the welfare state is in a dilemma, welfare pluralism blows a gust of fresh air to the social policy. It corrects the wrong understanding of too much emphasis on the state providing welfare; puts forward that multiple welfare providers, such as the state, families, market and voluntary organizations should share responsibility; establishes the structure of multiple welfare providers; provides important points of view for many models, for example, multiple departments replace the state to provide welfare.

However, the reflection of welfare pluralism also tells us that the government is still the main undertaker of social welfare as well as the only social institution that has the rights and resources and represents the public interests to carry out comprehensive actions. "Only the government (through taxes) can make citizens serve the public interests of the community, ensure that the excluded individuals receive assistance in times when having difficulty, ensure no man's welfare level is below the minimum standards, reallocate resources and promote equality. Scholars Gibson summarized the social services for elderly citizens of developed countries in the last few years, and pointed out the changes in the way to provide elderly services. Also, he pointed out that under the trend of welfare pluralism, the government strengthened the macro management functions, and provided services to non-profit organizations through vouchers, leasing, contracting and other marketing models. The government attached great importance to the development planning and service quality of social welfare. The functions of the government are mainly policy planning and the establishment of service supervision mechanism.[18]

[17]Ling [11].

[18]Zhou et al. [12].

Beyond doubt, the welfare pluralism of foreign countries is also a reference for the elderly mental security in our country. China has not yet had the rich economic foundation of Western welfare states, so we cannot copy the Western experience in the welfare model. Social welfare system of our country experienced the change from traditional security system of family security to the security system integrated by the country in the era of the planned economy. Later, in the process of transition from planned economic system to market economic system, a new social security system appeared. China has the traditional culture and value that emphasizes the family function, has the communities with strict organization structure, has or had the neighborhood relations of close interaction. Their role in the elderly mental support cannot be replaced. Welfare pluralism provides us with an elderly mental security theory model that we can use for reference: mental security must take the path of diversification. The main function of the government is to balance the roles of different elderly mental care providers to avoid excessive reliance on the government, giving full play to the role of the market, society, families and individuals in elderly mental security. The main functions of the government are macro planning, supervision and management. For example, in Australia, the federal government does not directly provide any services for elderly citizens, but develops policy planning, allocation of resources, formulation of service standards and supervision implementation. State government is also mainly responsible for the supervision and management of the project, while providing a small amount of care services for nursing homes, pension apartments and communities. Places under state level provide a number of services of pension apartments and communities, and play a supervisory role. The main providers of services are non-governmental sectors, including for-profit organizations and non-profit organizations. There are three types of family service providers: hospital and community service professionals, family caregivers and volunteers.

The aging process of the population in developed countries is a relatively slow process. All aspects of social life have a long time to prepare, so the social life has strong adaptability to the aging of population. However, our country is almost running into the aging society, with faster development and shorter conversion cycle, so the corresponding management systems and mechanism cannot meet the demands.

First, the existing institution setting of cause for the aging population is not conducive to the government's macro planning, supervision and management. Most countries which have entered an aging population period have established administrative institutions specialized in aging population affairs. For example, Aging Population Affairs Bureau was set up in United States Department of Health and Social Services to manage the work of cause for the aging population of the state, responsible for formulating policies, laws and regulations as well as providing legal services, medical services, transport services and life services for elderly citizens. The Bureau consists of assessment center, management center, plans and policies execution center, information communication and customer services office, welfare and community service center and other centers, and administers the first-level management institutions of 56 states and first-level management institutions of 655 regions. Our country does not have the department specialized in the elderly affairs management. As a result, the elderly affairs are under separate management of the Ministry of Health, the Ministry of Culture, the Ministry of Civil Affairs, the Ministry of Human Resources and Social Security, the Population and Family Planning Commission and other agencies. Committees on aging at all levels, from the central to the local, are defined as deliberation organs. Office of the National Working Commission on Aging is a public institution directly subordinate to the State Council with reference to civil service management, affiliated to the Ministry of Civil Affairs. It is a body affiliated to the Ministry of Civil Affairs Department, under the management of the Ministry of Civil Affairs. However, the Ministry of Civil Affairs itself is a member unit of the National Working Commission on Aging, so it is hard for it to coordinate with relevant member units of the National Working Commission on Aging in many affairs related to the work on aging. Though offices of committees on aging in various regions are working bodies under the National Working Commission, they have different functions: some are administrative departments or given administrative functions, some have both working body functions and administrative functions, and some belong to community organizations, resulting in hard performance of functions. For example, the elderly care security, medical care and education for elderly citizens, culture of elderly citizens and other work can only be scattered in different administrative departments. As a result, the department of work on aging should play a coordinating role among them. Nonetheless, because some local offices on aging are public institutions or associations of lower specifications, coordination is very difficult. The strength of the grassroots aging work is also very weak, especially in manpower. For example, nearly 70% of the counties (cities, districts) in Zhejiang Province have only one part-time worker working on aging. Most streets as well as villages and towns in Ningbo City have no special workers working on aging. In Henan Province, there are a total of more than 400 workers working on aging and Xuchang municipal government actually does not have full-time workers working on aging. Due to such institution setting, it is very difficult to establish a macro, comprehensive and scientific supervision and management mechanism.[19]

Second, there is a lack of relevant elderly care services, especially the evaluation system of mental services. In Japan, with the further intensification of aging, more and more elderly people need long-term care, but traditional family-based elderly care the elderly expect is increasingly difficult. In order to cope with the aging problems, Japan put forward a series of new welfare ideas. Influenced by these ideas and concrete policies, more and more elderly people receive elderly care services. Especially, after the implementation of the nursing insurance system in April 2000, non-profit service organizations have expanded the welfare objects. In the meantime, the joining of for-profit service providers further solves the problem of imbalance between supply and demand of elderly care services, forming a

[19]Research Group on Countermeasure of Chinese Population Aging: *General Report on Countermeasure of Population Aging in China*, August 2008.

diversified elderly care services market. In order to guarantee and improve the quality of elderly care services, Japan gradually began to study and introduce the service evaluation system from the 1980s. The implementation of this system not only strengthened the management of elderly care services, but also further improved the quality of elderly care services, promoting the continuous upgrading of elderly care services. In our country, elderly care services, especially the research on mental support services evaluation is still an undeveloped area, operating methods of private elderly care institutions are varied, and the law hasn't specified its position, which caused confusion to the service market and the service quality to be constantly questioned. Therefore, how to scientifically evaluate and regulate elderly care services and improve the quality of our country's elderly care services has become a top priority.[20]

(4) Create the cultural atmosphere of "respecting the elderly and paying attention to their mental needs". In the 2006 survey, most elderly citizens (66.7%) held that "more and more young people respect the elderly", only 13.3% elderly citizens didn't agree with this viewpoint, and another twenty percent (20.0%) said it was hard to say.[21] From the point of view of the longitudinal comparison, compared with the 2000 survey, the proportions of the elderly who agreed with this viewpoint in both urban and rural areas were on the rise; from the point of view of the longitudinal comparison, compared with the 2000 survey, the proportions of the elderly who felt lonely in both urban and rural areas had declined. This shows that the elderly in China had more and more positive and optimistic mentality, and reflects China's cultural atmosphere of "respecting the elderly" has improved, which is inseparable from some policies promulgated by the government to protect the interests of the elderly in recent years. For example, the government has taken various measures to facilitate the elderly people's clothing, food, shelter and transportation, providing them with free, discount or preferential policies in urban public transport, long-distance passenger transportation, rail, water route and air passenger transportation; law offices, notary offices, grassroots legal services and other social legal service organizations provide the elderly with lower-cost or free legal consultation and related services; promote the publicity of *Rights and Interests Protection Law for the Elderly* into communities and the countryside, combine legal education and moral education on respect for the elderly, promote the formation of socialist ideological and ethical view of respecting the elderly; carry out a wide range of educational activities with "respecting, loving and helping the elderly" theme among adolescents; organize regular sports activities for grassroots elderly people, develop sports and fitness programs suitable for physical characteristics of the elderly, creating conditions for spontaneous activities of the elderly; construct culture, health and community service facilities which are closely related to the daily life of the elderly, do well in the planning and

---

[20]Wang [13].

[21]Zhang and Guo [14].

construction of cultural entertainment sites, such as park, exhibition hall, museum and library, increase the elderly oriented construction projects and make them free or preferential to the elderly; set up some television programs for the elderly, such as *Sunset Glow* program on CCTV.

Singapore and Japan are two countries which are strikingly similar to China in the Confucian culture circle. In recent decades, in order to have a rapid development of industry, Singapore introduced Western capital and its culture, and focused on technology learning but ignored character training in education, so that the younger generation neglected the oriental culture and pursued material gain. As a result, the original Confucian pension culture was challenged. In the past 20 years, the Singapore government strengthened the construction of spiritual civilization. In addition to carrying out the courtesy campaign, since 1979, it has also held a "respect the aged week" each year. Through various activities, it encourages the public organizations and other organizations to care about the welfare of the elderly and encourage the elderly to participate in social activities. Singapore newspapers carried out "love other's elder as yours" education from time to time, proposing maintaining and carrying forward the oriental moral tradition through schools and families to cultivate a new generation of Singaporeans. Since 1986, all walks of life in Singapore had discussions, considering stipulating by legislation that children must care or support parents in their old age, or one of the children of the widowed elderly people must live with the elderly, to enjoy the inheritance tax preferential treatment and others. Singapore's Ministry of education has also announced that the ethical moral education of youth in secondary schools must be strengthened. The Singapore government has also strongly advocated three generations under one roof. The former Prime Minister Lee Kuan Yew pointed out that "three generations under one roof" is proven to be a system which can maintain the value of life by more than 3000 years of Chinese history. There are more people and less land in Singapore. The proposal of "three generations under one roof" not only carries forward the traditional elderly care culture and solves pension problem, but also greatly reduces living space, which is conducive to solving the housing problem. In order to cooperate with the government to promote "three generations under one roof", Singapore Housing Development Board has taken measures, such as allowing those who are going to live next to lineal relatives to participate in housing distribution by drawing lots in advance. Japan is a member of the Confucian culture circle. Because population aging is becoming more and more serious, culture of respecting the elderly has become a very important part of the community culture. Coupled with the government's propaganda and policy support, the community's helping the elderly service is in full swing development. All these remind people that the aging society has come, and we cannot ignore the existence of the elderly when doing everything. Thus, a new type of community culture, which needs to balance the needs of people of all ages, is produced.

Some typical examples which promote the good atmosphere of "respecting and loving the elderly" once appeared in some places in our country. For example, village cadres of Shicheng County in Jiangxi Province were evaluated in every half a year, whether they respected their parents was also an item. The village cadres who had unfilial behavior would be dismissed. When Shenyang North Railway Station Comprehensive Service Corporation recruited young workers, it followed the principle of "choosing filial workers". These young workers showed filial respect for parents at home and fulfilled their duties at work, which was widely praised by the society. Evaluation of cadres is one of the important responsibilities of the governments. The standards for evaluation of cadres become the goal of cadres' endeavor. At present, our country generally takes thought level, work ability, work experience, cooperation degree, professional knowledge and age as the standards of evaluation. Because they do not contain respect for the elderly and care for the elderly, thus it is difficult to truly form the good atmosphere of respecting and supporting the elderly in the whole society.

The government should vigorously advocate and promote the social atmosphere of "respecting the elderly and paying attention to the mental needs of the elderly" through media, education and design of some systems.

2. The position and role of market and non-profit organizations in the elderly mental security

Welfare pluralism is a theoretical paradigm to solve the welfare state crisis that emerged in the 1980s after classical liberalism and Keynes Beveridge paradigm. This paradigm advocates the diversification of sources of welfare, which neither advocates relying too much on market like liberalism, nor advocates fully relying on the state as Keynes did. Instead, it advocates that welfare should be shared by the government, the market, individuals, families and non-profit organizations.

The market is a very important pillar of welfare pluralism. The pension is an undertaking related to the people's livelihood. The government should bear the primary responsibility to take measures to make every elderly person be looked after properly. The fundamental purpose of the generation of the social welfare system is to make up for the deficiency of the market through the redistribution. It should not be related to the market again. However, social welfare is made up of two parts, namely welfare income and welfare services. Welfare income is given to the beneficiaries by the government, so each trading relationship does not exist. However, the government, the welfare service providers and the beneficiaries must all participate in welfare services in order to achieve the goal. The market can guarantee that the service items are produced at the lowest cost, and that the products produced are most needed by the consumers.[22]

Non-profit organization is a product of the development of western social system evolution. From a broad theoretical point of view, all for-profit social organizations which do not belong to the public sector can be said to be non-profit organizations. Due to the differences in life experience, health status, education level and so on, different elderly people have different lifestyles, life preferences and life demands. There are some conflicts between the services the elderly need and the services

[22]Zhang [15].

provided by the society, which lead to some people's disagreement on the formal care. The mental services for the elderly are informative and professional, which cannot be solved by ordinary occupation in the community. Non-profit organization, by virtue of its own characteristics, especially its non-profit and supportive characteristics, can get close to the service objects more directly and flexibly to form some kind of interaction, and can obtain higher social benefits at low cost, using its own resources and social resources to provide services, which eases the government's burden, reduces cost of government in social welfare expenditure, effectively satisfies the diversified and multi-level demands and guarantees the service quality. Especially, in the field of quasi public goods such as pension, the role and advantage of non-profit organizations is very obvious. Therefore, non-profit organizations play an important role in the elderly mental service.

In the late 1980s, economic globalization became faster, the competition between countries intensified, and domestic problems increased. Therefore, many countries carried out reforms on public sectors. Some public sectors were "marketized" and "socialized", which, on the one hand eased the national administrative and financial pressure, on the other hand, improved the efficiency of these sectors. Over the years, in the field of pension benefits and relief in China, the tendency of "socialization" or "marketization" was also very obvious. The country's services grew slowly. At the beginning of the 1990s, the services operated by the Civil Affairs Department had been exceeded by the facilities operated by the society or the community. Now, social services operated by individuals or the market have already shown up and flourished in some areas (such as Guangzhou and Shanghai). Therefore, elderly social groups, elderly social service organizations, elderly organizations with professional technology and self service, elderly cultural and sports organizations and other NGOs emerged.

All kinds of elderly groups and elderly service organizations held by these markets or non-profit organizations have played an important role in the elderly mental security. For example, China Association of the Universities for the Aged was founded in 1998. As a national social organization that organizes the inter-scholastic collaboration between universities and schools for the aged around country, as of 1998, it organized 1900 schools and universities for the aged with about 700,000 students; in order to activate the entertainment and sports activities for the elderly, all places generally established sports association for the elderly, the elderly choir, the elderly art team, elderly yangge team, Tai Chi Chuan class for the elderly, etc., actively carrying out various cultural and sports activities to activate the late life of elderly citizens; the elderly citizens organized various management organizations and various industry associations of self organization, self management, self service and self education, such as science and technology association for retired workers, association of retired teachers, association of retired doctors, association of retired lawyers. These professional associations managed by elderly citizens played an important role in promoting the elderly social interaction and enriching mental world of elderly citizens.

However, from a pure business point of view, the pension industry is not a highly profitable industry, but a labor intensive industry with high investment, high management requirements and high quality requirements that requires great enthusiasm and love. Throughout the world, few entrepreneurs who take profit maximization as the primary goal are really willing to invest a huge sum of money in developing the social welfare or pension industry. People from all walks of life who are enthusiastic about the pension industry, first of all, with a strong sense of social responsibility, passion for the cause of social welfare and love for elderly citizens, develop pension security, this sacred cause. However, pension service industry is a very realistic security and service system, which needs a lot of money and manpower investment as well as high management and service requirements. Much more is this the case with the mental service for the elderly industry. Some put a lot of money and talents, but people do not understand or are not keen on this service industry. Elderly citizens often do not want to say that they have psychological problems. They would rather take these problems as the physical and social conditions or the normal aging. Due to social discrimination, economic reasons, limited knowledge of mental disorders and lack of trust in mental health service professionals, the elderly are not willing to use mental health services. For example, many elderly citizens who are suffering from depression are willing to choose a common physician for treatment. These reasons lead to the great enthusiasm recession of the market or the society to set up the mental service for the elderly industry. According to the survey, 48.5% China's urban elderly people have a variety of realistic demands for pension services. 25.2% of them required domestic service, 18.04% of them needed care services and 13.79% of them needed diversion service by chat. However, the results of the survey show that the total satisfaction rate of China's current resident demand for pension service is only 15.9%, of which satisfaction rate of domestic services accounts for 22.61%, satisfaction rate of nursing service 8.3% and satisfaction rate of diversion service by chat 31.6%. Although the urban community home care service has a new development than before, it is far from meeting the needs of elderly citizens.[23]

5.3 Build a Comprehensive and Multi-level Elderly Care Mental Security System

According to the data of the fifth nationwide census in 2000, by the time point of the census in 2000, there were more than 130 million people who were over 60 years old all over China, which accounted for about 10.3% of the population all over China; the number of elderly citizens who were over 65 years old had reached 97 million, accounting for 7.0% of the population all over China. By the end of 2005, nearly 144 million people were over 60 years old all over China, accounting

[23]Office of the National Working Commission on Aging: *City Home-based Elderly Care Service Research in China,* website of the Office of the National Working Commission on Aging, December 23, 2008.

for 11% of the population all over China. As to the development trend of the aging of population by 2020 or the middle of the 21st century in China, the relevant departments and research institutions in China have made a lot of predictions, and the results are rather similar to them. The Population and Family Planning Commission in China predicts that by 2020, the population of elderly citizens between 60 and 65 years old and over 65 years old in China will reach 234 million and 164 million respectively, accounting for 16.1 and 11.2% of the total population respectively. By the late 2040s, the number of elderly citizens who are at these two ages will reach 430 million and 320 million respectively, accounting for 22 and 30% of the total population at that time respectively. By then, there will be an elderly citizen among 3 or 4 people. According to the classification standard of the United Nations, China now has entered the stage of aging population, thus entering the aging society. Rapidly increasing aging population, obvious trend of aging, the huge number of elderly citizens and getting rich before getting old become the main characteristics of the population aging in China. The huge number of the elderly group brings many adverse impacts on the development of the economy, society and other aspects of China in needs of elderly care, medical treatment, social services and other aspects. How to build a comprehensive and effective elderly care security system becomes an urgent and arduous task.

"Economic support, life care and spiritual solace" are three basic components of elderly care. Especially with the development of the economy and the improvement of the social security system, the spiritual needs of elderly citizens have been improving constantly. In China, more than 100,000 elderly citizens who are over 55 years old commit suicide every year, becoming the group with the highest suicide rate in China. The spiritual needs of elderly citizens can not be ignored in the comprehensive and effective pension security systems. Being influenced by the special cultural tradition and economic development level, the spiritual solace for elderly citizens comes from the informal support system primarily for quite some time. The family, relatives, neighbors and friends play very important roles in the spiritual solace of elderly citizens. However, with the development of the market economy, the society is increasingly open, so the professional and regional personnel flow strengthens little by little and the employment pressure and competition increasingly intensify These informal support systems are facing many challenges. With emotional foundation, the informal support builds proactive, almighty, flexible spiritual support. While the formal support system has its own advantages when disposing of the services which are related to social policies and laws and needs to combine a variety of resources and needs technology and knowledge. The formal support system and the informal one have different characteristics in the elderly mental security. Both of them are important support resources of the elderly spiritual solace. A perfect elderly mental security system should be multi-level and comprehensive.

(I) Formal support system—dominated by government with the participation of a variety of social forces

The government boasts implementing competence and policy measures which a series of informal support systems and other non-governmental organizations lack. The elderly mental security is a responsibility which the government can not shirk in the implementation of the national functions. The government is in a dominant position in the elderly mental security. The government should play unique roles in the legal policy making, the investment in qualified personnel and capital, the macroscopic supervision and management of the service system, the construction of the culture atmosphere of "respecting and cherishing elderly citizens and paying attention to the elderly spiritual needs."

(1) Make and improve relevant laws, policies and systems. Provide the social spiritual elderly care with positive policy support and reliable legal security policy. Fund the relevant countermeasure study and survey vigorously.

Our country has many laws, policies and systems on elderly citizens, but just a few of them are related to the elderly mental security and most of them are relatively abstract and of principle. "Law of the People's Republic of China on Protection of the Rights and Interests of the Elderly" has some contents of the elderly mental security, but these items are quite of principle and lack specific responsibilities. It isn't stipulated in detail that legal responsibilities shall be borne if obligation of mental support hasn't been fulfilled. In reality, it's always hard to execute laws. Presiding judge Lin Aihua of Adjudication Division for Elderly Citizens of Shanghai Jing'an District People's Court suggests that contents of mental support should be added into *Law on Protection of the Rights and Interests of the Elderly*: first, when mandatory standards are formulated, three-aspect contents of supporting and helping shall be overall stipulated and it shall be specified that spiritual solace is adult children's obligation to their parents. Specific explanation: "Adult children have the obligation of substantial support, spiritual solace and help in life to parents". Second, it's necessary to add standards to prohibit children from violating parents' personal dignity and often severely mistreating elderly citizens: "Prohibit the children from discriminating, abusing, humiliating and defaming elderly citizens, illegally intervening in the elderly participation in political or social activities, or other mental mistreatment behavior. Third, add authorized standards and explicitly stipulate: "When children conduct mental mistreatment or don't fulfill support obligation, parents have the right to stop children's mistreatment and ask them to apologize or satisfy basic mental and life needs".[24] Article 31 of *Regulations on Protection of the Rights and Interests of Elderly Citizens, Liaoning Province* which was brought into force on October 1, 2008, explicitly stipulates that supporters shall fulfill the obligation to the elderly spiritual solace and often extend greetings to and visit their parents if they don't live

[24]Hu [16].

with their parents. If national staff violate the above regulations, their units or higher authorities shall command a correction; if the situation is serious, administrative sanction shall be given or relevant legal liabilities shall be investigated. It's the first time that adding the elderly spiritual solace being supporter's obligation into local laws and regulations in the country. It's also the first time that the obligation of mental support and its legal liabilities are explicitly stipulated. This makes up for shortages of mental support. It is worth being accepted and popularized.

Understanding and knowing the elderly mental need and its change, law is the premise to discuss and solve elderly mental problems. Foreign developed countries have much mature experience which we should learn from, but because of our country's special cultural tradition and specific development phase, we can't totally copy the developed countries' modes. Only by making deep researches of our country's elderly citizens' mental situation and social, cultural and economic background can we make effective solutions. In the past years, there were more and more researches about the aging of population and problems related to elderly citizens made by scholars from all walks of life in our country and foreign countries. Material and data about elderly citizens' situation collected through surveys, interviews and observation become more and more comprehensive, accurate and systematic. Tracing surveys of Chinese rural and urban elderly citizens' situation organized by China Research Center on Aging in 2000, 2006 and 2010 provide us with much data to understand Chinese elderly citizens. However, by comparison, researches on elderly citizens' mental situation and mental security are still in the starting phase, and there are just a few relevant researches. Under special cultural and social backgrounds, what is the difference between Chinese elderly citizens' mental needs and those of the Western countries; what special roles Chinese families play in elderly citizens' mental support; what are the potential challenges; how is the prospect; how to build mental service evaluation system and other problems are in urgent need of investigation and research. The government shall give more financial assistance to this kind of survey and research.

(2) Strengthen capital investment and talent input in mental service system. Elderly citizens' activities and entertainment facilities in most villages and remote poverty-stricken areas are still very insufficient, so the government shall strengthen investment in these areas. In some urban newly-built neighborhoods, places for public benefit activities and facilities are commonly insufficient. On one hand, property developers go against national requirements and don't have public places and facilities and supporting facilities for elderly citizens' activities. On the other hand, they might have various profitable clubs and charge for a so high price that elderly citizens can't afford it. The government shall enhance supervision and build elderly service facilities in residence community plan like planning public facilities of hospitals and kindergartens.

By the end of 2009, there were 2.662 million beds only in various welfare institutions for elderly citizens in China. The total amount was just 2% over the amount of elderly citizens, which was far more less than the general rate of 5% to

7% in international society. The government should make investment to open more welfare houses for elderly citizens, nursing homes, care centers for elderly citizens and sanitariums for retired cadres. Besides, these welfare facilities should meet the elderly mental and spiritual requirements. Now a majority of such facilities are still characterized by welfare relief. The poor conditions and service can't satisfy the elderly spiritual demands. Elderly citizens will feel bored, disappointed and lonely without emotional exchange there. That's why most of them are reluctant to live in welfare institutions. For example, in Beijing, there are less than 2 beds for every 100 elderly citizens on average, but the vacancy rate is up to nearly 40%. Welfare institutions for elderly citizens run by the government should primarily aim at vulnerable groups and solve problems of senior citizens who have difficulties in life. For these elderly citizens, relevant authorities must have a strict evaluation system. Private-owned institutions can aim at people with high income and reasonably charge based on service level. The government supports and gives a certain amount of subsidy and carries out management and supervision.

Emotional communication and social communication are the elderly basic mental needs. To those elderly citizens who have life problems or live in poor and remote areas, the government should provide them with television, telephone, radio, mobile phone and other hardware facilities free of charge so that they can contact and know the outside world more and better. Make more TV programs for elderly citizens. If possible, specially make an elderly-oriented TV channel. Elderly citizens shall get a discount in communication expense so that they can communicate with the outside world or relatives more. In many provinces, elderly citizens can get a discount or take bus and enter park free of charge. This improves the elderly social communication, so it's worth being further developed and popularized.

The government should also invest more in talents in the elderly mental service system. Care workers' quality is the key to elderly care service level. The government should train on-post care workers for elderly citizens to improve their service quality, especially quality in mental service. The government should give subsidy to the service staff in training expense and carry out elderly service occupational certification system. We should learn from Netherland's policies. Netherland carries out elderly care service evaluation system, dividing care workers into four grades and nurses five grades. The outline of education training and textbooks are the same in the whole country. Diploma is valid in the whole country. Strict training system for care workers has been established. Only people with certifications can work as care workers in welfare institutions. Moreover, we should make efforts to cultivate professional talents of geriatric social work. Strengthen education of elderly service majors in full-time colleges and universities. Learning from experience of developed countries and regions, we should make favorable policies so that professional social workers and graduates majoring in social work of colleges and universities like working in elderly care service institutions. These geriatric social workers have professional ideas, approaches and skills, being capable of assisting elderly citizens in solving problems of physiology, spirit, emotion, money and so on, so that elderly citizens can continue to participate in social life and happily enjoy old age.

(3) Encourage other social forces to enter the elderly mental service system. The government should plan in a macro way, supervise and manage the whole elderly mental service system to guarantee the standard and quality of the market and services.

The elderly mental security is a public need that the government must respond to. However, because of the limit of abilities, the government must consider how to fulfill the country's responsibilities of guaranteeing citizens' security with forces of the market and non-governmental institutions. The government can authorize and outsource projects to and purchase services from the market and non-governmental institutions in terms of the supply system of the elderly mental security products. The government's main function is to plan in a macro way, supervise and manage the whole elderly mental service system to guarantee the standard and quality of the market and services.

Few profits are made in the elderly care service industry. The government should make favorable policies for the society to develop elderly care service industry. Establish an admittance system of an open, equal and standard elderly care service industry. More favorable policies should be made in terms of approval of land use, petty loan, business registration, tax, administrative fees, and so on. The government should lead and support all kinds of non-governmental institutions to provide elderly citizens with direct mental services, implement and improve the favorable policies in terms of tax and others. At the same time, the government also needs to gradually fulfill the policies of purchasing mental service from these institutions so as to advance its service resources and improve its service level. Mental service is a relatively new industry, so people know little about it and don't trust it. The government should adopt relevant policies to guide some elderly citizens to consume. In addition, we can make systems about industrial standards and criterions, industrial plans and admittance, service quality evaluation, and so on.

In order to meet requirements of planning in a macro way, supervising and managing the whole elderly mental service system, the government also needs to further complete the system and mechanism of the work on aging. In the world, most countries which have entered the aging phase have set administrative institutions specially in charge of affairs about aging. As a governmental administrative institution, the National Working Committee on Aging shall strengthen its function of overall planning, comprehensive coordination, administrative legislation and supervision of law enforcement on aging development. Further enhance basic-level management on aging affairs and cyber management and construction of the elderly service institution. Improve and complete aging work institutions and staffing in villages and towns (streets). Only in this way can we effectively plan in a macro way, supervise and manage the elderly mental service system to guarantee the standard and quality of the market and services.

(4) Build a social environment and an atmosphere of public voice that "the whole society should respect, understand and care elderly citizens and pay more attention to their mental needs." The government can take advantage of its influence and strength, through the media, education, systems and others, to

lead the whole society to give a full play to the Chinese traditional virtue of respecting and caring elderly citizens, mobilize and encourage the social forces to take part in activities of respecting and caring elderly citizens, and popularize the importance of the elderly spiritual solace so that the whole society can understand the elderly mental needs and try to form a good trend of caring, respecting and understanding elderly citizens. At the same time, the government can also, through all kinds of advertisement tools and educational approaches, lead elderly citizens to actively find happiness and strengthen emotional and social communication. With advertisement and education, more elderly citizens will understand happiness which is good for self-adjustment of physiology so that elderly citizens can stay healthy and live long; emotional and social communication can promote interpersonal harmony and help elderly citizens enjoy a happy life in old age among love. The active and favorable elderly care way is advocated. As Chinese traditional culture, fruits of "filial piety" should be reasonably accepted. Positively advertise and report typical cases of filial piety and supporting the old parents. One of the important duties of all the governments is to investigate cadres. Now, China generally regards thoughts, working abilities, working experience, cooperation degree, major knowledge and age as standards of evaluation. We can try to add the item of respecting and supporting elderly citizens into the examination system, strengthening the good atmosphere of respecting and supporting elderly citizens in the whole society in systems. In addition, in the education of primary and secondary school, the request of respecting and caring elderly citizens shall be explicitly made so as to form a social atmosphere of respecting and supporting elderly citizens. Besides, relevant laws shall be made for constraints. In fact, some East-Asian countries which are greatly influenced by Chinese traditional culture are also promoting filial piety. For example, the Korean government advocates "respecting and supporting elderly citizens and relatives." And Singapore carries out "law of filial piety."

II. **Based on emotion, informal support system creates a responsible, active, all-round and flexible life base—give play to mental solace function of primary groups centering on family**

1. Formulate some family policies to guarantee the most basic emotional exchange, family greetings and visits

In China, the specious and widely spread views in books about elderly citizens' situation in developed industrial countries believe that China's family elderly care style characterized by feedback is humanized and good and meets elderly citizens' needs, but family intergeneration relations in Western countries are extremely bad. They also believe that in Western countries, when children become adults, parents will live alone and are lonely and helpless. However, in fact, this isn't the real life of elderly citizens in those countries. Professor Grana Spitz of the State University of New York at Albany and others believe that in the US, family elderly care stresses more on emotional support for elderly citizens by the family; although they tend to live separately, they have formed the value of uniting the family and helping each family member, and thus children and parents can help each other.[25] It has been proven by practice in developed countries that with economic development and constantly perfect social security, family's function of supporting elderly citizens financially and in life becomes worse, but family's function of giving elderly citizens mental solace is still vital and irreplaceable. We should formulate some family policies to support the family's function of mental solace and guarantee the most basic emotional exchange between family members and elderly citizens. For example, in terms of dwelling, the young people who are willing to live with elderly citizens have the right to get benefit from some policies. For example, people living with their parents can get financial help such as discount in loan, discount or free of charge in areas where parents live, have priority in using some public facilities, have a deduction in personal income tax, etc. Based on current situation of China's family relations and trend of being independent, the close but separate compound dwelling should be built in urban areas so that two generations can live separately. The way of "the old living with the young" in Japan can be learnt from. Their features are two entrances and exits and two independent kitchens. In this way, the young generation and elderly citizens can take care of each other and family conflicts can be avoided. The government should greatly develop transportation and communication and introduce preferential policies to support and promote the contact between children and parents. As there are many holidays in China, we should encourage the young generation to visit their parents during holidays, instead of tourism for increased percentage of economic growth. The Double Ninth Festival is a traditional festival for elderly citizens in China. However, it seems that it's very quiet during every Double Ninth Festival although it's very bustling during other festivals. The government should encourage the young generation to call their parents by introducing preferential policy on communication fees, give half a day off so that they can stay with their parents, or carry out some preferential plans on family-style activities in social service during the festival. The system of family-visit holiday should be changed. There are many unreasonable plans in the present-day family-visiting holiday. A couple needs to visit each other's parents. That is to say, they need to visit more family members. Besides, in terms of common sense, after marriage, their parents are older and need to be taken care of and visited. However, in the current family-visiting system, people have family-visiting holiday once a year before getting married, but they have family-visiting holidays once every four year after getting married.

We should also broaden the resources of family mental support, understand the great potential and function of elderly care given by spouse, elderly care given by daughters, elderly care given by grandchildren and elderly care given by relatives, especially daughters' mental support function. Relevant policies should be formulated to encourage these resources for mental elderly care. For example, in urban demolition, try best to maintain the original relative communities and build

[25]Zhang [17].

"network-style" families. Females play an important role in mental elderly care, so relevant polices should be formulated to help them to smoothly become such an important role. In one-child family with a lack of elderly care resource, family welfare and subsidy should be carried out. "Virtual family elderly care system" proposed by some scholars is worth being encouraged and supported by policies. Those families which don't have blood relationship but good interpersonal relationship can help each other in the form of a family for the sake of interest to jointly overcome challenges of life. It's also an alternative way of the shortage of elderly care resource and a new type of "family elderly care way" in a certain sense.

2. Build a new type of neighborhood relationship

China's proverbs or stories such as "Neighbors are more important than dwelling's location," "Good neighbors can be compared to treasure," "A distant relative is not as good as a near neighbor" and "Mencius mother moves her home three times to live close to good neighbors to better her son's education" show the importance of neighbors. However, currently, there are many tall buildings in cities of China. The cold iron doors lock city dwellers' mind and even break the happiness of getting along well with neighbors. Neighbors become strangers who are close but far as well. Some communities hold some meaningful activities such as "neighbor festival" and make "convention for neighbors" to build a new type of neighborhood relationship. However, because of communication difficulties arising from heterology of urban communities, the sense of insecurity arising from high mobility, broadness of social interpersonal circle and faster life pace arising from popularity of private car and network, it's still a long way for the establishment of a new type of neighborhood relationship to walk. Since the 20th century, the acceleration of urbanization not only makes cities a major stage for Western countries in economy, but also makes urban communities an important object of academic researches- will traditional communities disappear or change? For example, according to researches, Stacey and Fischl learnt that permanence and illiquidity are the important factors in forming worker-level communities. Personal, occupational and living mobility and range in workers' level are relatively small, so it's good for the horizontal contact between relatives, neighbors and friends. Only by enhancing the researches in different features and advantages in China's different communities can we formulate effective countermeasures.

3. Create growing and meaningful elderly personal life

Individual is also an important role in welfare pluralism. Domestic and foreign experiences of solving problems on elderly citizens show that "elderly citizens' self-supporting and self-reliance" does have great meanings. "If you want to get help from others, you should help yourself by yourself first" can be said an external and time-tested law of success. Just as the Western proverb goes, "God helps those who help themselves". The Chinese proverb "Heaven helps those who help themselves" refers to the same point. Usually, as a mysterious cultural instruction, "self-esteem" will ask elderly citizens to try their best to be independent and support themselves. For example, in China rural areas, most elderly citizens won't stop working until they're too old to work. Such a "natural employment system" basically determines "retirement" or not according to workers' physical strength and life needs.[26]

James A. Autry pointed out in *The Spirit of Retirement*: "Years after retirement should be the most beautiful time in life. The heavy burden of making a living is gone. What is in front of you is the great time for personal development and mental growth. It's the real time belonging to you. It depends on whether you can hold and enjoy this awesome journey".[27] People should think of the later years positively. It's far more than senescence and death. Have colorful personal life based on your hobbies such as travelling, joining in various amateur associations, and taking part in family and social activities. The country should formulate different policies to support active aging such as establishing different amateur associations, organizing various activities and giving discount to elderly citizens in tourism so as to encourage elderly citizens to take part in family and social activities by creating different opportunities.

Family elderly care has a deep foundation in China and plays an important role in ancient and present time. However, to those elderly citizens who excessively rely on families and totally entrust the happiness in the rest of their life to their children, we should guide them in changing the concept of childbearing "bringing up children for the purpose of being looked after in old age" so that they won't excessively rely on family members and regard their investment in their children and economic burden and obligation upon minor children as inescapable social obligations and children's inalienable rights. In this way, they can create their growing and meaningful personal life in old age.

Practice tells us that, in fact, formal and informal support cannot support the mansion of elderly citizens' life quality independently. For this, Western sociologists pointed out "Theory of Shared Function" and "Balance Theory of Coordination". The former combines formal and informal support organically. The latter requires constantly finding the balance. Only by seriously analyzing features and functions of formal and informal mental support and combining them together can we create a practical and reliable security system to improve elderly citizens' life quality.

---

[26]Mu [18].

[27]James [19].

References

1. Wu Cangping, and Guangning Mu. 1996. *Actively Achieve Healthy Ageing, Guangming Daily*. 30 Apr 1996.
2. Zhang Youqin. 2001. *Comparative Study of the Social Support Network in Urban and Rural Areas—A Case Study in Xiamen, Sociological Study*, Issue 4 of 2001.
3. Li Jianxin, and Fengyu Zhang. 1997. *Research on Mental Health of Urban Elderly Citizens and Relevant Factors, Chinese Journal of Population Science*, Issue 3 of 1997.
4. Zhou Yuping et al. 2008. *Aging Social Work*, 130. China: Intellectual Property Publishing House.
5. China Research Center on Aging. 2003. *Analysis of One-time Simply Survey Data of Urban and Rural Elderly Population in China*. China: China Standard Publishing House
6. Zhang Xiti, and Ping Guo. 2010 *Blue Book of the Aging of the Population and the Situation of Elderly Citizens in China*, 226. China: China Society Press.
7. Zhang Xiti, and Ping Guo. 2010 *Blue Book of the Aging of the Population and the Situation of Elderly Citizens in China*, 240. China: China Society Press.
8. Zhang Xiti, and Ping Guo. 2010 *Blue Book of the Aging of the Population and the Situation of Elderly Citizens in China*, 207. China: China Society Press.
9. Hu Xiying. 2006. *Mental Support, Easy Said Than Done—For the 10th Anniversary of the Issuing and Implementation of the Protection of the Elderly Rights and Interests, Friend of Old Comrades*, Issue 10 of 2006.
10. Pan Jinhong (2009) *The Government's Pension Liability Under Multiple Fields of View, Nanjing Cadre Institute of Population Management*, vol. 25, no. 2 of 2009.
11. Ling Xiaoping. 2008. *Dilemma and Choice of Gerontological Social Work in China, Social Observation*, the 7th Issue, 2008.
12. Zhou Yuping et al. 2008 *Gerontological Social Work*, 135. China: Intellectual Property Publishing House.
13. Wang Lifang. 2010. The Evaluation System of Japan's Elderly Services and Its Inspiration to China's Elderly Care Services. *Journal of Zhaoqing University* (7).
14. Zhang Xiti, Ping Guo. 2010. *Blue Book of China's Ageing Population and Conditions of the Elderly*, 207. China: China Society Press.
15. Zhang Liangli. 2006. *To Deal with the Aging of the Population—Construction and Planning of Social Pension Service System*, 2006. China: Social Sciences Academic Press.
16. Hu Xiying. 2006. *Mental Support, Easier Said Than Done—For the 10th Anniversary of the Issuing and Implementation of the Protection of the Elderly Rights and Interests, Friend of Old Comrades*, Issue 10 of 2006.
17. Zhang Minjie. 1994 *Comparison between Chinese and Foreign Ways of Elderly Care and Improvement of China's Way of Elderly Care, Sociological Studies*, Issue 4 of 1994.
18. Mu Guangzong. 2005. *Thinking About Chinese Policy for the Elderly, Market and Demographic Analysis*, supplement in 2005.
19. James A. Autry. 2010. *The Spirit of Retirement*. Trans. Wenli Cao. China: SDX Joint Publishing Company.

Chapter 6
Study on the Management System of Modern Pension Security Systems

Dasheng Shan

Since the implementation of the reform and opening-up policy, especially since the 21st century, China has increased its capital investment in pension security, quickened its pace in establishing its social pension insurance system and social pension service system, and a pension security system that covers all citizens has experienced full-length development. However, compared to the goals and requirements of ensuring the elderly receive security in old age, the construction of the current pension security system is still in its early stages and can hardly satisfy the gradually increased and diversified social demand of the elderly in the context of population aging and residential consumption structure upgrading, and a lot more work needs to be done. In the long run, to further quicken the pace of establishing the pension security system, it requires not only more capital investment and a faster pace of institution building, but also a reforming and perfecting of the management system to promote the establishment of a comprehensive, coordinated, sustainable and long-lasting mechanism. This chapter reviews the history of the reform and development of the Chinese pension security management system, analyzes the major contradictions and disadvantages that exist in the current system, learns lessons from typical countries and proposes basic ideas and policy recommendations for further reforming the Chinese pension security management system.

D. Shan (✉)
Development Research Center of the State Council,
Room 324, #225, Chaonei Street, Dongcheng District,
Beijing 100010, People's Republic of China
e-mail: shandasheng@126.com

Y. Wang (ed.), *The Development of Security and Whole Care System for the Aged in China*, https://doi.org/10.1007/978-981-10-6481-4_6

6.1 Concept and Basic Principles of the Pension Security Management System

6.1.1 Concept of the Pension Security Management System

Generally speaking, the pension security system in the industrialization phase was mainly composed of the social endowment insurance system and the social pension service (or pension social service) system, and the two major systems respectively solved economic security and service security problems. A social pension insurance system refers to a social security system in which the country and society provide corresponding income revenue assurance to guarantee the basic living conditions for a worker who reaches the working age limit of being released from his or her working obligations or retreating from his or her working position due to a loss of working ability, with the purpose of strengthening the worker's ability to resist the risks of old-age. The social pension service system refers to the network and the assorted service standards, operating mechanisms and supervising systems. The network was formed to go hand in hand with economic and social development with the purpose of satisfying the elderly's demands for pension services and promoting the elderly's living standards. It is aimed at all the elderly, provides facilities, organization, talented personnel and technological factors including life care, rehabilitation care, mental support, emergency aid, community involvement and many others.[1] With the change in social structure, the growth of the economy and the elderly's increasing demands, mental support for the elderly has gradually become an important part of pension security and a socialized pension mental security system needs to be set up.[2] However, due to the restrictions of various aspects, the institutional building in this respect has not been systematic or comprehensive and is still under exploration. This is not specifically analyzed here.

In order to manage complicated pension security affairs, an entire set of comprehensive management and service systems needs to be established. Pension security affairs need to be planned, organized, led, coordinated, controlled, supervised, etc., in accordance with laws and procedures and via certain ways, means and methods.[3] The total factors of pension security, including management functions, institution structuring, subordinate relationships, the division of power, the management system and other factors, form the pension security management system. Adapting to the state legislative system and administrative management system, the pension security management system includes the legislative level, administrative supervisor level and the business handling level, each of which performs different functions.

[1]*Plan for Building Social Pension Service System (2011–2015)* (released by the General Office of the State Council of the People's Republic of China [2011] No. 60), 2011.

[2]Shi Jinqun: *Research Report of Special Studies on Spiritual Security of the Elderly.*

[3]Zheng [1].

From the perspective of legislation, this primarily means that legislature formulates regulations on the scope and subject of pension security, basic conditions for enjoying security, capital sources, payment standards and methods, and management methods as well as liabilities, rights, obligations and other aspects of relevant subjects within pension security. Pension security is greatly policy-based, closely related to the interests of all relevant parties and involves many pairs of legal relations, so, as the pension security legal system is continuously perfected, from the perspective of maintaining the normal operation of the pension security system, certain judiciary institutions should be set up to issue corresponding criminal and civil penalties to crimes that occur during pension security's operating process.

From the perspective of the administrative supervisor level, the government's functions in the pension security system include necessary capital support, the establishment and perfection of the pension security system, guaranteeing the normal expenditure of relevant handling institutions, supervising such links as handling, fund management, use and others, providing basic public pension services, carrying forward supervision for the pension service industry, establishing capital effective use and risk prevention mechanisms for pension security to promote the legislation of pension security and other functions. Most countries have set up a pension security management institution in their administrative management sequence. Its functions are set up in accordance with corresponding laws and regulations and it exercises its administrative management rights in accordance with laws. In terms of institutional setup, there are different methods including unified management by a single department, simultaneous management by two departments, simultaneous management by several departments, etc.

From the perspective of the business handling level, the social pension security system and the social pension service system cover different business contents in accordance with their respective division of work. Two types of management, including social pension insurance public services and social pension insurance internal affairs management are included in the social pension insurance system. The former is the pension insurance management institution's function of providing services outwardly and means directly providing services such as information consultation, insurance participation registration, application, fee payment, modification, evaluation, payment, annual examination, reimbursement, appearance, etc., to organizations and individuals who have or have not bought the social pension insurance and require social pension insurance services, the purpose being to provide convenient, efficient and quality social pension insurance public services for the insured. The latter refers to the pension insurance management institution's internal management function of pension insurance affairs with its core being pension insurance fund management. Pension insurance fund management can be divided into five links including budget management, collection management, finance management, investment management and payment management in accordance with the procedures. Budget management refers to planning for the revenue and expenditure of the pension insurance fund; collection management refers to checking and ratifying the capital base and pushing all financing parties to pay the pension insurance fund according to regulations in accordance with relevant laws, regulations and policies on pension insurance; finance management refers to the daily management of pension insurance of different types in accordance with the financial system from different sources of pension insurance financing; investment management refers to putting the accumulated pension insurance fund into the capital market for preservation and appreciation; payment management refers to paying the pension fund to beneficiaries in accordance with relevant laws, regulations and policies on pension insurance. As for providing social endowment insurance public service and fund management, the business institutions involved shall at least include endowment insurance management institution, endowment fund management institution, social service institution, etc. The social pension service primarily includes home-based pension services, community pension services and institutional pension services. Home-based pension services refer to the service model whereby the government and society rely on the community to provide life care, household management services, rehabilitation care, mental comfort and other services for the elderly at home. Community pension services are a crucial support for home-based pension services that include two functions including community day care and home-based pension support and are mainly aimed to provide services for a family in which nobody can take care of the elderly during the daytime or where the family members are unable to take care of the elderly. Institutional pension services realize their fundamental pension service functions via facility construction and focus on realizing such functions as life care, rehabilitation care and emergency rescue.[4] The bodies of the social pension service system include the pension institutions and pension service institutions. A pension institution refers to the location where accommodation and support is specially provided for the elderly. A pension service institution refers to the service organizations, locations, facilities and others that provide life and value development for the elderly (a pension institution is regarded as an accommodation and support institution for the elderly in counting pension service institutions in Shandong Province). Because social pension services cover a wide spectrum and pension institutions and pension service institutions have not been scientifically classified in reality, the social pension service institution is applied to uniformly indicate the two categories of institutions in the text.

6.1.2 Basic Principles of the Pension Security Management System

To establish a scientific and improved pension security management system to promote good health and coordinate the operation of the pension security system is

[4]*The '12th Five-year Plan' for the Construction of a Social Pension Service System* (Draft for seeking opinions).

a goal that society expects to reach. To achieve this goal, the pension security management system should stick to and abide by the following principles.

6.1.2.1 Legalization Principle

From a comprehensive survey of the establishment, development and operation of pension security systems around the world, a pension security system is established on the premise of standard lawmaking, its mandatory nature is ensured by the rigidity of laws, and this is showed by standardizing the rights and obligations of all parties, standardizing the behaviors of all parties, giving the government the authority to interfere in pension security, etc. The legalization principle of the pension security management system includes two aspects: ① The pension security management institution setup and functions should be based on corresponding laws and regulations, relevant laws and regulations should have clear regulations on this, i.e., the nature, legal status, responsibilities and authorities, internal institutions, staff formation and so on should be clearly defined and regulated in detail so as to provide a legal basis for the government to perform pension security management and service functions comprehensively. ② The pension security management institution must be operated in accordance with laws, i.e., the management institution can only exercise its rights within the designated range of responsibilities to realize the legalization of institution, function, formation and working procedure. In short, the legalization of the pension security management system requires the pension security system to practically follow the principles of having laws to abide by, firmly adhering to laws, enforcing laws strictly and investigating into illegal affairs in aspects including institution setup, functional equipment, work operation and so on.

6.1.2.2 Scientific Principle

The scientific principle refers to determining the management functions, institutions and management methods of the pension security management in a scientific and reasonable manner in accordance with the internal requirements of the administrative management system and the general rules of management science. ① Scientifically determine the government's functions in pension security management. In accordance with the requirements of constructing a service-oriented government, a responsible government, a constitutional government and an incorruptible government, the relationship between the government and the market, the government and society, and the government and citizens should be correctly handled and attention should be paid to solving the problems of over interference, misplacement and absence of the government. ② In accordance with the requirements of unity of rights and responsibilities, reasonable division of work, scientific policy decisions, smooth implementation and powerful supervision, the pension security management functions of governments of all levels and all departments of the governments should be scientifically divided in order to establish a mechanism where the policy-making power, implementation power and supervision power restrict and coordinate with one another. ③ In accordance with the principles of simplification, unity and efficacy, the pension security management institution should be reasonably set up and the division of work for all departments and on all levels should be rationalized in accordance with the management level and scope. ④ To optimize the pension security management method, the government should improve the scientific and democratic policy-making mechanism, realize professional, standard and informationalized management and carry forward a performance management system and a supervision and accountability system.

6.1.2.3 Socialization Principle

The socialization of management services is an internal requirement of the pension security system. The contents of socialization of pension security management services include the following: ① Socialization of financing. Pension security capital comes from various aspects including companies, individuals, the government, society and other parties rather than being solely undertaken by the government, companies or individuals (it is not company security or family security). ② Socialization of organizational institutions. Break the identity limit of rural and urban areas, regions and companies, establish a socialized institution which is independent from enterprises and public institutions that provide all sorts of pension security services for the insured and carry out localized management. ③ Socialization of management subject. In addition to the special pension security management institutions, various subjects including the community, families, etc., are also involved in the management and supervision of pension security.[5]

6.1.2.4 Efficiency Principle

The efficiency principle refers to learning from modern management ideas, management organizations, management means and management methods under the restriction of limited resources by means of optimizing the management system so as to promote the pension security system's effective operation and allow the system to achieve long-term sustainable development. The efficiency principle is represented in the following aspects: ① A clear distinction of responsibilities among management institutions. ② All management links and management institutions are connected well, operating smoothly with the system running healthily. ③ The operating costs (including short-term costs and long-term costs) are relatively low, which is convenient for securing the subjects and improving the degree of satisfaction of the secured subjects and helps to increase the earnings of the

[5]Song [2].

insured and society.[6] ④ On the premise that the pension security capital and fund are ensured to be secure and complete, improve the operating efficiency to obtain preservation and appreciation to the greatest extent. ⑤ Management resources (mainly referring to human resources) are brought into full play.[7]

6.2 Reform and Development of Chinese Pension Security Management System

Since the founding of new China, great changes have taken place to the pension security system, different pension security systems and corresponding management systems that respectively adapt to them have emerged in different historical periods.

6.2.1 The Pension Security Management System in the Command Economy Period

Throughout the command economy period, with laborers taken as central, a labor insurance system was established for enterprise staff, a retiring system and a resigning system was established for staff of organs and public institutions, and "those without the source of income, the ability to work or a legal support obligor", including the unemployed, would receive social assistance from the Civil Affairs Department via the sub-district. In rural areas, ordinary agricultural laborers were still mainly dependent on the family for endowment and the village collective would provide guarantee in five aspects for those in need, including the old, the weak, the solitary, the widowed and the disabled,[8] forming a "five guarantees" support system with Chinese characteristics.

The labor insurance system was the core of the pension security system at that time, and treatment included the treatment for the injured and disabled, the dead, the diseased, the old, childbirth treatment and supporting directly-related family members treatment and others persons. As the top supervising organ of the national labor insurance business, the All-China Federation of Trade Unions coordinates the processing of national labor security business, supervised the implementation of policies related to the labor insurance business by all local subsidiary trade union organizations and trade union organizations of all industries, and the trade union

[6]Wang [3].

[7]Zheng [4].

[8]Including food guarantees, clothes guarantees, fire guarantees (fuel), education guarantees (kids and teenagers) and burial guarantees, and such contents as "accommodation guarantees", "medical care guarantees" and others have also been added with the development of the "five guarantees" system in rural areas.

organization of all levels should set up a labor insurance department. The Ministry of Labor was the top supervising organ of the state labor insurance business, and it was responsible for carrying forward labor insurance regulations, examining the implementation of the country's labor insurance business, labor administrative organizations at all levels supervised the payment of labor insurance funds, examining the implementation of the labor insurance business and dealing with appeals related to labor insurance incidents. In order to simplify the government structure, the Government Administration Council released *Notice of Transferring the Labor Insurance Business Work to Trade Unions for Unified Management* in May 1954, the enterprise labor insurance business in charge of the labor department would be gradually transferred to trade unions of all levels for unified management. The labor insurance work was specifically carried forward by enterprises, a labor insurance committee was generally set up in an enterprise, labor insurance committee members should be appointed in the workshop and a labor insurance manager should be appointed for each team and group. The labor insurance fund should be fully paid by the enterprise, and 70% of the fund should be deposited into the account of the basic-level committee of the enterprise's trade union, with the remaining 30% in the account of the All-China Federation of Trade Unions, the labor insurance fund flows among all levels of the trade union and the supervising trade union organization was responsible for coordination and regulation, showing certain social coordinating characteristics. Staff of state organs and public institutions were not included in the labor insurance system. Instead, an independent system for them had been established which was primarily in charge of the internal affairs department to begin with, transferred to the human resources department later, revoked when the human resources department was repealed afterwards and then managed by the internal affairs department again.

According to the prevailing system at that time, the labor department was only responsible for managing the pension security work of urban workers (who were employed), the social assistance and social welfare for "those without the source of income, the ability to work or a legal support obligor" in urban areas and "households enjoying the five guarantees" in rural areas were in the charge of the Ministry of Internal Affairs (modified into the Ministry of Civil Affairs afterwards), so there were capital guarantee as well as the service security provided by state-run pension institutions. A social welfare service institution under the direct supervision of the Ministry of Internal Affairs and civil affairs departments of all levels should be established to adopt the solitary elderly who had no one to depend on, no working ability or normal economic income. Private-run welfare institutions were basically prohibited during this period. As a collective welfare course in rural areas, the "five guarantees" support has always been organized and carried forward by the village collective organization (Agricultural Production Cooperative at first, changed to the production brigade after the People's Communization in 1958 and then changed to the production team after 1962). In order to solve the problem for the elderly who were unable to take care of themselves, nursing homes had been set up in some places to gather those who enjoyed the "five guarantees" support to support them together, gradually forming the "five guarantees" supporting mode combining concentrated support and scattered support with the latter being the major one factor. In order to solve guarantee problems for many disabled soldiers, sacrificed and diseased soldiers and their relatives, regulations on preferential treatment and placement had also been issued by the state, implementing the method of granting one-off pension to relatives of revolutionary martyrs, relatives of soldiers who were sacrificed while on duty and relatives of soldiers dying of diseases by the county, municipal, municipal district people's government in the residence of the relatives.

During the Cultural Revolution, enterprise business activities suffered from serious disruption, the united collection, management and expenditure system of the labor insurance fund could hardly continue, along with the fact that the trade union stopped its activities during this period, causing some disorder in labor insurance work. The Ministry of Labor was dismissed in 1969 and merged into the Labor Bureau of the National Planning Committee, the Salary Bureau which was mainly in charge of labor insurance was changed to the Salary Group (the Welfare Group), most carders of the Ministry of Labor were decentralized to local organs, basically paralyzing the labor insurance institutional framework. In 1969, the Ministry of Finance required all state-run enterprises to stop withdrawing labor insurance funds, and labor insurance was coordinated and taken charge of by the enterprise administration, realizing thorough self-management of the enterprise. In 1968, the Ministry of Internal Affairs was also revoked and many welfare service enterprises were forced to consolidate or were revoked. In March 1972, the State Council of the People's Republic of China called upon such departments as the Ministry of Finance, Ministry of Public Security, Ministry of Health, National Planning Committee and others for negotiation. They studied businesses which were principally in the charge of the Ministry of Internal Affairs, proposed the opinion of division of management with the Ministry of Finance being responsible for the assistance, preferential treatment and other work, while the National Planning Committee and Ministry of Labor were responsible for the work related to the resignation and retirement of staff of state organs.

6.2.2 Restoration and Reform of Pension Security Management System

Back in 1975, Deng Xiaoping returned to lead the work of the central government, and the work which was suspended during the "Cultural Revolution" started to recover. According to Notice on Adjusting Organizations Directly under the State Council (Guo Fa [1975] No. 153) issued by the State Council on September 30, 1975, the Labor Bureau of the State Development Planning Commission was renamed the State Administration of Labor which was in the charge of the State Development Planning Commission as an organization directly under the State Council. In 1978, the Ministry of Civil Affairs was set up, and the work of the former Department of the Interior before the "Cultural Revolution" was restored.

The personnel of the competent authority, social relief and social welfare work of the Ministry of Civil Affairs were made clear. In the same year, the Organization Department of the Central Committee of the CPC set up the Bureau of Retired Veteran Cadres. Work institutions of retired veteran cadres at all levels were gradually set up. Some national ministries and commissions with more retired veteran cadres also set up work organizations of retired veteran cadres. In October 1978, the 9th National Congress of Workers decided to restore the labor insurance work, and trade union organizations at all levels were rebuilt in succession. In 1979, the State Administration of Labor set up the Insurance & Welfare Department, and labor departments throughout China successively set up the Insurance & Welfare Office (Section). In 1980, the Personnel Bureau, a government agency of the Ministry of Civil Affairs, merged with the Demobilized Army Cadre Emplacement Leading Group of the State Council, and the State Personnel Bureau was set up directly under the leadership of the State Council. On March 14, 1980, the State Administration of Labor and All China Federation of Trade Unions issued *Notice on Reorganizing and Strengthening Labor Insurance Work* to require labor departments and the federation of trade unions at all levels to strengthen the leadership of grass-roots labor insurance work, thus restoring the administrative systems of labor insurance at all levels gradually. In 1982, the government carried out institutional reform. The State Administration of Labor, the State Personnel Bureau, the Science and Technology Cadre Bureau of the State Council and the State Organization Commission were merged, and the Ministry of Labor and Personnel was set up to be in charge of the Insurance & Welfare Bureau to comprehensively manage social insurance and employee welfare. In the same year, it was clearly ruled that friendless and homeless people without source of income in society would be in the charge of and helped by departments of civil affairs.

In 1984, throughout China, labor departments and PICC began to carry out the social overall planning of pension insurance in state-owned enterprises and most of urban collective enterprises respectively,[9] and trade unions also carried out the employee mutual security project in part of enterprises and institutions. In addition to local overall planning, some industries tried to carry out the industrial overall planning of pension insurance.[10] In the process of labor system reform, some places set up labor service companies which were affiliated with labor departments and took on more work of the social overall planning of pension insurance. After the volume of business increased, independent social insurance companies were set up. In 1986, China carried out labor system reform. It was ruled that social insurance specialized agencies under labor administrative departments were in

[9]During the process, PICC started to manage the pension insurance of workers in part of urban collective enterprises.

[10]From 1987, the State Council allowed 11 industries, namely railway, electricity, coal, water conservancy, post and telecommunications, transportation, non-ferrous metals, construction (China State Construction Engineering Corporation), finance, civil aviation, oil and gas to manage the pension insurance of their workers separately in succession.

charge of the retirement and pension work of contract workers,[11] the collection of retirement and pension funds, the payment of retirement and pension fees and the organization and management of retired workers. In 1987, retirement fee overall management committees at all levels were set up in succession. Offices were located in departments of labor and personnel, and most of them worked in the same place with administrative offices to achieve the combination of politics and affairs. Since the new system hadn't been formed, the management service work was still in the charge of enterprises primarily.[12] In order to solve the difficulties brought about by the inconformity of social insurance agencies in the actual work, in 1988, the Ministry of Labor decided to name social insurance agencies under labor departments in all provinces, cities and counties "Social Insurance Business Administration" collectively. It was a business unit, and its duties were ruled. The development of social insurance agencies was gradually standard.

Although labor departments played an increasingly important role in social insurance, when social insurance reform was carried out in government departments and public institutions, other ownership units and the countryside, labor departments encountered new problems, such as the repeated debates on the administrative power over pension insurance of collective ownership enterprises between PICC and labor departments after the reform, the debates on the administrative power over rural social pension insurance between departments of civil affairs, PICC, labor departments and agricultural sectors, and the debates on the administrative power over pension insurance of government departments and public institutions between personnel departments, labor departments and departments of civil affairs later. There was once a tendency to set up social insurance agencies separate from labor departments locally, which caused the vigilance of labor departments. For example, Jilin, Tianjin and Shenyang successively set up social insurance agencies separated from labor departments, which was criticized by labor departments. On September 10, 1988, the 6th meeting of the State Commission Office of Public Sectors Reform chaired by Premier Li Peng decided that social insurance was in the charge of the Ministry of Labor. The Ministry of Labor standardized names, nature and duties of social insurance institutions under labor departments in all provinces, cities and counties.[13] During the government institutional reform in 1988, the Ministry of Labor and Personnel was cancelled, and the Ministry of Labor and the Ministry of Personnel was newly set up to be in charge of the retirement work of enterprises and government departments and public institutions respectively.

During the period of "the 7th Five-Year Plan", China began to explore the establishment of accumulative rural social pension insurance system and made it

---

[11] *Notice on Reforming Four Stipulations of the Labor System of the State Council* (Guo Fa [1986] No. 77); *Interim Provisions on the Implementation of the Labor Contract in State-owned Enterprises* (Guo Fa [1986] No. 77); *Interim Provisions on the Unemployment Insurance for Employees in State-owned Enterprises* (Guo Fa [1986] No. 77).

[12] Meng [5].

[13] Feng [6].

clear that the Ministry of Civil Affairs was in charge of it. The Ministry of Civil Affairs newly set up the Rural Social Pension Insurance Office (the Rural Social Insurance Department) and the Rural Social Pension Insurance Service Center of the Ministry of Civil Affairs. During this period, the pension security management function of departments of civil affairs for senior citizens "without the ability to work, a source of income or a legal supporter and caregiver in urban areas" and "households enjoying the five guarantees" in rural areas didn't change significantly. After it was gradually difficult to implement the system of the five guarantees based on collective economy, in 1994, *Regulations on Support Work of "Households Enjoying the Five Guarantees" in Rural Areas* was formulated. It was ruled that support costs for "households enjoying the five guarantees" were primarily provided by rural collective economic organizations and disbursed from village public accumulation funds and town overall plan funds to adapt to the new changes of rural financial management after the implementation of household contract responsibility system. Some places also drew some costs from enterprises of villages and towns to support "households enjoying the five guarantees" and establish elderly care homes. The old management system of public pension agencies like welfare houses and elderly care homes was still used. The Ministry of Civil Affairs issued *Interim Measures for the Management of Rural Elderly Care Homes* and *Interim Measures for the Management of Social Welfare Institutions* to standardize the management system of pension agencies. It is important to note that in order to alleviate the contradiction between the increasingly prominent aging population and senior care services, in the mid-1980s, the Ministry of Civil Affairs put forward the development thought of socializing social welfare.[14] Social welfare institutions were allowed to use rest beds to provide paid services for necessary social members and try to introduce social capital to expand the supply of senior care services. In order to adapt to such new situations, the Ministry of Civil Affairs strengthened the industrial management function of senior care services.

6.2.3 Establishment of the New Pension Security Management System

Since the mid 1980s, China explored the reform of the old security system with the security of traditional units as the main body to establish a social security system independent of enterprises and public institutions. In 1991, *Decision on the Reform of the Pension Insurance System for Enterprise Employees* issued by the State Council made clear the establishment of the system combining the basic pension

[14]In November 1984, the Ministry of Civil Affairs held the meeting of exchanging experience in reform and rectification of national urban social welfare institutions in Zhangzhou and put forward that social welfare programs should be further transformed to the collective system of the state, the collective and the individual to further transform from relief to welfare, from support to support and rehabilitation and from closing to opening (hereinafter referred to as "three transformations").

insurance, enterprise supplementary pension insurance and employee personal saving pension insurance. Specific business of the basic pension insurance, enterprise supplementary pension insurance was in the charge of social insurance agencies under labor departments at all levels, and employees chose agencies for personal saving pension insurance voluntarily. Social insurance management agencies were non-profit business units. Local governments at all levels set up the Pension Insurance Fund Committee to guide and supervise pension insurance fund management. Offices were located in labor departments. It also clearly put forward that the overall planning of basic pension insurance funds should be gradually transferred to provinces from cities and counties.

The 3rd Plenary Session of the 14th Central Committee of the CPC made it clear that the direction of the reform of social security management system was to establish a unified social security management organization, and social security administrative management and social insurance fund management should be separated. In order to adapt to the said requirement, the former Insurance and Welfare Division of the Ministry of Labor was divided into the Social Insurance Division, the Social Insurance Business Administration and the Social Insurance Research Institute. The Social Insurance Division was an administration department which comprehensively managed social insurance and welfare of enterprise employees in China and was in main charge of the formulation, organization and implementation of reform plans for pension, medical treatment, industrial injury and birth and employee welfare as well as the coordination of interior social insurance administration, fund management and other business. The Social Insurance Business Administration, entrusted by administrative departments in charge, was responsible for managing national urban social insurance business and fund management work and guiding local governments to carry out handling and management work, thus separating administrative units from public institutions in social insurance. The Social Insurance Research Institute primarily researched policies and theories of social insurance to provide decision-making consultation. According to the principle of separating administrative units from public institutions, independent social insurance agencies were set up generally in all places to carry out social insurance management service work.

In this period, pension insurance work was given priority to in social insurance work. Social insurance agencies were mainly responsible for raising, adjusting and managing funds. Due to partial payment and allocation and enterprise distribution of pension insurance, much management service work was still borne by enterprises. In 1997, the State Council issued *Notice on Problems about the Transfer of Basic Pension Insurance for Enterprise Employees from Provincial Overall Planning and Industrial Overall Planning to Local Management*, which solved the long-standing contradiction in the separation of industrial overall planning and local overall planning and basically realized the localized management of basic pension insurance.

At the beginning of the socialization of enterprise insurance, due to the pressure from government departments and the rule-breaking operations of social insurance agencies, social insurance funds were embezzled seriously. In 1996, the State

Council issued *Decision on Strengthening the Management of Extra-budgetary Funds*, requiring that social insurance funds should be managed based on the management measure of extra-budgetary funds and incorporated into the special financial account of social security. Financial departments started to get fully involved in the management of social insurance funds. In 1998, in order to strengthen the management of funds, the separation between revenue and expenditure was implemented in basic pension insurance funds for enterprise employees. Audit departments took great pains to investigate the embezzlement of social insurance funds. In 1998, the Audit Office set up the Department of Social Security Audit independently, and local audit bodies at all levels also set up social security audit institutions to further strengthen the social security audit.

Compared with the rapid development of the urban pension insurance system, the pilot rural social pension insurance system which began in 1986 was not smooth. Policy makers thought development conditions were not mature, so it was rectified. In 1998, the government carried out institutional reform. Social insurance for urban employees in the charge of the Ministry of Labor, social insurance for government departments and public institutions in the charge of the Ministry of Personnel, rural social insurance in the charge of the Ministry of Civil Affairs, social insurance overall planned by industrial sectors and medical insurance in the charge of health departments were managed by the Ministry of Labor and Social Security collectively. In 2000, the Social Insurance Business Administration under the Ministry of Labor, the Management Service Center of Rural Social Pension Insurance under the Ministry of Personnel, the National Management Center of Public Medical Care Affairs under the Ministry of Health, central state organs of the Ministry of Personnel and its social insurance management centers for public institutions in Beijing were merged into the Social Insurance Business Management Center under the Ministry of Labor and Social Security to be responsible for the comprehensive management of national social insurance work. Most of overall planning areas nationwide had set up social security agencies at provincial, city and county levels affiliated to the Ministry of Labor and Social Security to manage basic pension insurance of urban employees. During this period, in order to solve the sudden peak of unemployment, especially the unemployment in state-owned enterprises, the Chinese government adopted the "two guarantees", the "three security lines" and other measures. Labor security management service agencies were set up on urban streets to further strengthen the handling ability of labor security grassroots units.

In 2000, the State Council put forward the solution to improve the urban social security system, asked to practice individual accounts of basic pension insurance, gradually improve the overall planning level and transfer to provincial overall planning, and put forward to establish a social security system independent of enterprises and public institutions with diversified funding sources, standardized security system and socialized management services. Among them, "socialized management services" meant that social security funds under the planned economic system were transferred to be managed by social institutions from the closed management of enterprises and public institutions, which accorded with the international practice on pension security management. In 2002, the Central Committee of the CPC and the State Council issued *Notice on Doing a Better Job in the Re-employment of the Laid-off and the Unemployed* to specifically require that urban and rural grassroots organizations should undertake the management and service work of retirees of enterprises and public institutions. Streets and villages and towns with heavy tasks could set up or determine the agency in charge of labor security business and employ professional service staff in the community.

To cope with the potential pressure of fund payment brought by population aging in the future, from 2000, China began to build national social security funds for strategic reserve and set up the National Council for Social Security Fund. In the same year, the State Council decided to carry out market-oriented operation management of enterprise annuity. To standardize the management of enterprise annuity, in 2004, four ministries and commissions, including the Ministry of Labor and Social Security, issued *Trial Measures for Enterprise Annuity Fund Management*, and in 2011, the newly revised *Measures for Enterprise Annuity Fund Management* was issued.

In 2002, the former Ministry of Labor and Social Security began to implement the "Golden Insurance Project". Social insurance information network was gradually extended to streets, communities and service organizations and service outlets, and the handling mode was being innovated. In June 2003, the general office of the Central Committee of the CPC and the State Council forwarded *Opinions of Actively Promoting Socialized Management Service Work of Enterprise Retirees which was* jointly formulated by 13 departments like the former Ministry of Labor and Social Security, greatly promoting the socialized management service work of enterprise retirees. In May 2009, the Standardization Administration of China allowed the Ministry of Human Resources and Social Security to establish the National Technical Committee of Social Insurance Standardization to be mainly responsible for pension insurance, unemployment insurance, medical insurance, industrial injury insurance, maternity insurance and other social insurance services as well as the evaluation and management of the standardization work in this field. Members were from government departments, scientific research institutions, institutions of higher learning, industrial associations and user units.

Since the 16th National Congress of the Communist Party of China, China balanced urban and rural development, and the rural pension security system was rapidly developed. In 2006, the State Council issued the revision of *Regulations on Supporting Rural Households Enjoying the Five Guarantees*, and thus rural collective support was transferred to financial support and the management and service work of supporting households enjoying the five guarantees was further standardized. When rural social pension insurance (commonly referred to as "the old rural insurance") stagnated due to rectification and before the new unified rural social pension insurance system was issued, some departments took the lead to establish the pension insurance system for certain groups in the countryside. From 2004, led by family planning departments, for rural single-child families and two-daughter families, the reward and support polices for rural family planning families were implemented. In 2007, special pilot support system was provided for single-child families in which the child was disable or dead. Driven by organization departments, some provinces and cities also established the pension insurance system for village cadres. Health departments in some places established the pension insurance system for rural doctors, and water conservancy departments established the pension insurance system for reservoir immigrants. Local labor departments at all levels also established the pension insurance system for landless peasants and peasant workers based on actual local situations. From 2009, driven by social pension branches, part of the country started to pilot the new rural social pension insurance. During the "12th Five-Year Plan", this system would achieve full coverage. The labor and social security departments at all levels undertook the management service work of the new rural insurance. In March 1997, the Office of Military Insurance of the Chinese People's Liberation which was organized jointly by three headquarters of the People's Liberation was set up in Beijing, which marked that the military insurance system in China was formally established.

In 2010, China began to implement social insurance fund budget, which changed the system in the past that social insurance administrative departments prepared and financial departments at the same level reviewed social insurance fund budget and actual budget drafts into the budget system that social insurance agencies or (tax departments) prepared budget, departments of human resources and social security audited and summarized, financial departments audited, the people's governments at the corresponding level examined and approved and social insurance agencies implemented. The actual budget system that social insurance agencies prepared actual budget drafts of annual social insurance funds, departments of human resources and social security audited and summarized, financial departments audited, the people's governments at the corresponding level examined, approved and submitted to financial departments and departments of human resources and social security at the higher level.

6.2.4 Present Situation of the System

6.2.4.1 Legislative Level

The pension security policy system includes laws, regulations, rules, normative documents and other different levels. According to the present legislation jurisdiction, the State Council, according to the development needs of China's pension security cause, organizes trial pension security reform, evaluates and summarizes the mature experience of pension security practice and promotes pension security legislation drafts to the National People's Congress or the Standing Committee. After the National People's Congress or its standing committee deliberates and approves, such legislation drafts can be issued and implemented. The State Council can also formulate relevant pension security regulations according to the pension security laws. Departments in charge of the pension security work can formulate regulations and normative documents based on work needs and pension security laws and regulations. Local governments and related departments can propose local legislation drafts to the local people's congress according to legislative authority, so as to issue local rules and regulations. Provinces, autonomous regions, municipalities directly under the central government and people's governments of large cities can formulate rules and regulations according to laws, administrative regulations and local laws and regulations.

Standing committees of people's congresses at all levels hear and review special work reports on receipts and disbursements, management, investment, operation, supervision and inspection of social insurance funds by the people's government at the corresponding level, inspects the implementation of *Social Insurance Law of the PRC* and exercises the power of supervision according to law.

6.2.4.2 Administrative Level

According to national economic and social development plans at different periods, governments at all levels put forward development plans for pension security to promote the development of pension security cause. At the same time, departments in charge of pension security should also formulate several plans, including medium and long-term plans, specific work projects, annual key points, special work plans and other different forms.

At present, the Ministry of Human Resources and Social Security is responsible for the policy-making, supervision and management of basic pension insurance of organs, enterprises and public institutions and the new rural social pension insurance and participates in making the national social security fund investment policies. The Ministry of Human Resources and Social Security and departments associated with pension security are pension insurance departments and rural social insurance departments. All levels of administrative department for labor security, under the leadership of the government at the corresponding level, carry out the work, make local pension insurance policies and guides, supervise and inspect local pension insurance work and the business of social security departments. The Ministry of Human Resources and Social Security sets up the Supervision Office of Social Insurance Funds to supervise national social insurance funds. Institutions in charge of the supervision of social insurance funds under administrative departments of human resources and social security at or above the county level are responsible for the supervision of social insurance funds within their respective administrative areas. The main contents of the supervision of social insurance funds are as follows: the implementation of management laws, regulations and state policies of social insurance funds; the implementation and actual budget of social insurance funds; the collection, expenditure and balance of social insurance funds.[15] Administrative departments of labor security at the county level or in cities divided

[15]*Administrative Measures for Social Insurance Funds* (No. 12 order of the Ministry of Labor and Social Security of the People's Republic of China).

into districts are also responsible for the labor security supervision of employers' participation in various kinds of social insurance and payment of social premiums. Supervision and law enforcement ways like daily patrol inspection, special investigation into reports of the masses, yearly check of labor security and special inspection are adopted.[16] Administrative departments of labor security has the administration punishment power to deal with behavior in violation of labor security laws, regulations or rules.[17] Social insurance agencies, entrusted by administrative departments of labor security, can participate in the investigation related to the collection of social premiums. The Ministry of Human Resources and Social Security is responsible for preparing relevant policies for the management of enterprise annuity funds. The Ministry of Human Resources and Social Security supervises enterprise annuity funds. Business supervision departments of enterprise annuity trustee organizations, account managers, custodians and investment managers supervise business activities in accordance with their respective responsibilities.

The Ministry of Civil Affairs is the department in charge of social welfare and social relief. It is responsible for making development planning, policies and standards for social welfare and social relief formulating management measures for social welfare institutions and guiding the security of elderly rights. The Ministry of Civil Affairs and departments associated with pension security aim to promote social welfare and charity. The departments of civil affairs under the State Council are responsible for guiding the management of national social welfare institutions. The departments of civil affairs under the local people's governments at or above the county level are in charge of the business of social welfare institutions, manage, supervise and inspect social welfare institutions, develop setup planning for social welfare institutions and perform the functions of examination and approval and inspection.[18] Departments of civil affairs, according to the regulations, supervise and manage elderly care institutions. Departments of civil affairs under the people's governments at or above the county level also regularly inspect the work of elderly care institutions annually.

Local party committee organization departments, departments of retired veteran cadres and government departments of human resources and social security carry out macro guidance, supervision and inspection of the service management work

[16]*Supervision Regulations on Labor Security* (No. 423 order of the State Council of the People's Republic of China).

[17]According to *Supervision Regulations on Labor Security*, if employers conceal the total wages or staff members when declaring payable social premiums to social insurance agencies, administrative departments of labor security instruct them to correct and collect penalties 1 time above yet 3 times below the concealed wage amount. If they gain social insurance treatment or social insurance fund expenditure by cheating, administrative departments of labor security instruct them to return and collect penalties 1 time above yet 3 times below the cheated amount; in case of crime, they shall be investigated for criminal responsibility according to law.

[18]*Interim Measures for the Management of Social Welfare Institutions* (No. 19 order in [1999] of the Ministry of Civil Affairs).

for retired cadres in respective regions. Departments of retired veteran cadres, together with relevant departments, are responsible for the implementation of political treatment for former members of the leadership and retired leading cadres within the scope of cadre management authority in respective regions, departments and units.

The National Working Commission on Aging is a deliberation and coordination agency in charge of national aging work under the State Council. With the Vice Premier of the State Council serving as the chairman, its member units are composed of 26 national sectors, and a deputy director of respective national sector serves as the member. Its main responsibilities are: ① research and make development strategies and major policies of aging cause, coordinate and push related departments to implement aging development plans; ② coordinate and push related departments to do a good job in the protection of the elderly rights and interests; ③ coordinate and push related departments to strengthen macro guidance and comprehensive management of aging work and carry out various activities helpful for old the physical and mental health of elderly citizens; ④ guide, supervise and check the aging work of various provinces, autonomous regions and municipalities directly under the central government; ⑤ organize and coordinate major aging events of the United Nations and other international organizations in China. The Commission has offices to deal with daily work. Governments at all levels in provinces (autonomous regions and municipalities directly under the central government), places (cities, states, and leagues), counties (cities, districts and banners), villages and towns (streets) establish working commissions on aging and agencies, and specially-assigned persons of village (neighborhood) committees are responsible for the aging work.

In addition to competent departments, financial departments, audit departments, development and reform departments also perform their administrative functions in accordance with relevant laws and regulations. At the central level, the Ministry of Finance is responsible for the management of central fiscal social security expenditure, formulates the financial management system of social security capital (funds), takes on the financial supervision work of social insurance funds and prepares drafts and actual drafts of central social security budget jointly with relevant departments. The Department of Social Security is the Ministry of Finance is in charge of social security.[19] Offices of special commissioners are set up locally to be responsible for the financial supervision of social security within their respective jurisdictions. Financial departments at all levels also participate in the supervision of social security funds by budget management, management of receipts and disbursements, management of balance of funds, financial subsidies and tax incentives. Pension insurance funds are incorporated into the special financial account, separate revenue and expenditure and fixed sums are for fixed purposes. Development and

---

[19] *Notice on Printing and Distributing the Regulation on Internal Institutions with Main Responsibilities and Size of Personnel Force by General Office of the State Council* (Guo Fa Ban [2008] No. 65).

reform commissions at all levels are responsible for perfecting policy recommendations for the coordinated development of local social security and economy and coordinating the development of pension security cause and major issues and policies in reform. Audit organs are responsible for the audit supervision of social security funds, and carry out business guidance and supervision for the internal audit of social security funds. The National Audit Office is the supreme authority of the state audit, and the Department of Social Security Audit under it is in charge of social security funds. The National Audit Office sets up offices of special commissioners in various places to be responsible for the audit supervision of social security funds within their respective jurisdictions. Government audit institutions at various levels carry out the dual management system of the government at the corresponding level and the national audit institution at the higher level, and audit services are led by the audit institution at the higher level.

The financial departments, administrative departments of social insurance and audit institutions under the State Council also supervise the receipts and disbursements, management and investment operation of national social security funds.

The division of labor of each department in the preparation, examination and approval, implementation and final settlement of social insurance fund budget is as follows. In overall planning areas, social insurance agencies prepare budget drafts of social insurance funds and the department of human resources and social security at the corresponding level examines, approves and summarizes. Upon examination and approval by the financial department, the financial department and the department of human resources and social security submit to the people's government at the corresponding level for examination and approval. In case social premiums are collected by tax authorities, budget drafts of social insurance fund income are prepared by social insurance agencies and tax authorities. The financial department and the department of human resources and social security give an official written reply to the budget approved by the people's government at the corresponding level, and social insurance agencies implement. Social insurance agencies regularly report to the department of human resources and social security and the financial department at the corresponding level. Social insurance fund income budget gives an official written reply to tax authorities and social insurance agencies. Tax authorities regularly report to the financial department and the department of human resources and social security at the corresponding level. At the end of the year, in overall planning areas, social insurance agencies prepare budget drafts of social insurance funds and the department of human resources and social security at the corresponding level examines, approves and summarizes. Upon examination and approval by the financial department, the financial department and the department of human resources and social security submit to the people's government at the corresponding level for examination and approval. In overall planning areas, after drafts and actual drafts of social security budget are submitted to the people's government at the corresponding level for examination and approval, they should be submitted to the financial department and the department of human resources and social security at the higher level. The financial departments and the departments of human resources and social security at the provincial level submit drafts and actual drafts of social security budget of respective provinces (districts and cities) to the people's government at the corresponding level, the Ministry of Finance and the Ministry of Human Resources and Social Security.[20]

6.2.4.3 Business Agency Level

At the national level, the Social Insurance Management Center subordinate to the Ministry of Human Resources and Social Security undertakes the responsibility of guiding the social insurance agencies, being responsible for the comprehensive management of enterprise basic pension insurance, medical insurance, industrial injury insurance, maternity insurance, rural old-age insurance and other aspects. The center is a full funding public institution directly affiliated to Ministry of Labor and Social Security of the People's Republic of China. Most regions in the overall planning in China have established social insurance agencies affiliated to the Ministry of Human Resources and Social Security at provincial, city and county levels, and set up branches and service outlets as needed, forming the organization system and service network with agencies at all levels as the backbone, the banks and all kinds of fixed service agencies as the support and community labor and social security platform as the basis, which forms a set of management systems and service process. There are as many as over 7400 social insurance agencies at all levels, more than 30,000 street and township labor and social security offices. The basic operating expenses and administrative expenses of the personnel funds and the handling of social insurance for the pension insurance agency should be guaranteed by fiscal institution at the same level in accordance with the provisions of the state. For enterprise retirees, socialized management service work should be executed. After enterprise workers handle emeritus formalities, the work and the original enterprise are separated. Socialized payment of pension payment should be implemented. The personnel are transferred to city street and community office for localized management, and the community service organizations provide corresponding services.[21]

The enterprise annuity corporate trustee, account manager, trustee and investment manager carry out the management of enterprise annuity fund related tasks in accordance with the *Labor Law*, *Trust Law*, *Contract Law*, *Securities Investment Fund Law*, *Enterprise Annuity Fund Management Approaches* and other laws and regulations.

---

[20]*Opinions of the State Council on the Implementation of the Budget of the Social Insurance Fund* (Guo Fa [2010] No. 2, China issued by the State Council.

[21]*Notice on Forwarding "Opinions on Actively Promoting Socialized Management Service Work of Enterprise Retirees" by the Ministry of Labor and Social Security and Other Departments of General Office of the CPC Central Committee and General Office of the State Council* (Zhong Fa Ban [2003] No. 16).

The National Council for Social Security Fund is the management and operation organization of the national social security fund, responsible for the investment operation of the national social security fund assets, a public institution directly subordinate to the State Council which is under the direct leadership of the State Council and accepts the supervision of the State Council or the department authorized by the State Council.[22] The national social security fund is constituted of the fund and equity assets of state-owned shares reduction, the central government budget allocations, fund raised in other ways with the approval of the State Council and the investment returns, which is dedicated to the supplement of social security expenditure and fund adjustment. The Ministry of Finance, in conjunction with the Ministry of Labor and Social Security, worked out policies related to social security fund management operation to supervise the social security fund investment operation and the trusteeship situation. Securities Regulatory Commission and the People's Bank of China supervise the business activities of investment manager and trustee of social security fund activities in accordance with their respective functions and powers. The street (township) labor and security affairs agencies carry out socialized management service work of retirees, including: basic social pension payment; socialized management of personnel files and basic information of social enterprise retirees; establish self management and mutual service organizations for retirees; transfer organization relationship for the party members among enterprise retirees, organize them to participate in the street (township), community party organization activities; provide social insurance query service; master the living and transfer situation of enterprise retirees, carry out pension qualification work; assist the community health service agencies in providing medical services for enterprise retirees; organize the enterprise retirees to carry out cultural and sports activities; assist retirees' families in applying for mortuary grant-in-aid and survivor allowance.[23]

Most government departments and public institutions set retired cadres management institutions whose functions include: implement guidelines and policies of the work of retired cadres, formulate the measures for the implementation of the Department; organize retired cadres to read, study files and participate in political activities; be responsible for arranging services, such as health care, livelihood welfare, recuperation and vehicle for retired cadres; cope with the funeral and the aftermath of the funeral for retired cadres in conjunction with the relevant departments, be in charge of the inspection and guidance for the work of retired cadres of directly subordinate units, and forwarding and conveying related documents; assume the matters concerning the work of retired cadres assigned by the Department.

The national social insurance information system is planned by the country with unified standards and norms, and is constructed jointly by the people's governments

[22]Statute of the National Council for Social Security Fund.

[23]*Notice on the Implementation of Zhong Ban Fa [2003] No. 16 Document of the Ministry of Labor and Social Security* (Lao She Bu Fa [2003] No. 20).

at or above the county level, whose business is guided by information integrated management department. The social security card is under unified planning of the Ministry of Human Resources and Social Security with unified standards and norms, and is issued by the local labor and social security departments. By making overall planning of the social security card construction, establishing a unified national standard quasi norm system, developing unified key management system and other management systems, the Ministry of Human Resources and Social Security ensures the steady and orderly local social security card construction.

In addition, coordinate regional people's governments to establish the Social Insurance Supervision Committee composed of representative of the employing unit, representative of the insured, the union representative, experts and others. The committee organizes persons concerned to inspect the implementation of social insurance policies, laws and regulations, supervise the budget and final accounts of social insurance fund, collection, management, adjustment, investment operations and payment, put forward advice and suggestions on the work of social insurance, and exercise social supervision according to law. The Social Insurance Supervision Committee may engage accounting firms to conduct annual audit and special audit of the social insurance fund's income and expenditure, management and investment operation, and makes public the results of the audit.

6.3 The Main Problems of the Pension Security Management System in China

China's pension security management system has been greatly improved in the past few decades, and it is obviously in the direction of more legal, social, scientific and efficient development. However, the existing pension security management system cannot be said to have been very sound and has a lot of problems.

6.3.1 The Degree of Legal System of the Pension Security System Is Low, and the Pension Security Work Mainly Depends on the Policy Regulation and Administrative Means

In the process of promoting the construction of the pension security system, China put more management functions and responsibilities on administrative organs to coordinate the interest relations of all aspects in a centralized way, which formed the management pattern that pension security policies are made by administrative departments, agency management right belongs to administrative departments and the right to supervise is also under administrative departments. In this case, the legislature failed to play its due role, judicial organs are basically absent in the field of pension security. Such a system is related to the dependence of government management on the original system path, the need of reform and transformation of the pension security system as well as the fiscal responsibility borne by the government.[24] Since the beginning of the 21st century, China's legislature significantly accelerated the progress of the pension security legislation, and the legislative transparency and the degree of participation by all people also continues to increase. However, from an overall point of view, the pension security is still a weak area of legislation, and the operation of the pension security affairs still relies on the relevant provisions, guidance and departmental regulations of the State Council as well as the local regulations in a large extent. The administrative organs play a substantial role, and lead the reform and development of the pension security system. It should be said that this system has certain rationality. It is suitable for China's unique political system. Because the pension security in China itself is "crossing the river by feeling the stones", and the external environment of the pension security includes population, economics, finance, politics and other factors and is changing rapidly, it is difficult to formulate stable, normative, and standardized laws. More importantly, it is an unprecedented undertaking to establish pension security system of universal coverage in such a developing country as China. However, the earlier system formation also provides the local governments with operational space to continue to explore the pension security system that adapts to China's national conditions. This is beneficial to arouse the local enthusiasm and facilitates the implementation of the work. Fix the unstable reform and development projects or those with larger dispute in local pilot areas, which avoids big repetition and potential turmoil caused by the sudden reform in full swing. However, after the pension security system is gradually perfected, this system is more and more difficult to adapt to the needs of the healthy and standardized operation of the pension security system, and its disadvantages are from many aspects. First of all, the operation of pension security affairs still relies on the policies developed by administrative regulations, departmental rules and administrative institutions to a large extent. The absence of the law led to greater uncertainty of the system. Second, the administrative organs are in the leading position in decision making, but are not subject to effective supervision and control in the process of decision making as they are highly closed. No matter how fair and impartial the system is, it is still difficult to achieve the goal of equal participation of other subjects of responsibility in the system construction. Again, compared with the public, the administrative organs have more adequate information and more specific knowledge, so they have more conditions to hide and distort the relevant knowledge and information. By information superiority when formulating policies, the competent department requires expanding the various powers of the department as far as possible consciously or unconsciously to reduce or avoid the responsibility the department should bear as much as possible. As a result, it is difficult to ensure the policy of impartiality and credibility. Finally, under the big background of different regional economic and social development status differences, pension security

[24]Zheng [7], pp. 41–42.

regulations and policies developed by the local governments are of low level and ineffective, In addition, the difference in regions is so large that mutual recognition is unavailable, making the policy "fragmentation" phenomenon more serious.

6.3.2 The Slow Transformation of Governmental Functions and the Macro Management Function and the Cause Holding Function Hasn't Been Separated Substantially yet

In the period of planned economy, as all economic and social organizations managed all kinds of social affairs by administrative means, management activities and functions related to the elderly security were complicated and varied, but they were almost included in the administrative system of the government. The administrative management and career management weren't separated, making the government bear excessive and unrealistic functions and responsibilities. This was the inevitable product of the political and economic system at that time. In quite a long period of time after the Reform and Opening-up, people once tried to solve this problem. However, as we applied the one-sided experience of economic system reform mechanically and simply and inappropriately learnt and used the foreign models and practices for reference, the government's role in the elderly security experienced a process of wide-range withdrawal or reduction. The lack of the elderly security system once became an important factor restricting the vertical deep propulsion of the economic system reform and threatening social stability.

In other fields, the government also continues to follow the management mode in the period of planned economy, causing the offside and dislocation of functions. In the process of promoting the public welfare undertaking management system reform, including the elderly security, China always stresses that these two functions, namely governmental "management" and "holding" should be relatively separated and makes clear that the government is mainly in charge of policies, rules and supervision, realizing the separation of administrative units and institutions and the regulating function and the holding function. In addition, the social security system reform has long made clear the principle of separation of pension insurance fund management and administrative management. However, from the actual situation, the fund management agencies (agencies) aren't completely separated from the administrative departments. Compared with the model in the period of planned economy, the government centralized management model of setting, management, holding and supervision by the government hasn't been established although the pension insurance system which was carried out separately in the past is done by the government organs in a unified way.[25] As a result, after the scale and workload of the pension insurance agencies are gradually expanded, much energy of the

[25]Zheng [7].

competent department is put into the management of the agencies' operation and other micro affairs and the macro supervision function it should mainly bear is neglected or weakened. Moreover, due to the improper transformation of governmental functions, survivals of old systems and deeply-rooted influence of the department culture, such a closed system is also very likely to form profit-maximizing departmental functions. Especially, the chain of departmental interests (administrative units and institutions aren't separated) arising from the administrative subordinate relationship between the competent department and agencies further strengthens the departmental interests. In the chain of multi-layer principal-agents, as the social security agencies always have greater autonomy, opportunism, decision-making errors, bureaucracy and other problems may exist inevitably and various rent-seeking activities may happen easily. The agencies in charge of the pension insurance fund management and supervision institutions in charge of fund supervision belong to the administrative department of labor and social security. It always fails to carry out fair and effective supervision in terms of fund irregularities. Due to the low-level overall planning, the pension insurance fund is always regarded as the local fund, departmental fund or governmental fund. The situation of holding back and misappropriating the fund exists in some local areas. There are various subjects of fund irregularities in various areas, such as social insurance agencies, financial departments, local governments, labor and security departments and tax authorities.

In the social elderly care service system, due to improper transformation of functions, the obstacles and constraints of the original system hasn't been completely broken yet. The competent department is the policy maker as well as the sponsor and manager of the elderly care service industry. It manages the large-scale directly-affiliated units (besides the elderly care institutions, the civil affairs competent department also manages associations and social service institutions in funeral, interment, marriage, social welfare, social relief and other aspects all the time, having a wide scope of business). Much of its energy is put into the management within the system and the function of social management and public service facing the whole society is restricted. Objectively, many areas have formed the reform idea of separating the administrative units and institutions and management and holding in the elderly care service management system and carried out some explorations, but they haven't found an effective way to achieve it. In reality, the state-run elderly care institutions still manage in accordance with the administrative organs' model, lack an effective incentive mechanism and the reasonable distribution system and have no enough power and vitality of self development, so they can't response to the increasingly diversified and personalized elderly care service needs in time and cause a serious structural contradiction. First, in the shelter of the competent department, the state-run elderly care institutions take part in the elderly care service market activities with the private elderly care institutions. Many state-run elderly care institutions mainly take in elderly citizens with ability to work or rather good economic conditions rather than those "without ability to work, a source of income or a supporter or caregiver", causing the dislocation of service objects and deviating from the functional positioning of state-run elderly care institutions; second, they also enjoy the favored support of public resources and greatly distort the elderly care service market.

6.3.3 The Administrative Duties, Responsibilities and Limits of Authority of Various Departments in the Government Overlap, and Competent Departments Have Insufficient Limits of Authority to Make Overall Plans for Pension Security Cause

In the initial stage of the reform, labor departments, personnel departments, insurance companies, industrial management department, the Civil Affairs Department, the trade union system are involved in pension insurance management, thus forming a very decentralized management system and causing a lot of problems. During the "7th Five-Year Plan" period, China had asked social security institutions to manage social insurance, social welfare and social relief work, and make plans and arrangement. Decision making departments also tried to establish an independent social security department, but there was no result. However, the local governments tried unifying social security management institutions. For instance, Guangdong Province, Hainan Province and Shanghai City have set up social insurance commission to manage social security, accumulating a lot of useful experience. In every reform of government organization, a proposal of establishing a large social security department was put forward later, and the social security management system has become centralized gradually. In particular, the establishment of the Ministry of Human Resources and Social Security in 2008 marks the administration of social pension insurance in the department level has been united. (Within the department, it is still divided in accordance with the urban and rural areas, government departments and public institutions, and enterprise). Of course, whether we should further centralize the pension security management, especially pension service industry management and other functions and let one department perform them is a problem worth considering. Now it seems that this integration is not necessary, but there is the disadvantage of decentralized power in different systems (social pension insurance, social elderly care services), so it is hard for competent departments to centralize the right to supervise and criticize effectively.

At present, the most prominent problem is that basic pension insurance is managed according to the groups. Seemingly, unified management of pension insurance system at the competent department level has been achieved. However, due to the residues of old systems and the lag of the related supporting reform, competent departments don't manage the retirement system and pension insurance of various people. The pilot pension insurance system reform in public institutions is confronted with great resistance in practice and almost halts. It is mainly because the social security department is in charge of pension insurance in public institutions, but the coordinated efforts of all kinds of affairs involving the reform of pension insurance in public institutions is weak. At present, pension insurance system in public institutions involves at least finance departments, departments in charge of social undertakings, institution preparation departments and so on. Quite a few leaders and cadres in public institutions are still managed by the Organization Department. In addition, the reform of various social undertakings (education, health, science, technology, etc.) and public institutions (personnel, classification, wages, etc.) has been promoted on a full scale. The reform of the pension insurance system of public institutions will damage the interests of financial departments and departments in charge of social undertakings. If the relevant supporting reform is not well organized timely, it will be hard for competent departments to manage the pension insurance business in a centralized, effective and unified way. In the field of social elderly care services, the Civil Affairs Department is obliged to promote, supervise and manage the construction of social elderly care service system with policies. However, preferential policies for various elderly care institutions involve the Civil Affairs Department, Labor and Social Security Department, Education Department, Health Department, Industry and Commerce Department, Taxation Department and other departments, so it is hard for the Civil Affairs Department alone to plan and coordinate them. All departments cannot form a cohesive force, which has a direct impact on the sound development of elderly care service industry. It is pointed out that the direct management of public elderly care institutions the Civil Affairs Department has long used is not necessarily applicable to the supervision and management of the elderly care service industry. Actually, now civil affairs departments at all levels, health departments, fire departments and communities supervise social elderly care institutions. The Civil Affairs Department supervises and manages elderly care institutions through the report approval and the review of the procedure standard in principle, and it doesn't know much about the actual situation of institutions. Besides, in the process of monitoring and evaluation, multiple management and mutually making excuses often appear.

6.3.4 The Responsibilities and Duties of Governments at Different Levels Are not Clear, and the Central Government and Provincial Governments Have Insufficient Overall Planning for the Pension Security Cause in Their Regions

Any government cannot arrange all its activities, so governments at different levels must be set up to implement the administrative management. However, it is very difficult to define the functions of governments at different levels. In the early years of New China, unit security was implemented (collective security was implemented in rural areas). Under the financial system of unified collection and allocation of funds by the state, it is sort of a social overall planning mechanism. The "Cultural Revolution" destroyed the function of social overall planning, and pension insurance evolved into business insurance. After the reform and opening up, China began to rebuild the social overall planning mechanism of pension insurance and gradually increased the overall planning level. When system of tax distribution in the fiscal system was established in 1994, China was exploring the reform of pension security system. There was no clear duty and it was impossible to definite the right of governments at all levels over social security expenditure. New pension insurance system was established later, but the responsibility and right of governments at all levels were not divided clearly. In particular, the responsibility of governments at all levels was not clear. As for the pension security expenditure and compensation, there was no unified and fixed system.[26] In particular, the central government and provincial governments had insufficient overall planning for the pension security cause in their regions. The overall planning level of pension insurance fund directly showed that the relationship between the central government and local governments was not clear. The overall planning of pension refers to the system in which social insurance agencies collect, manage, regulate and use pension insurance fund in a centralized manner in a certain administrative region. Generally speaking, pension insurance system follows the "law of large number". With the support of large population, pension insurance and security fund can be used in a larger scope. It enhances the pension insurance's mutual aid function, and unified payment base, proportion of payment and treatment standard can be achieved, thus eliminating the barrier of cross-regional flow of labor and people. It improves the overall planning level and reduces the management link and management level, thus reducing the cost of management.

Due to the unbalanced development of different regions and restriction of administrative management system and government management level, in a very long historical period, mutual assistance and aid of pension insurance were always realized within limits (mainly taking administrative area as unit and adapting to administrative management system) rather than be realized in the whole society (regions with funds of mutual aid within the boundary are called overall planning regions). Since the reform, China holds a clear attitude toward the improvement of the overall level of pension insurance. The "12th Five-Year Plan" and the *Social Insurance Law* explicitly require the achievement of basic pension insurance in the whole country.[27] However, besides the fact that Beijing, Tianjin and Shanghai

[26]Yang [8].

[27]In 2007, *Notice on Related Problems in Provincial-level Overall Planning for Promoting Basic Pension Insurance for Enterprise Workers* issued by the Ministry of Labor and Social Security requires that provincial-level overall planning to unify the basic pension insurance system, the basic pension insurance payment, the basic pension insurance treatment, the basic pension insurance fund use, the basic provincial-level fund budget and the basic pension insurance services regulations.

implement the unified collection and allocation of funds by the state in the whole city at present, only a dozen provinces and municipalities carry out the provincial-level adjustment fund system while other provinces and municipalities carry out the prefecture-level overall planning. Nationally, there are at least 600 prefecture-level overall planning regions. More than 90% of them take the county and municipal level overall planning as the basis. Different payment rates, treatment standards and specific management modes exist in all overall planning regions. Overall planning regions of the current system are excessively scattered. As a result, they fail to develop the most important social and mutual aid functions effectively. Elderly care burden problems with uneven joy and sorrow become quite prominent among regions and the mutual aid function of social insurance is weakened seriously.[28] Urban and rural division, occupation division and regional division of social security system not only stop labor mobility, but also cause "cream skimming" phenomenon and increase management cost.[29]

Division management of pension insurance causes different regions to carry out their own pension insurance policies and break the unity and seriousness of pension insurance system. Due to the rigidity of pension insurance, the scattered pension insurance funds will easily become vested interests and won't be utilized once it forms in the lower overall planning level. Ratcheting effect will also occur as a local government has the autonomous right. In other words, once the local government gets some power, the central government will have difficulty in taking them back totally, which makes the unit making requests for many times to get more privileges. The local government asks the central government for financial support as the main channel to solve the problem of fund shortage there. In this case, partial pensions are greatly surplus and the central government gives some regions the massive transfer payment simultaneously. The division management of pension insurance stops the unified completeness of labor market and contorts the relationship between the central and local governments. Though the Ministry of Human Resources and Social Security issued the transfer and renewing measures for the basic pension insurance relation of urban workers in recent years and solved transfer and renewing problems of pension insurance relation in the process of labor flow to a certain degree, but unnecessary management cost of social insurance agencies was increased. With the further acceleration of labor flow and increasing development of urbanization, this management mode of one-to-one transfer in agencies of different regions can't remove the drawbacks of division management fundamentally.

[28]Wang [9].

[29]Zhu [10].

6.3.5 Management System and Operation Mechanism of Pension Insurance Funds Haven't Been Finalized and Serious Dissension Still Exists in Some Major Issues

(1) In the course of fund collection. China's social pension insurance fee collection is first taken charge by social insurance agencies completely. In the late 1990s, due to the economic system transformation, structural adjustment and the influence of financial crisis in Asia, some difficulties occurred in social insurance fee collection.[30] To boost fund collection and payment, some places allowed taxation departments to collect social insurance fees. In this case, some different places formed two modes that social insurance fees were collected by taxation departments and social insurance departments. In 1999, Provisional Regulations on Collection and Payment of Social Insurance Fees put the institutions to collect social insurance fees under provincial and municipal people's governments. Before then, around social insurance fee collection system, there was some violent conflicts and dissension in theoretical cycle and practical work departments. So far, there is no any unified understanding. That two departments are allowed to collect social insurance fees simultaneously in a country is a rare precedent internationally and also difficult to adapt to the intrinsic requirement of the finalization, stability and sustainable development of social insurance system.

---

[30]In the late 1990s, the situation that enterprises didn't pay social insurance fees was quite serious. At that time, many departments took action to enhance the collection and payment of social insurance fees. As stipulated, enterprises with the arrears of over one million yuan should be monitored by provincial-level labor security departments directly; for enterprises with the arrears of over 10 million yuan, the provincial-level ministry of labor and social security should report the situation to the Ministry of Labor and Social Security for the record and the Ministry of Finance, Economy and Trade Department and other relevant department. After a survey, if an enterprise was verified to have the ability to pay the fees but didn't fulfill obligations, it should be punished according to stipulations of *Provisional Regulations on Collection and Payment*. Moreover, the following administrative measures could be taken: the labor security administrative department should make it public via news media; Securities Regulatory Institution wouldn't approve the listing of the enterprise; the administrative department of enterprise leaders would involve the payment conditions of state-owned enterprise in the important content of enterprise leaders' performance evaluations. Those legal representatives and chief accountants who didn't pay the fees according to law couldn't be promoted, appraised and elected to be progressive personnel or get any year-end bonus. Moreover, they should get a warning in the arrear year. If they still refused to pay the fees, they would be dismissed from their posts and get punished; furthermore, Industrial and Commercial Administrative Department wouldn't examine and approve their sole proprietorships, let them found a new enterprise by sole proprietorship or joint stock, or examine and approve their branches or extend business scope. See this in *Notice on Relevant Problems in Cleaning and Collecting the Arrears of Social Insurance Fees Issued by the Ministry of Finance of the State Economic and Trade Commission in the Ministry of Labor and Social Security* (Lao She Bu Fa [1999] No. 36).

(2) In the course of fund daily management. Nowadays, some major issues such as pension insurance agencies are administrated from level to level or vertically, institutions are set by social insurance integration or insurance types respectively, and personnel are managed as civil servants or public service workers, are still short of scientific analyses and arguments. Systems in each place are completely different as well as the understanding of both internal and external system. Besides the *Labor Law* of 1994, there are no other laws and regulations specifically stipulating the functions and institution setting of social insurance and fund agencies. In practice, the nature, names, levels, functions, financial resources, payment, worker status, processes, internal control rules and other aspects of social insurance agencies across the country lack uniformity, authority and independence. Moreover, they change frequently.[31] Seen from the line of business of five social insurance types, they can be handled simultaneously; some insurance types are individually handled by agencies; some special accounts for funds of different insurance types are set in different commercial banks; some social security institutions and other social service institutions share responsibility for a part of business respectively. After China adopted the pension system mode of partial accumulation integrating social overall planning (pay-as-you-go) with individual account (funded accumulation), since some senior workers didn't have individual accounts to accumulate funds, the social overall planning funds in the new system are insufficient to pay the pension promised under the traditional system. Under the circumstance without any clear and definite channels for compensation, funds of individual account and social overall planning should be managed together and borrow the funds of employed workers' individual accounts to meet the need of current payment.

(3) In the course of fund investment. Currently, pension insurance funds of China's accumulation system include the national social security funds and social pension insurance accumulation funds. As pension reserve fund, national social security funds are invested directly or in an entrusted way by the National Council for Social Security Fund. So far, the investment goes well. Especially, it underwent the impact of the international financial crisis in 2008. However, there are still some problems, such as the law environment needs to be improved, the found resources are uncertain and increasingly unstable, and the investment channels are too narrow. There is still a very long way to go to establish a sound governance mechanism. Due to the very low level of overall planning, social insurance agencies established by governments at all levels take charge of the investment management of accumulation funds of social pension insurance. These departments lack talents and knowledge about financial investment and are short of intrinsic motivation to maintain and

[31]Take Guangdong Social Insurance Agency as an example. It had five names successively in last over 20 years, namely "Social Labor Insurance Company", "Social Insurance Business Office", "Social Insurance Management Office", "Social Insurance Fund Management Office" and "Social Insurance Fund Management Center".

increase the value and the ability to participate in the market competition because of the operation based on administrative management mode. Moreover, lacking market-oriented investment channels, they can only perform the investment management function by purchasing and holding national debts and deposit in bank. In this case, the investment management of funds in individual accounts works inefficiently and the goal of funded accumulation system can't be realized. Therefore, individual account system seems to be nominal.[32] After the case of Shanghai social insurance, the investment restriction of funds in individual accounts is increasingly tightened. Although the risk of administrative embezzlement reduces, the risk of depreciation increases. In December 2006, central financial aid funds in basic pension insurance individual accounts of enterprise workers of nine experimental provinces were entrusted to the National Council for Social Security Fund for investment and operation. The National Council for Social Security Fund promised a quite favorable yield rate and doesn't charge any management and operation fees. Relevant fees should be verified and appropriated by the central budget. However, the investment operation of the real funds and new funds in individual accounts in local places are quite conservative, so they can only be saved in banks or used to purchase national debts.

(4) In the course of fund supervision. At present, China's pension insurance funds are mainly supervised by competent departments and cooperatively supervised by the Ministry of Finance, Audit Committee and relevant departments, and peripheral inner-party supervision, discipline inspection and supervision, supervision of public opinions and so on.[33] As a whole, it is a kind of quite scattered supervision system and has some drawbacks: ① Without clear responsibility. Many supervisors supervise an object simultaneously, which causes "irresponsibility of all supervisors" and makes it hard to investigate responsibility. ② Repeated labor. It makes supervision behavior of different subjects similar. ③ Lack of supervision efficiency and authority. ④ Without good coordination. To boost fund supervision and solve supervision coordination problems, most provinces and cities founded social security supervision committees. However, most of them are inattentive coordinating institutions and fail to give full play to comprehensive coordination.[34]

---

[32]Wang [9].

[33]*Notice on Strengthening the Supervision and Management Work of Social Security Fund* issued in 2002 (Lao She Bu Fa [2002] No. 12) gives labor and security departments, finance departments, auditing departments, taxation departments and postal departments at all levels and the branches and sub-branches of People's Bank of China great powers of supervision and investigation.

[34]For example, in 2005, Kunming carried out a pilot inspection and research into the bank deposit interest of medical insurance fund on behalf of the Supervision Committee. As the inspection involved finance departments and banking departments and other departments, it is rather difficult to carry it out.

6.3.6 Management System of Elderly Care Service Industry Is Imperfect and Management Modes of Social Elderly Care Service Institutions Are Still so Extensive

The direct management modes of state-run elderly care institutions formed under competent departments for a long time can hardly adapt to the supervision and management of elderly care service industry. Scientific and reasonable management system of elderly care services hasn't been established yet.

(1) A lot of factors affect the development of private elderly care service institutions, resulting in insufficient opening up of industry,[35] the low level of social involvement and the inflexible operating mechanism.[36] China has issued documents, making provisions about the tax, land, credit, water, electricity and others of private elderly service institutions in principle, but it is still difficult to implement national policies, because departments involved and local authorities did not develop operational measures and approaches. Meanwhile, some public elderly care institutions, which enjoy state subsidies and are designed mainly for elderly citizens who have low income and are hardly accepted by the market, directly participate in market competition and don't figure out whether the elderly citizens who move in are poor or not, hindering the development of private elderly care service institutions. Competent departments offer insufficient industry guidance and management, so private elderly care service institutions are basically in a state of laissez faire. They lack industry guidance, supervision and self-regulation. Industry organizations copy the pattern of government departments, and it is hard to communicate with peer organizations and report the appeal to the leaders. As China has not promulgated industry service standards, the regulation of the service and the handling of the dispute are not standardized.
(2) Competent departments' main functions of setting planning, examination and approval, providing annual inspection, assessing institutions, providing industry development service are relatively weak. Perfect industry management system has not been established. Some problems exist. For instance, the elderly care service system is not perfect; there is a lack of construction standards, service specifications, industry access, self-discipline in institutions and market supervision. China's elderly care service market is not fully developed. In particular, the construction of home-based elderly care and community-based elderly care system is still in the initial stage. In recent years, social elderly care services also gave birth to mutual-support elderly care, "housing reverse mortgage loans", tourism elderly care, "migratory bird" elderly care (an elderly care way of which elderly citizens migrate from one place to another according

[35]The most prominent symbol is that, compared with the huge demand for elderly care services, the current supply capacity is seriously insufficient.
[36]Hu et al. [11].

to the change of seasons), remote elderly care and other different elderly care ways.[37] It is expected that there will be more and newer business models in the field of social care services as elderly citizens' consumption structure upgrades. However, the government's ability to manage the industry is very weak, and the access, supervision, service specifications and standards of industry management are not perfect.

(3) Due to the lack of top-level design and overall planning, China has not classified the elderly care institutions according to the different levels of care. As the development of China's elderly care service industry is still in the initial stage, most elderly care institutions still belong to mixed type elderly care institutions. They are not so professional and the adopted elderly citizens vary from those who can take care of themselves to those who have to stay in bed for a long time. They even include elderly citizens with mental illness or mental deficiency. The management and service in elderly care institutions are often in disorder. Elderly care institutions are classified according to the administrative level, the size of facilities and the forms of ownership rather than the functions. Such classification is not scientific and reasonable. According to the function of elderly care institutions in China, the care degrees of elderly citizens in elderly care hospitals (also known as elderly nursing hospitals) governed by the health department and apartments for elderly citizens governed by departments of civil affairs are obviously different. The common social welfare homes and elderly care homes have no functional positioning, and the elderly citizens they adopt include those who can take care of themselves, those who stay in bed for a long time and even those who need palliative care. In these elderly care institutions, nursing is classified into a few grades, namely special nursing, grade one nursing, grade two nursing and grade three nursing according to the nursing degrees elderly citizens need, so as to implement management based on departments or districts. No special elderly care institution adopts elderly citizens who need specialized nursing and grade one nursing.[38] It is hard for this unclassified and extensive management mode to meet people's multi-level and diversified elderly care service needs.

(4) The elderly care service resources are divided by urban and rural areas, departments and units, so they haven't been effectively integrated. Due to the segmentation of the government functional departments, it is hard to share resources in departments of civil affairs, health departments and other departments. Even the development in regions and urban and rural areas is not balanced. Due to the lack of scientific planning, in the same area, people wait in line in some elderly care institutions with few beds left while few persons go to some elderly care institutions with vacant beds. The newly built elderly care institutions are located in the outer suburbs where the traffic is inconvenient, so the bed vacancy rate is high. Too many high-grade elderly care institutions are

---

[37] Research Group of the Capital Institute of Social Economical Development [12].

[38] Gui [13].

built, and elderly citizens cannot afford them, so there is an oversupply of such institutions. Some of the elderly care institutions are distributed too densely and unreasonably, which also causes a vicious competition. Some elderly care institutions pursue "hotel" design and service, so there is a lack of family atmosphere. Some people even confuse elderly care institutions with residence for elderly citizens, which is taken advantage of by illegal real estate businessmen to cause chaos in the market.

6.3.7 The Social Management and Service System Is Relatively Weak, and Social Management Has not yet Been Applied to Some Retired People

Geared to the establishment of social pension insurance system, China has made significant progress in the social distribution of pension and social management services for enterprise retirees since the reform. Social insurance agencies have established a social security system and service network which is mainly relied on agencies at all levels, supported by banks and all kinds of fixed service organizations and based on community labor and security platforms.[39] However, the social management and service system is relatively weak, and social management has not yet been applied to some retired people.

(1) The management and service network is still relatively weak. After the implementation of the social pension distribution, the employers no longer take on the daily work of managing social insurance objects. Enterprise retirees receive social management services from the streets and communities. However, there is no special department for elderly citizens in the streets and communities, so elderly citizens are basically unmanaged.
(2) Much attention is paid to the management but less to services. The handling modes and processes of the pension insurance in each place are diverse. Pension insurance agencies design process primarily according to administrators' business practices. It makes it easy for administrators to manage it, but it doesn't reflect characteristics of serving the insured. For example, a number of agencies are set up and the handling window only provides single function, which makes it easy for administrators in the agency management, but it reduces the quality and efficiency of handling services. To handle a business, the insured unit and the individual often have to go to more than one department and wait before a number of windows and ask for many approvals.

[39]Meng Zhaoxi: *China's Social Insurance Handling Management and Service Industry Develops Rapidly since the Reform and Opening up in the Past* 30 *Years*, China *Labor and Social Security News*.

(3) China's social security information construction is still in its initial stage. Quality management and standardization construction of data are backward, and the basic data collation and statistical analysis are rather extensive. Traditional manual operation and the low-level stand-alone management make it difficult to meet the needs of hundreds of millions of business handling.

(4) The traditional pension insurance agency mode with stable employment group as the object, supported by employers and run by the local authorities obviously cannot adapt itself to the challenges of the strong population mobility and the diverse employment modes. As the mobility of people in urban and rural areas, regions and industries is increasing, a number of problems like remote elderly care, transfer and continuation of pension insurance appear. Lifetime accurate record, settlement and payment of hundreds of millions of the insured must be guaranteed, and the management and service capacity is obviously insufficient.[40]

(5) From 1978, government departments and public institutions and enterprises began to perform different retirement systems. Now the retirement salary of senior technical personnel in the enterprises and the similar personnel in public institutions is different greatly, which results in social injustice, affects the rational flow of labor and restricts the self-reform in government departments and public institutions. In the reform, the relevant departments had been aware of this problem, so they carried out the reform and deployment a long time ago. However, due to various factors, the basic pension insurance system for employees in government departments and public institutions has not been changed. As a matter of fact, at present, the most prominent problem is not the dual-track system for employees in government departments and public institutions and enterprises. Instead, the problem is, there has not been a unified and socialized pension fund raising system and management and a service system for retirees in the internal pension insurance system of government departments and public institutions. Take the Central Committee as an example. Now each department, committee, office and bureau separately sets up retired management agencies to give retired cadres' political treatment, manage retired cadres' ideological and political work, offer retired cadres material amenities, help retired cadres with special difficulties and carry out elderly education and recreational and sports activities.[41]

---

[40]He [14].

[41]*On Further Strengthening the Work about Retired Cadres in the New Situation* (Zhong Zu Fa [2008] No. 10).

6.4 Reform Ideas and Policy Recommendations of China's Pension Security Management System

China is experiencing the largest and fastest aging process in the history of the world. To achieve the goal of providing senior citizens with a sense of security, we must speed up the establishment of a sound pension security system. A scientific and reasonable management system should be established to perfect pension security system, which needs to be carefully designed and scientifically planned. Under the condition of socialist market economy, how to manage and organize pension security affairs is a new subject. As the government lacks experience in social management, it must fully absorb and learn from the successful practices and experience in the world. But it should also be noted that government agencies in developed countries are established based on the mature market economy. It is in the context of power separation and supervision and checks and balances, which is greatly different from the situation in China. The names of the pension security agencies or organizations are almost the same, but they may function very differently in different countries, which makes it difficult to compare the management systems in different countries. Due to the unique political system in China, we cannot completely copy the model of other countries, so we must establish a pension security management system that is suitable for the existing political system. This means that the establishment of a scientific and reasonable pension security management system cannot be achieved overnight, and the reform will not be finished at one go.

6.4.1 Actively and Steadily Push Forward the Legislation of Pension Security

With the gradual formation of social pension insurance system and the rapid development of social elderly care service system, the goal principle, system structure and policy measures of pension security in China have gradually been clear, and time is ripe for accelerating the construction of the legal system of pension security. As the public opinion organ, the state legislature shall be responsible for the final decision of pension security system, so it can gather public opinion the most widely, reflect the ruling party's pension security policy and proposition, ensure broad participation of all social strata, establish authority, stability and reliability of the system and serve as a prerequisite of supervising the operation of pension security system as a judicial authority.[42] Due to the strong rigidity of law, we must adhere to positive and steady principles when promoting the construction of the legal system.

---

[42]Zheng [15], pp. 68–76.

(1) The basic objective is to establish a pension security laws and regulations system with a clear goal, a complete system and wide coverage. There are three different ideas about pension security legislation.① According to the principle of high intensiveness, formulate a highly comprehensive pension security law to regulate all kinds of major affairs of pension security. ② Formulate a basic law for elderly citizens. Under the basic law, formulate special laws, including *Social Insurance Law of the PRC* to regulate pension security affairs. ③ According to the needs of sub projects of elderly pension security, formulate a lot of equal, parallel and cooperative laws respectively to standardize a certain kind of pension security affairs. At present, China has promulgated laws and regulations such as *Adoption Law*, *Law of the PRC on Protection of the Elderly Rights and Interests*, *Law of the PRC on Protection of the Rights and Interests of Consumers*, *Basic Norm for Social Welfare Institutions for Elderly Citizens* and *Interim Measures for the Management of Social Welfare Institutions*. Every place has promulgated and implemented similar local regulations and rules. In particular, *Social Insurance Law of the PRC* has been promulgated, taking a key step in the legalization of pension security. It becomes the most important law to guide the construction of pension security system in China. What kind of legislative model should be taken in the next step is worth studying. From the historical perspective, in the early stage of social security system construction, legislation is generally in a decentralized form, but the excessively dispersed legal system construction is not conducive to the normal operation of the system.

In the long term, "the decentralized social security laws should be integrated and brought together as soon as possible, and the law should be drafted in the most clear and reasonable language."[43] It is an inevitable trend that the decentralized legislation will be further developed to the modestly intensive legislation and then intensive legislation. In 1994, China adopted integrated legislation in the legislative plan of *Social Insurance Law*. That is, formulate a comprehensive and unified social insurance law model. *Social Insurance Law of the PRC* covers all kinds of social insurance systems, including pension insurance and satisfies the principle of intensive legislation. Given the fact that pension security covers a wide range, it is difficult to manage it. Different items are strongly independent from each other, so it is not necessary and practical to regulate all of pension security affairs with one law. At this stage, it is acceptable that each law regulates a sub-system of pension security of the same nature or of the similar nature.[44] What comes next is to address the problem that there is a lack of laws in the field of social elderly care services and elderly spiritual security. Formulate laws of the most urgent and ripe project. The top priority is to issue Management Regulations *for Elderly Care Institutions* and revise *Law of the PRC on Protection of the Elderly Rights and Interests* as soon as

[43] International Labor Organization [16].

[44] Zheng [17].

possible. For mature pension security projects, speed up the construction of the legal system to ensure that competent administrative departments perform their duties in accordance with the law.

(2) Actively and steadily realize the transformation from administrative legislation to legislature legislation. The top-down reform of pension security system is an internationally common practice. That is, the state first develops programs and legislates for them, and then implements them. However, the national conditions of China is that there is a huge gap between urban and rural areas as well as regions with profound planned economy, a large population and a low level of economic development. To establish a standard and universal pension security law is indeed a test for legislators. For example, after the draft of *Social Insurance Law of the PRC* is open to the public for opinions, all sectors in the society have a fierce confrontation and disagreement on all aspects of this law. In this case, if we hastily legislate before the conditions are ripe, and copy the foreign legal provisions, it will also depart from China's national conditions, and even pose obstacles on the reform and the establishment of a new system. Given the current situation of local partition of China's elderly care security system and the historical tradition of administrative leadership, it is impossible to realize the transition from the legislation by administrative organs at all levels to the legislation by state. The basic idea is to gradually transfer the decision-making power of local governments at all levels to the central government, and then gradually transfer the decision-making power to the legislature. The decision-making process of the legislature shall be carried out under the supervision of the public, and the legislature shall fully solicit opinions from all walks of life in the form of public hearings and other forms of democracy before making decisions. It must be pointed out that, letting the legislature perform the right of final decision does not mean that the government doesn't take charge of pension security system and assume direct financial responsibility. The government will still be the main provider and promoter of the pension security act, but the establishment and revision of systems must be passed by the legislative process before taking effect.[45]

(2) Strengthen the role of the legislature in supervising the judiciary authorities. Let the People's Congress and its Standing Committee assume the responsibility of deciding pension security system and supervising the budget, and judiciary authorities assume the duties of judicial supervision. Establish a system in which the people's congresses and their standing committees at all levels supervise and evaluate pension security system. Pension security is closely

[45]According to a comparative study of social security institutions in 50 countries, the regulations of social security are generally proposed by the administrative departments in charge, passed by the legislature after consultation procedures, and implemented by the administrative departments. Enterprises, trade unions and other social service organizations primarily promote, consult and supervise them. See *The Enlightenment of International Social Security System for Us* by Zhou Hong, *Guangming Daily*, November 17, 2008.

related to the vital interests of the members of the society. Whether the dispute caused by pension security can be properly resolved is related to the normal operation of pension security system. The administrative procedures in pension security system are not enough to solve the pension security disputes, so judicial authorities need to step in. In developed countries, judicial authorities' supervision of the social security is based on laws and regulations, and they are responsible for the corresponding criminal penalty and civil punishment for the crime that appears in the operation process of pension security, playing a deterrent role. The International Labor Organization points out in *Looking ahead to the 21st Century: the Development of Social Security* that "people shall have a clear right to complain about the way they are treated by any social security institution. If the complaint within the scope of the social security institution itself is not satisfied, it shall be submitted to an independent competent authority."[46] Some of the countries in the world set up a special social insurance judicial organ to try social insurance dispute cases through the trial in the special court. The social security court has been set up in the people's court in some places of China. In 1994, China's *Labor Law* stipulated that the social insurance dispute was part of labor disputes. Social insurance disputes shall be solved according to the relevant provisions in *Regulations on the Settlement of Labor Disputes*. In 2001, the Ministry of Labor and Social Security promulgated *Regulations on the Social Insurance Administrative Disputes*, breaking the restriction that social insurance disputes belonged to labor disputes, expanding social insurance parties' channels of right relief. *Social Insurance Law of the PRC* further stipulates the resolutions on social insurance disputes comprehensively. It is advised to further deepen the reform of the judicial system and the working mechanism, actively cultivate and train legal workers of social security at all levels. The people's courts at all levels shall have judges who are familiar with social security affairs, and actively explore and set up special court of social security, so that judicial authorities at all levels can be really responsible for supervising pension security.

6.4.2 *Promote the Government's Functions of Supervision and Management and Provide the Transformation of Basic Pension Security*

Pension security management system is the important content of the administrative management system. In a sense, various present problems in pension security management system are the specific reflections of the defects of administrative

[46]International Labor Organization [16].

management system in the field of pension security. Just like the reform of the administrative system, the key and core of the reform of the pension security management system is the transformation of government functions.

(1) Separate administrative management from pension insurance fund management to practically strengthen the fund supervision function of the government. Due to the difference in factors like the function area of administrative power, the effect degree of social power, the growth of social integrity and the allocation of social resources, conditions of pension insurance fund management are different in various countries. According to the power types of management governments, from the administrative power and industrial self-discipline and management degree, it's roughly divided into government management mode, private marketization management mode, social autonomy management mode and the mixed system combining government guidance with social autonomy. All these management modes have their own advantages and disadvantages. China has early established the principle of separating the social security administrative management from fund operation. Although it has not been carried out completely, the principle that the government shouldn't directly manage the social security fund has been recognized publicly.[47] The social security agencies monopolized by the government have been increasingly challenged by private organizations, so a lot of social security agencies start to try to put a part of business that private organizations can provide more effective services out to contract. Since the 1990s, some places in China have started to entrust social security business to professional commercial insurance companies, thus accumulating rich experience. However, different from the commercial insurance, the basic pension insurance has higher requirements for safety. It is doubtful whether the private marketization management mode can be popularized widely. Actually, few people hold such a view. In view of the disadvantage that the government intervenes in the fund management too much, Chinese theoretical circles strongly require the autonomy management of social

[47]In terms of the definition of the social security agency, according to the survey of the research team (Yang Yansui, Chen Yangdong, Bao Danru, etc.) of *Study on Capacity Construction of Social Insurance Agencies*, the results of the questionnaire by the staff of social insurance institutions in 2004 showed: 71% of people chose it as "one of government departments", 18% "a public institution directly under the government", and 10.5% "a government statutory institution (or a monopolistic commonweal organization)". The results of the questionnaire by the staff who weren't from social insurance institutions in 2007 showed: 86% thought social insurance institutions badly needed to be improved as the government executive department. The results of the questionnaire by officials above the division level in 2008 showed: 44% thought the social insurance institution should be the government executive department. See also *Summary of the Interim Report about the Study on Capacity Construction of Social Insurance Agencies* (Exposure Draft) on the website of School of Public Policy and Management of Tsinghua University by the research team of *Study on Capacity Construction of Social Insurance Agencies*.

security fund in recent years.[48] Social autonomy management means that non-governmental organizations are in charge of the fund management business, while the government is only in charge of lawmaking and providing guidance and supervision for the policy implementation. An independent institution consisting of representatives from the government, employees and employers is in charge of the fund management, and can be equivalent to a foundation, a board of directors, a committee and a council in terms of the nature.[49] Undoubtedly, the social autonomy management indeed has some unique advantages. However, according to international experience, services for the independent social security fund management will depend on the sound legal environment, mature society and strict government supervision.

Even Germany with the most mature social autonomy management has strengthened the macro-management of social insurance autonomous institutions in recent years in order to remove some disadvantages of autonomy management. However, the practice of some transitional countries shows that under the circumstances of the immature market, imperfect supervision systems, and a lack of the accountability system and transparency, the bigger gap between public service institutions and the government will probably lead to a series of serious consequences.[50] Considering the autonomy management can't represent the international trend,[51] and has higher requirements for the governance mechanism, at least it's improper to adopt the method of autonomy management in China in a short term. It's best for China to follow the path dependence, and it doesn't need to introduce it for attraction.[52] We advocate that the mixed system combining government guidance and social autonomy will be also adopted in the future fund management. The

---

[48]Germany is a typical country with social autonomy management. The pension insurance system of autonomy management and negotiation under the government supervision is adopted in Germany. Instead of being attached to any governmental departments, pension insurance organizations refer to the social autonomous self-management institutions autonomously managed by the Representative Committee consisting of representatives of staff and employers. These autonomous institutions include the German Pension Insurance Union, the German Pension Insurance Union of Miners, Railroad Workers and Seamen and so on. Pension insurance agencies achieve self-management through respective standards and norms formulated by industrial associations. And the pension insurance comes into operation based on the agreement reached by both autonomous parties through negotiation. The division and cooperation of the government and numerous autonomous organizations are as follows: Firstly, it avoids the problems that the excessive regulation of the government leads to low efficiency. Secondly, it brings the business superiority of professional organizations into full play to reduce the asymmetric information. Thirdly, the interest relations in all respects should be comprehensively coordinated through the independent negotiations between autonomous organizations.

[49]Sun [18].

[50]World Bank [19].

[51]Jia [20].

[52]Zheng [21].

fund management institutions should be independent of administrative management institutions and also be operated in accordance with the forms of standard public institutions.[53] Public service units should be separated from the government in the basic pension insurance. As the relatively independent institution is under the leadership of the administrative departments and keeps a certain distance from competent departments, the fund management institution carries out relevant management activities on behalf of the interests of the insured. And those fund management institutions operating in accordance with the forms of public institutions should strengthen the fund supervision function of the government. The government supervision is the core of the pension insurance fund, the internal need of the government as the leader and public manager of the pension insurance system and also the internal need of the pension insurance system as the public goods. In terms of the goal of the pension insurance system, the government guarantees that the pension insurance system should normally operate on the path of the legal system and norms, including the daily and financial supervision of the pension insurance implementation institutions. The former is to guarantee the normative operation of the pension insurance system and the effectiveness of delivering and implementing policies and to correct the phenomena of failing to meet the standard, while the latter is to guarantee that the collection, management, payment and operation of the pension insurance fund should conform to the legal system and norms.[54]

(2) Separate the management function of the elderly service industry from the function of organizers to promote the transformation of government functions to the mode of providing basic public elderly care services. Under the condition of socialist market economy, it's impossible for the government to meet the elderly care service needs of all elderly citizens. However, some elderly citizens with relatively low income in the society always get the basic public elderly care services with the support of the government, which is the foothold for the government to establish public elderly care institutions. And the public elderly care institutions should insist on meeting people's basic living needs, basically provide free services or low-cost and cheap services, mainly guarantee the basic

[53]The research team led by Professor Yang Yansui thought that the social insurance institution, as the executive institution of the service-type government, should take responsibility for the insured of social insurance and also be entitled to be entrusted with national public welfare projects (social insurance) management, subcontract and entrust projects, purchase services and implement supervision on behalf of the government. That is, it also serves as the trustee of public benefit. See also *Summary of the Interim Report about the Study on Capacity Construction of Social Insurance Agencies* (Exposure Draft) on the website of School of Public Policy and Management of Tsinghua University by the research team of *Study on Capacity Construction of Social Insurance Agencies*.
[54]Zheng [16], pp. 80–81.

elderly care service needs of "elderly citizens without ability to work, source of income, supporter or caregiver, households enjoying the five guarantees" in rural areas, "those without the ability to work, a source of income or a legal supporter and caregiver" in cities, and special urban and rural disabled elderly citizens who have low income, suffer from dementia and so on, and try to realize the equalization of basic public elderly care services in urban and rural areas.[55]

The current public elderly care institutions with nonstandard management and poor services should be entrusted to social forces for management and operation. Public elderly care institutions should be specified to play a fundamental, demonstrative and radiation-driven role among various elderly care institutions with the ownership nature and the mechanism that the government purchases elderly care services should be actively explored at the same time. The management mode of the existing public elderly care institutions should be innovated to reform the internal management and operation systems, change the previous mode of directly establishing elderly care service institutions, actively explore various realization forms of separating public service units from the government, the government administration from commune management, the management from enforcement and the state-run from the private-owned,[56] confirm the independent corporate position of the public elderly care institutions and establish the scientific and reasonable management mechanism of public elderly care institutions. Competent government departments should practically perform supervision and management functions in the elderly care service industry like development planning, qualification admittance and service supervision, mainly focus on formulating relevant policies and development plans, and switch to the policy and regulation management from the management of specific affairs, the industry management from the administrative subordinating management, and the indirect management from the direct management. Competent government departments should also scientifically and rationally allocate the elderly care service resources and implement localized management in the whole industry.

[55]Public financial security has been implemented in supporting "elderly citizens without ability to work, source of income, supporter or caregiver, households enjoying the five guarantees" in rural areas. The equalization of basic public elderly care services in urban and rural areas has made solid advances. However, rural elderly care homes are still defined as rural collective welfare institutions. As a result, the management system of elderly care homes needs to be straightened to sort out the relationships of "being sponsored by towns and villages", "being supervised by the civil affairs departments" and "the assistance of the departments".

[56]Exploration in this aspect has been carried out well in Beijing and Shanghai, and the basic thought is that the competent departments retreat from the micro-management of elderly care institutions, strengthen the functions of macro-management and industry supervision, organize social welfare centers to uniformly manage various social welfare institutions, improve the industry associations and strengthen the industry guidance and self-regulation.

All elderly care institutions, regardless of their ownership, investors, subordination relationship and operational nature,[57] should be uniformly planned, admitted and supervised by the local civil affairs departments.

(3) Handle the relationship between pension security and public finance well. The pension security system in China has a very complicated relationship with public finance. During the planned economy period, the pension security fund is mixed in the financial system of unified collection and allocation of funds by the state, while the Ministry of Finance provides support according to the investing modes of public institutions. After the reform and opening up, the socialized pension insurance financing mechanism and ways to provide market-oriented elderly care services have been established, but the Ministry of Finance still bears very important responsibility for investment (including paying the reform cost, making up for the pension financing gap, paying the pension of government departments and public institutions, guaranteeing the development of the elderly welfare cause, providing reduction and exemption support of taxes for elderly care institutions) to show the public nature of the government finance. Meanwhile, the Ministry of Finance comprehensively intervenes in the supervision of pension security fund. Since the basic pension insurance fund was integrated into the financial special account in 1997 to implement "separate management between revenue and expenditure", the Ministry of Finance started to perform relevant financial supervision functions. In 1998, China officially proposed establishing the public finance framework and required the Ministry of Finance to comprehensively manage extra-budgetary funds including the social security fund. The social insurance fund has been put into trial use since 2010. Overall, the public finance has obviously strengthened the grass-roots miscellaneous function and macro-control role of the social pension insurance fund and enhanced the investment security and police guidance functions of social elderly care services. The problem is that after the social pension insurance fund and social elderly care service system that were outside of the public finance system for a long time have been gradually included in the scope and management of the public finance system, how to draw a clear distinction between the two conditions, how to ensure the responsibility of the public finance in the pension security system, and especially how to establish scientific and stable fiscal pension security investment system, have become problems that call for immediate solutions. We think, in terms of the relationship between pension

[57]According to different sponsors, China has three kinds of elderly care institutions: one is organized and invested by the government, registered in the organization management departments of institutions of the people's government above the county level and managed by the civil affairs departments with the pubic institution nature. One is organized by social forces and registered in the civil affairs departments with the private non-enterprise nature. The last one is organized by social forces and registered in the industrial and commercial departments with the enterprise nature. Elderly care institutions with the public institution nature and those with the private non-enterprise nature are generally called social welfare institutions of elderly citizens.

security and public finance, the following several principles should be adhered to: ① The national finance should provide necessary investment for pension security, including the management and operation fees for social insurance administrative agencies and public elderly care institutions, the social insurance payment responsibility for staff of government departments and public institutions and fiscal subsidies to pension insurance (mainly the compensation for the historical contribution of old staff),[58] provide financial subsidies for a certain project or special people of pension security (such as the subsidies of the financial departments at all levels to new rural social pension insurance, and the security for rural "elderly citizens without ability to work, source of income, supporter or caregiver, households enjoying the five guarantees" and urban ones "without the ability to work, a source of income or a legal supporter and caregiver"), and provide reduction and exemption support of taxes for various supplementary security projects and social elderly care service institutions. ② The national finance should perform the supervision function comprehensively, gradually bring various pension security funds into the financial revenue system, establish the social security budget system and supervise the financial and accounting activities of various pension security institutions. ③ The pension insurance fund should keep a distance from the national finance to realize self-development and avoid the pension insurance system's great impact on the stability of the national finance when a major financial crisis appears in the system.[59]

6.4.3 *Establish Moderately Centralized Pension Security Supervision and Management System and Coordination Mechanism with the Integration of Power and Responsibility*

After the government functions are gradually transformed into supervising, managing and providing basic pension security, the current government structure must be optimized to form the scientific and reasonable government management system of pension security. From the transverse aspect, according to reform requirements of the national administrative management system, "the large department system" should be promoted to make competent departments of pension security perform complete and powerful industry supervision and management functions and establish enough authority. Meanwhile, the accountability mechanism should be established according to the requirements of the integration of power and responsibility.

---

[58]Zheng [7], p. 42.

[59]Zheng [16], p. 39.

(1) Establish the administration and supervision system according to the principles of centralized supervision and administrative accountability to form the pension insurance supervision system that is led by the supervision of competent departments, assisted by the supervision of special institutions and supplemented by social supervision. The social insurance fund is a kind of special public fund. There is a principle-agent relationship between the insured and management institutions of the pension insurance fund. In order to ensure the safe and effective use of the fund, the pension insurance supervision system must be established. The government supervision is the core of the pension insurance supervision system. Currently, the government supervision and management system focus on the supervision of the pension insurance competent departments, while the Department of Finance, the Audit Department and relevant departments assist in the supervision. As a whole, it is a relatively decentralized supervision system. Such a system has certain rationality because the operation of pension insurance is very complicated and involved in numerous links and a lot of profession business. A single institution is short of relatively comprehensive skills of supervision technology and business to difficultly supervise the pension insurance fund comprehensively. Therefore, numerous institutions should be bestowed with equal supervision responsibility and the business and professional advantages of various departments should be made the best of to improve the supervision efficiency. More importantly, under the circumstances that the functions of pension insurance competent departments have not been transformed completely, establishing mutual supervision and interaction mechanism of various departments and units can ensure the safe and efficient use of the fund.[60] Besides, under the current framework of the administrative management system, some departments naturally have professional supervision functions of the pension insurance system. Surely, such a decentralized supervision system will also inevitably have the disadvantages of uncertain responsibility and poor coordination. Under the circumstances that various departments implement supervision at the same time, their supervision actions of almost the same nature will cause unnecessary waste and overlapping duties, thus easily leading to the phenomena of shifting responsibility and scrambling for merits. The decentralized supervision and management system with many participants should be changed to moderately centralize the government supervision, which helps define the supervision responsibility that the fund supervision departments should take to form effective accountability.[61] Change the current situation that the Department of Finance, the Audit Department, post offices and banks have the right to supervise the pension insurance fund to relatively centralize the supervision function of the fund in

[60]The Editorial Committee of *Financial Supervision of Social Security Funds*: *Financial Supervision of Social Security Funds*, China Financial and Economic Publishing House, 2005, p. 68.

[61]Zheng [22].

the competent departments. Meanwhile, enough authority should be established to form the administration and supervision system that is led by the supervision of the competent departments and assisted by the financial and auditing supervision and establish the overall coordination mechanism between different administrative supervisors. It should be pointed out that the moderately centralized supervision and management don't mean to hand the supervision duties of financial and auditing institutions over to the competent departments of social insurance, but specify that the pension insurance competent departments should be responsible for pension insurance supervision to ensure their sufficient fees, independent staffing right and compulsory supervision right. Meanwhile, enough authority should be established to overall coordinate the forces of all aspects to comprehensively supervise and manage pension insurance. The advantage is that it helps to specify the supervision responsibility that the supervision departments should take to form the effective accountability. The departments of finance, audit and supervision still need to supervise relevant pension insurance affairs and parties involved within the scope of legal duties.

(2) Establish the centralized and unified management system of the elderly care service industry. On the whole, the construction of China's elderly care service system is in the early exploration stage. And the privately-run elderly care service institutions are especially in the initial development stage generally with a small scale. However, the construction of the elderly care service system generally lacks the strong support of policies, which is related to the currently decentralized government management system. Although the civil affairs departments are in charge of the construction of the social elderly care service system, the civil affairs departments play a limited role in promoting policies because the elderly care services are related to the departments of land and resources, financial planning, housing and urban construction, health education, culture and sports, public security and fire control. Even as the competent department of the industry, the civil affair department still has isolated, divided and incomplete management functions of the social elderly care service system with the objects of all elderly citizens and contents of life care, rehabilitation nursing, spiritual solace, emergency rescue and social participation. The civil affairs departments focus their management on the institution services, but lack the overall planning, supervision and management in the welfare services of community-based rehabilitation, culture, entertainment and housekeeping needed by elderly citizens. And some functions don't belong to specific competent departments. At present, the civil affairs departments have been in charge of the construction and management of elderly care institutions (Office of the National Working Commission on Aging is in charge of the construction and management of privately-run elderly care institutions in few places), while Office of the National Working Commission on Aging is in charge of the most work of home-based elderly care services. The competent departments in charge of the professional training of elderly care services are still

undetermined.[62] And now, the departments in charge of the construction of the elderly care service system haven't been clearly determined. The civil affairs departments, as well as the departments of aging, are in charge of it in various places. No department administrates some elderly care institutions without a business license. For the sake of promoting the sound development of the elderly care service industry, it is suggested to moderately centralize the functions which are dispersed in administrative departments such as organization and personnel departments, institution preparation departments, development and reform departments, finance departments, price departments and other departments about the management of elderly care service industry. Such functions should be exercised by the civil affairs department collectively, so that the pension insurance competent departments truly become the administrators of pension service industry management. Ensure that it has adequate financial support, independent personnel power and coercive supervision authority. Hence, civil affairs departments gradually manage elderly care service institutions et al. levels, home-based elderly care service and community-based elderly care service. Make it clear that civil affair departments and the departments of aging are in charge of vocational training of elderly care services (nursing). Give the civil affairs department the right to issue professional qualification certificate of elderly care service training and the right to examine and appraise vocational qualification certification (under the supervision and management of human security departments). Civil affairs departments and education departments cooperate to set up elderly care services (nursing) major in colleges and universities to cultivate high-level management and technical personnel experts in elderly care services.[63] A complete administrative body is the organic unity of function, power and responsibility. In modern government theory, more and more attention is paid to the principle of the integration of power and responsibility. That is, the allocation of power and responsibility should be the same, and the administrative body should exercise power according to law through accountability. Therefore, after competent departments centralize power, accountability mechanism for civil affairs departments should be established, which is the basic goal of the reform of government institutions in China.

(3) Adjust to elderly citizens' diversified and comprehensive needs, and build a high-level elderly security coordination system. Period of the "12th Five-Year Plan" is a vital period that China moves forward to the high-income phase from the upper and middle-income phase and achieves the goal of finishing building a moderately prosperous society in all aspects. In this development phase, elderly citizens' social needs gradually show features of diversification, multi

[62]Human security departments are only responsible for non-profit vocational training. Currently, industrial and commercial administrative departments only have the right to examine and review the enterprises' application for computer training. As for other training, documents from competent departments are needed.

[63]Li [23].

levels and overall release. Besides, all kinds of social needs mingle with and influence each other. For example, to solve the problem of long-term care and nursing of elderly citizens, it's a necessary trend to build elderly citizens' long-term care and nursing insurance system. This, meanwhile, involves social pension insurance system and social elderly care service. In the future, the elderly care service will fully extend from basic life care to medical treatment and health, assistant tools allocation, spiritual solace, law service and emergency aid. Although projects in these aspects haven't entered the system construction phase, they have become quite big problems in some areas. The above security projects are related to more governmental departments, so the management becomes more complicated. Judging from China's development phase, the pension problems are related to all families and form the most basic issue on the people's livelihood. In addition, population of elderly citizens and relevant problems will influence the economic and social development situation as a whole day after day.[64] International experience shows that aging of population phenomena is a long-period event. We must do the top-level design and advanced arrangement well. In the world, most of the countries which have entered aging have set up special administrative institutions in charge of aging affairs, and valued making an overall plan in general and from a macro perspective to gather all government functional departments' forces to study development strategy of coordination among population and economy and society, major issues, priority fields, intervention keys and expected effects. It's worth being further studied whether China should gather management of elderly security, appoint only one department to perform elderly security management function scattered in different departments and build a large department system of elderly security. In theory, the complete elderly security should include projects such as basic necessities of life, medical care, study, education, fitness, entertainment, emotional solace, law consultant, life aid and participation in the society. All projects should be independent to a certain extent. Management ways and means vary from different projects. It's unnecessary for one department to be only responsible for all of it. At present, China's capital security and service security of pension security are respectively managed by social security departments and civil affairs departments. Working Committee on Aging performs the function of overall planning and coordination. As a whole, they work well and don't have any bitter conflicts and clashes. However, we still need to make well preparations for comprehensiveness and

[64]The most direct influence is that aging of population significantly changes both Chinese labor market structure and Chinese consumer demand structure and has a long and deep impact on Chinese economy in this way. In recent years, policy issues on pension security are beyond the scope of social construction and people's livelihood security day after day, and most of them have been involved in the Party and national overall situation of working. For example, they are closely related to the elderly care service industry development and changes in the economic development way; accumulated pension insurance fund is related to capital market development; the pension insurance reform of public institutions is related to the Party's policy toward intellectuals, etc.

wholeness of pension security system, actively break strict limits of the current departments' division of work, and enhance coordination and cooperation between different security projects and different departments so as to clear up potential conflicts and problems from systems in the period of implementing the pension security system. Luckily, now we have such a work system of Working Committee on Aging. We'd better further complete system and mechanism of working on aging, and based on development of situation, enhance its function of making an overall plan, comprehensively coordinating, administrative legislation and supervision of law enforcement upon work on aging so as to perform the function of the large department system.[65] Meanwhile, we should further strengthen basic-level aging work management and service network construction to provide institutional support in overall response for aging and elderly citizens' living quality improvement.

6.4.4 Build the Scientific and Reasonable Pension Insurance Fund Management System

China gradually establishes the basic pension insurance system mode of combining social coordination and personal account. Although there are many problems about implementation, from the perspective of reform steadiness, management cost and social affordability, we should try our best to maintain the current system and mode's continuity and steadiness. It's unwise to go back to accounting on the cash basis or implement the full fund accumulation system. We'd better not start over or start from scratch. After all, the pension mode itself of combining social coordination and personal account is relatively flexible, so we can adjust the structure in the framework of the current system as needed. No matter how the future basic pension insurance system will be adjusted, we must understand clearly that social coordination fund and personal account fund should be managed separately in different management systems.

6.4.4.1 On the Collection and Payment System of Social Insurance Fund

According to relevant researches, there isn't a unified social insurance collection system mode in the world. Social insurance institutions do collection on the basis of fees, or taxation departments do collection on the basis of tax, or independent

---

[65]Large department system mainly solves the coordination problems between departments arising from departments' function division so as to increase the management efficiency. Large department system can turn the relationship between departments into internal-departmental relationship, thus avoiding departments' potential long discussion, argument and even conflicts of interest and ensuring the government's work.

institutions do collection, but collection is mainly carried out in the first two ways.[66] In general, all countries set up collection institutions based on their national conditions and historical traditions. China also needs to seek how to build a scientific and reasonable social premium collection system on the basis of the concrete national conditions. There are mainly three views: first, start to collect social insurance tax; second, continue to implement social insurance payment;[67] third, it's mixed financing, that is, to raise employers' social coordination capital in the way of social security tax, and to raise individuals' capital in the personal accounts in the way of social premium.

We believe that we should mainly consider the feature of the current government functional departments' division of business to determine the collection system of China's social premium. We should do our utmost to take advantage of relevant institutions' business strength of the current departments' work division structure. In theory, tax collection and fee collection are means of raising capital, so it's not a question having a theoretical answer that which department should do collection, but the test of business division and technical means to different departments. In the current administrative management system, tax authorities' business advantages are quite obvious. Having set up a strict system of supervising the source of taxation, taxation departments clearly know situation of enterprises' production, operation, salary level, change of personnel, financial revenue and expenditure, and common residents' income. In general, the objects of tax collection and social premium have a high coincidence. The taxation departments at all levels have good collection facilities and perfect network of declaration of dutiable goods. They can provide the insured with comprehensive good-quality services, thus saving learning and management cost. Also having strong rigidity of law enforcement, taxation departments can make hard penalization for illegal behaviors. It's worth being noticed that China has determined the goal of national overall plan of basic pension. The national overall plan of basic pension means that this part of social coordination has had features of taxation and technical difficulties don't exist. Personal account fund doesn't have features of taxation, but taxation authorities can also authorize other departments to collect tax so as to give play to scale effect.[68] In this way, agencies can free themselves from fund collection and spend more energy on management and service.[69] In recent years, the international events that big faults of pension

---

[66]Similar and relevant researches also point out that, in 28 countries which have the independent social security system of OECD, 17 of them manage social security taxes and dues collection through independent social security institutions (or similar institutions). In the rest 11 ones, national social security taxes and dues collection and common tax collection business are united. See Zhang [24].

[67]Zheng [25, 26].

[68]Some provinces implement separate collection of overall planning and personal account (the overall-planning fund is collected by local taxation departments and the personal account capital social insurance agencies). We don't think this is necessary, but it can be regarded as a mode of transition.

[69]Lü [27].

payment records led to turbulent political scenes have warned us that the basic work with an emphasis on personal account management is related to social steadiness, and, if agencies still spend more energy on fund collection and cause the incomplete record of social security rights and interests, it may have a big political risk.[70]

6.4.4.2 On the Relationship Between Pension Agencies and Other Social Insurance Agencies and Social Service Institutions

According to a comparison of relevant international researches, agencies in charge of different projects of social security seem relatively separate in general. Agencies in charge of unemployment project have obvious separation; due to the similarity of annuity management, projects of pension and industrial accident are commonly dealt with in a united way; projects of medical treatment and projects of pension and industrial accident tend to separate management. In countries implementing the common medical treatment system, projects of medical treatment are generally managed separately. In countries with the social medical insurance system, project of medical treatment has a certain trend of jointly operation, but it's not a common phenomena.[71] In various social security projects, medical security is highly professional. A good few of countries set up independent medical insurance agencies. Many regions of China also set up medical insurance agencies and pension insurance agencies separately. According to experience, the separate establishment is good to gather manpower and energy to specially do work of pension insurance and medical insurance, thus improving the efficiency.[72] Surely, industrial accident

[70] *Social Insurance Law of the PRC* stipulates that social insurance agencies should build files for employers in time, completely and accurately record the social insurance data such as personal information, payment and others of the insured, and keep the original documents of registration and declaration and accounting documents of payment and settlement well. According to *Administrative Procedure Law*, if mistakes of the social security information harm citizens' social security rights and interests, the burden of proof can be shifted. *Regulations on Social Security Business File Management* (*Trial*) in 2009 explicitly stipulates the responsibilities of staff in social insurance agencies if they lead to file missing, damage, leakage of the unit or personal information, and other results.

[71] Yang and Sun [28].

[72] Wang Dongjin, the Former Vice Minister of the Ministry of Labor and Social Security, pointed out at the national medical insurance work forum held on February 27, 2001, that the local government should decide whether medical social insurance agencies such as medical insurance ones and pension insurance ones should be established in a separate way or a unified one based on realities. No matter whether it's handled by separate agencies or just one social insurance agency, as long as it is managed by social security departments, it fits the spirit of "gathering five insurances into one". *Returning Letter of Building Urban Employees' Basic Medical Insurance Agencies* (Liao Lao She Han [2001] No. 25) of Liaoning Province pointed out: "At present, Dalian, Jinzhou, Liaoyang, Huludao, Chaoyang, Yingkou, Panjin and other cities of our province have set up independent medical insurance agencies. These cities reflect that it goes better after medical insurance agencies are established separately. It simplifies procedures and gathers manpower and energy so that they can specially handle medical insurance work and make a contribution to local social steadiness".

insurance, maternity insurance, and nursing insurance which may be set up in the future, can also be dealt with by medical insurance agencies in a united way. However, all social security projects are quite different from one another, but they have a certain coincidence in business procedures. From the perspective of procedure of the social security fund management, financing management, current management and control of expenditure are mainly included. Their difference mainly lies in current management and fund expenditure part, and they share similarity and homogeneity in the financing part. To reduce management cost, all social security funds should be collected by taxation authorities only in the future.[73] It is also necessary to point out that only one institute can't satisfy all needs of various security objects and various handling services of the security system, but in the basic level, we should try our best to gather all social security affairs on one platform. We'd better set up basic-level pension insurance agencies with other public service institutions together and provide urban and rural residents with "one-stop" services. To put it simply, the so-called "one-stop" service is just that service suppliers provide targeted customers with a service platform with unified population through Internet, so that customers can get overall services by accessing the unified portal.[74]

6.4.4.3 On the Management System of Personal Account Fund

China has already explicitly pointed out that during the period of the "12th Five-Year Plan," provincial overall plan of urban employees' basic pension insurance should be implemented in an all-round manner to achieve national overall plan of basic pension. Just operate social coordination fund in the form of tax. In principle, the current fund surplus is deposited in banks or used to buy national debts without other investments. As to personal account fund whose pension insurancc is fully accumulated, due to the fund's huge amount and long accumulation period, its value must be maintained and increased by being used for capital market investment. Currently, there is nearly no dispute in investing personal account fund in trust and achieving multiple portfolios, but there are still different views on whether provincial social insurance agencies or the central government should be the investor of personal account fund. At present, the advantaged view is that provincial social insurance agencies should manage the investment of personal account fund, and build an investment decision committee led by leaders of provincial governments and with members of directors and relevant experts, whose main function is to analyze and study major issues such as investment strategy, directions, selection of investment institutions and income distribution during the

[73]German pension insurance, unemployment insurance and medical insurance capital are collected by medical insurance organizations in a unified way and then are allocated to various insurance kinds management institutions according to the collection amount.

[74]National Information Security Engineering Center [29].

investment process of personal account, and then relevant departments handle organization and implementation. Some researches oppose this. They think that the fund investment management level and profit opportunity are significantly different in different provinces, so the earning rate will be very different inevitably. It's hard to preclude administrative intervention if the personal account fund is controlled by different provinces, and it can't achieve full competition in the whole country as enterprise annuity which is totally of marketization does. It'll easily become "public fund" which is actually controlled by different provinces. In the situation that it's hard for the central government to have centralized supervision and unified compensation system is incomplete, local governments' investment constraint will be further loosened, and a quite big social risk is hidden.[75] We think that problems of motivation and supervision of the personal account fund management system need to be solved at the same time. Levels of main investors are relatively low, so we can give play to all-level governments' activeness of using the personal account fund well. However, local governments should have a high ability of fund management. There are many points of risk and it's hard to supervise the fund. The eight provinces, which have launched a pilot project of the third group of personal account implementation at the end of 2006, have handed the central finance's subsidy of the implemented account capital over to National Council for Social Security Fund to centralize the fund for management and investment, and has entrusted all accumulated personal account funds to National Council for Social Security Fund for management. We believe, because of the regional disparity in China, there are great differences between different regional system construction and governmental management ability. It's very hard to implement nationwide centralized management of personal account fund in a short time. The urgent priority is, based on the fact that the basic pension moves towards national overall planning, to hand over the personal account fund to the province by level as soon as possible so that provincial social insurance agencies can manage and operate it in a unified way and carry out provincial business accounting and give money to different levels. In line with this, provinces manage all-level social insurance institutions in a vertical way, and leaders of all-level social institutions are appointed and dismissed by provincial social security bureaus.[76] We can learn from international experience to set up pension insurance offices in a country's level and their branches are provincial social insurance agencies. At the same time, as accumulated fund, personal account fund should have flexible investment ways, and it's necessary to introduce in factor of marketization, but there must be strict control.[77] Collection and payment of

[75]Zheng [30].

[76]Shaanxi Province has made a vital step forward in this field.

[77]In 2009, China Insurance Regulatory Commission issued *Notice on Affairs Related to Trying out Pension Security Mandatory Administration Business*, which allows pension insurance companies to try out pension security mandatory management, set up the independent pension security mandatory management fund, and give full play to advantages of business insurance in actuarial technique, account management, investment management, risk management, annuity payment, and so on.

social insurance fund are enforced by the country. Its investment request should be different from enterprise annuity after all. What's more, China's capital market isn't totally standard yet and principles such as honesty and credibility haven't been set up, so it's impossible to hand all capital over to financial institutions of marketization for management. Therefore, we must build risk prevention and supervision system at the same time. Problems of risk control need to be solved through multi-level guarantee systems. Fund supervision are mainly achieved by enhancing the government's functions. At the same time, the right to know, right of participation, right of expression and right of supervision of pension insurance fund of the insured units and individuals should be guaranteed as beneficiaries. We should build a system to release the pension insurance fund information and agencies should regularly announce information of pension insurance fund income and expenses, management and the pay of the insured to the society. Build a system of pension insurance fund supervision carried out by relevant departments' representatives of the government, employers' representatives, labor unions' representatives and relevant experts. In the current phase, National Council for Social Security Fund has intervened in the personal account fund investment management. As conditions become mature gradually, personal account fund will finally be managed by the country in a centralized way. By then, National Council for Social Security Fund can fulfill functions of investment and management fully. Since National Social Security Fund has the nature of sovereign wealth fund, so its function needs to be further confirmed to determine a scientific and reasonable management system based on this. As to enterprise annuity, we should mainly change the governmental function from the excessive control to the effective control of enterprise annuity.

6.4.5 Improve the All-Around and Scientific Management System of the Elderly Care Service Industry

How to deal with the aging population became an important agenda in the whole work of the CPC in the 5th Plenary Session of 17th Central Committee of the CPC. It was proposed to give priority to develop the social elderly care services, cultivate and expand the elderly care service cause and industry. Predictably, the elderly care service industry will develop rapidly in some period in the future. The establishment and improvement of the management system of the elderly care service industry must be accelerated.

6.4.5.1 Achieve the Industry-Wide Management of the Elderly Care Services as Soon as Possible

As the elderly care service industry is a very broad concept, the departments of civil affairs should earnestly supervise and manage the elderly care service industry as competent departments, including all the social elderly care service institutions, community services, home-based services in the scope of the industry supervision and carrying out the local management. The departments of civil affairs should make overall arrangements with reasonable layouts and unified management to avoid the redundant construction for all kinds of elderly care service institutions. Encourage the elderly care institutions to carry out the industry self-discipline management to promote the healthy development of the elderly care service industry. In the so-called industry management, the government functional departments and the corresponding industry groups are the management subjects and the policies and coordination are the main service contents with the measure of indirect management and the producers of the similar products cover the whole society.[78] There are some explorations of this aspect, including the Community Service Industry Association founded in Pudong New District, Shanghai in 1993, China Association of Social Welfare founded in 2001 and the Self-disciplinary Convention of the National Private Elderly Service Institutions and Service Organizations Related to the Elderly Citizens sponsored by China Silver Industry Association. However, since China is a power-oriented and administration-oriented country for long, there isn't a tradition of self-discipline and autonomy of the industry associations and the industry self-discipline management must rely on the social morality, professional ethics and the establishment and implementation of the credit system, while the relevant social basis is very weak in China. Besides, the public elderly care institutions and the practitioners work in accordance with the administrative manner for long and the autonomous elderly care institutions and service market are far from being formed, so we still have a long way to go to implement the self-discipline management. At this stage, we should adopt a management system of relying on the supervision and management of competent administrative department primarily with the combination of the self-discipline management of the industry and the self-management of institutions.

6.4.5.2 Strengthen the Industrial Management Functions of the Competent Departments of Civil Affairs

(1) Make plans for the layout and implementation of the elderly care service institutions scientifically with unified plan and rational distribution. Make plans which are prospective and operable and can reflect the regional characteristics for the elderly care service institutions and the layout of the facility construction

[78]Tang [31].

of the elderly care services according to the development speed of the aging population and conditions of the elderly care service needs. Classify the elderly care service institutions according to the functions with rational layout based on the differences of the functional orientation and the medical intervention degree.

(2) Improve the open, equal and standard elderly care service industry access system as quickly as possible. Further standardize the standards, conditions, hardware facility configuration and the technical standards of the establishment of elderly care institutions. Control the access strictly and intensify the supervision function. Control the qualification review and annual assessment strictly. For the elderly care service institutions founded without the approval of the departments of civil affairs, the relevant departments should not provide them with check-in services of legal person registration.

(3) Strengthen the daily supervision of the elderly care service institutions. Improve the policies, standards and plans of the elderly care service industry and refine them step by step. Improve the specific supervision measures and enrich the supervision power as quickly as possible.

(4) Improve the policy support system which can promote the development of elderly care services. Help social forces found elderly care institutions with funds in many forms, including construction allowances, bed allowances, population allowances and general allowances. Improve the preferential policies for elderly care service institutions in aspects of land supply, capital investment, reduction or exemption of tax, financial aid, social finance, water, power and heat supply, free services, etc.

(5) Strengthen the construction and supervision of qualified personnel of elderly care services. Develop the elderly care service post professional standards and operation specifications as soon as possible. Carry out systematic professional training under professional education and improve the specialization level of the service team. Establish the professional qualification system of the elderly care nursing workers step by step. Conduct professional skill appraisal (assessment) work of the elderly care nursing workers. Strengthen the daily supervision of the elderly care service practitioners.

(6) Explore effective management ways for the home-based elderly care and community-based elderly care. Encourage the establishment of the elderly care service allowance and service quality evaluation systems in the places with good conditions. As the home-based elderly care and community-based elderly care involve many aspects and their management is complicated, the departments of civil affairs should earnestly fulfill their functions, including comprehensive coordination, business guidance, supervision and inspection and the supervision of the government's funds for elderly care services, develop the specifications of the community care services and the home-based elderly care services, standardize the service links and service procedures and developing the assessment criterion. As the leading units of the construction of harmonious community, the departments of civil affairs should perform the comprehensive coordination function, arouse the enthusiasm of all walks of life and various

units within areas under its jurisdiction, coordinate the internal existing service facilities of it, including elderly care, health and recreation and sports, integrate the unused service resources of the communities, put them into use in public and provide the elderly citizens with timely and comprehensive services.[79] Reform the elderly care home appropriately and make it a comprehensive and multi-functional social welfare service center in the locality. Make overall plans for the forces in all aspects to establish the community-based elderly care service information platform. Build unified archives for elderly citizens who accept the services and the personnel who provide them with services.

(7) Construct and manage the elderly care institutions respectively according to different classifications and levels.[80] Divide elderly citizens who live in elderly care institutions into different levels. Provide elderly citizens in different levels with different service contents. For example, provide elderly citizens who can not take care of themselves with rehabilitation nursing services primarily and provide those who can take care of themselves with daily care services, etc.

6.4.5.3 Establish a Classification Management System for the for-Profit and Non-profit Social Elderly Care Service Institutions

In the process of the expansion of the elderly care service resource orientation, the preliminary basic framework of the elderly care service system in China is established. The preliminary classification management system is established according to different natures of private non-enterprise units and enterprise units. The investment system of the elderly care service industry in China has changed ultimately. Now the elderly care institutions are invested by diversified subjects, including the country, collectives, enterprises, associations, individuals and the foreign capital from unitary investment of the country and collectives. The business pattern becomes diversified step by step. Now the property and legal status of the private elderly care institutions can not be distinguished and classified clearly, so people confuse the for-profit private elderly care institutions with those non-profit ones in reality. The classification management of the social elderly care service institutions means to manage the for-profit private elderly care institutions with those non-profit ones which have different kinds of organizational property

---

[79]Yuetan Sub-district in Xicheng District, Beijing City makes contact with the service security departments of the state organs (over 10 units, including the NDRC, the Ministry of Finance, the SARFT and the MOR) in Xicheng District, starts to try to share the resources (including the dining hall, the rehearsal hall of the activity center, the auditorium, exhibition rooms and the stadium) with the residents in the community. There are cooperation plans for the resource sharing between the social units and communities in all the 15 sub-districts in Xicheng District. See *The Auditoriums and Dining Halls of Some Central Ministries and Commissions are Opened for the Residents in the Community*, *Beijing Daily*, February 12, 2011.

[80]Research Team of the Capital Institute of Social Economical Development [32].

respectively, make their own legal property and their consistent internal management systems and external supervision systems clear and implement different finance and taxation policies and supportive policies. The classification management of the for-profit and non-profit social elderly care service institutions is a fundamental system related to the healthy development of the elderly care service industry. There have been clear systems and stipulations of the classification management in the healthcare industry. The private education is also exploring and making experiments of the classification management of the for-profit and non-profit private elderly care institutions. The classification management is a new topic for the elderly care service institutions. We do not have a set of clear assessment criteria and management systems of the non-profit organizations, finance and taxation policies and property right systems which encourage and standardize the non-profit elderly care institutions, asset management systems and accounting audit systems which accords with the international conventions. All signs indicate that the elderly care service industry in China faces a historical period of development and transition. China proposed to set up a social elderly care service system with perfect systems, perfect organizations, appropriate scale, good operation, excellent services, perfect supervision and sustainable development in 2015. In the face of the critical moment of the transition, we should explore a classification management system of the for-profit and non-profit elderly care institutions and pave the way for healthy development of these two kinds of elderly care service institutions instead of repeating the past indistinct policies and measures of expediency. The non-profit elderly care service institutions primarily provide the service objects in the non-welfare elderly care service institutions and the ordinary elderly group of ordinary economic conditions with elderly services. Their elderly care fees are borne by individuals. After the assessment of corresponding elderly care needs, the government will provide them with appropriate allowances. The for-profit elderly care service institutions provide elderly citizens with good economic conditions with elderly care services primarily in a marketable charging way. The elderly care fees are borne by individuals or families.[81]

6.4.6 Continue to Promote the Pension Security Management and Service Socialization

Management and service socialization are the basic requirements of the pension security system. With the issuing of *Social Insurance Law of the PRC* and other laws and regulations, the management and services of the pension security in China

---

[81]Opinions of Promoting the Development of the Elderly Care Service Cause of Hangzhou Municipal Committee of the CPC and Hangzhou Municipal People's Government (Shi Wei [2010] No. 24).

enter into a new stage of law-based administration and law-based handling. It is required that we should continue to promote the social management and services.

6.4.6.1 Strengthen the Construction of Pension Insurance Agencies

(1) Strengthen the construction of pension insurance agencies. Establish a stable talent training, evaluation, usage and incentive mechanism of pension insurance agencies. Introduce professional social workers into grass-roots social insurance agencies. Improve the professional registration, post setting and salary management systems and other systems for social security workers. Improve the support system of the Ministry of Finance for the management and services of the grass-roots agencies. Establish the long-term support mechanism of the Ministry of Finance supporting the social security management services. Explore and establish a mechanism of the personnel allocation and fund security related to the business volume. Besides, solve the problems of personnel shortage of the agencies and backward management service measures through entrusting and purchasing services and other ways.
(2) Integrate the resources of the grass-roots agencies. Provide different kinds of security objects with efficient and convenient pension insurance public services intensively. Break the current division situation in which the agencies are carved up layer by layer and extend to sub-districts and even neighborhood committees and village committees. Establish a comprehensive social security service network in urban sub-districts and rural townships and carry out the decentralized and subsided service forms to achieve the handling service functions organization of service function extending to the sub-districts and communities, narrow the service radius, improve the service accessibility and achieve the nearby handling of pension insurance affairs. Set up comprehensive service windows. All the social insurance handling business achieves the "one-stop" handling. Treat the service objects in a just and fair way. Provide them with standard and unified pension insurance. Accept the supervision of society and the service objects actively. China has put forward a goal that the national unified social security cards will be issued to cover 60% of the population of China during the 12th Five-Year Plan. Expand application field of the social security cards to achieve the information resource integration and process connection among the social security business. The settlement of all social insurance business can be finished with a social security card. Explore the way to apply the social security cards in civil administration, health and other public service fields.
(3) Strengthen the data and information management of pension insurance. Build a multi-channel and all-around information system of public services of human resources and social security. Build the individual account into an important carrier and information platform with records, transformation, investment and benefit distribution based on the individual account payment and the records of

the rights and interests. Pay special attention to data quality, fulfill the idea of "records, services and security for the whole life" and collect and provide authentic social insurance data. Strengthen the internal information statistics of the agencies and cooperation of the business departments. The information integrated management institutions should perform the business guidance function for the agencies. As the social insurance handling system possesses personal information of huge potential commercial value, the agencies[82] should assume the obligation of protecting the right of privacy of the public. Explore the ways to manage the human resources and all kinds of information resources, equipment resources and technical resources of the social security uniformly. Build a unified information platform of public services which extends down to the sub-districts (towns) and communities (villages).

(4) Actively promote the standardized construction of the elderly care insurance handling. Achieve the standardization and normalization of the handling management services step by step. The social insurance has many procedures, so it has standardized operation conditions objectively. The social security cause in China has entered a new stage in need of the standardized management and services. The standardization of the handling services means a process of achieving the service quality with targets, the standardized service measures and the procedural service procedures so that the first-rate services can be obtained by making and implementing the service standards and the application of the standardized principles and measures.[83] In 2009, Standardization Administration of the People's Republic of China approved *Social Insurance Service General Principles* and *Equipment and Facility Requirements of the Social Security Service Center,* two national standards of the Social Insurance Standardization Technical Committee of China. There's also good exploration in local authorities. For example, Shaanxi Province has made three levels of standard system tables, namely the province, city and county, which cover business, management, integrity, information and other fields and unify the names of the institutions, image recognition, archive management, professional terms, service specifications and other aspects all over China. Wuxi City has carried out the social insurance handling mechanism innovation and standardization work under the "five insurance into one." Tianjin City has established the service management system specifications. Shanghai Medical Insurance Center, Huai'an Social Insurance Center in Jiangsu Province and other institutions carry out ISO9001-2000 QMSC and so on. Combined with the international current standardization situation and development trend, we shall absorb the local practical experience and ideas actively to accelerate the improvement of the standard system of the social insurance.

[82]Hong Kong Octopus Company (the Octopus cards issued by the company have multiple consumption functions and record cardholders' personal information. Nearly everyone in Hong Kong has one Octopus card) sold nearly 2 million customers' personal information to the third parties for business promotion, getting profits of over 40 million yuan. WWW.CHINANEWS.COM.

[83]http://ww.calss.net.cn/n1196/nl211/nl545/2126539.html.

6.4.6.2 Strengthen Social Management and Service for Retirees

(1) Broaden the scope of social management and service for retirees. China's original system of social management and service for retirees is formed in the process of deepening the reform of state-owned enterprises step by step. Responsibility of management and service for retirees of enterprises and public institutions are mainly assumed by rural and urban basic-level institutions, and it has a certain limitation.

We should bring retirees of non-public economic institutions such as private-owned enterprises and privately- or individually-owned businesses in social management in an all-round manner so as to carry out community management.

(2) Build departmental coordination system of social management and service for retirees. Social management of enterprise retirees is a systematic project related to the whole society. *Outline of Sustainable Development Action in the Early 21st Century of China* printed and distributed by the State Council in 2003 includes the realization of social management and service of social insurance as the important content of social development. To adapt to such a new request, all-level governments shall build interdepartmental coordination system to join forces of different departments.
(3) Encourage retirees to improve their own living quality through self-management and mutual-help services. Community is the carrier of enterprise retirees to carry out the social management and service.
(4) Actively push forward social management and service of retirees of government departments and public institutions. At present, the governmental public servants and most staff members of public institutions also basically carry out the system that the national finance and the units "are responsible for all". In recent years, the reform of pension insurance system of public institutions which is promoted according to the mode of combining social coordination and personal account meets a great obstacle in the process of promotion and nearly stops. Based on the current situation, government departments and public institutions' pension insurance system reform has formed articulation point and conflict combination point of reform of government institutions and public institutions and other kinds of reforms. What happens to it may affect the whole. It is related to many aspects and deep levels of interests. In prediction, it still needs a long time of fierce theoretical disputes to finally confirm the reform direction.[84] However, this doesn't mean that no achievements are made in the current government departments and public institutions' pension insurance reform. The urgent priority is to make real achievements in social management and service. Based on the international situation, as to most of pension insurance of national public servants (generally equal to staff members of government departments and public institutions of China), independent centralized

[84]Chen and Wang [33].

management institutions are established in the central authorities and management institutions and agencies are established in regions. In most countries, governments set up unified and social agencies and keep them independent from supervision and management institutions in the principle of "separation of management and function performance."[85] Learning from foreign practice and based on levels of administrative management, we can establish pension security bureaus of all-level government departments and public institutions, to centralize functions of retirement management institutions which are subordinate to government departments and public institutions currently. Those functions will be performed by pension security bureaus of government departments and public institutions in a unified way. After leaders and cadres who hold certain positions in government departments and public institutions, management can be implemented according to original channels.

References

1. Zheng, Gongcheng. 1997. *On the Social Security Path with Chinese Characteristics,* 497. Wuhan University Press.
2. Song, Xiaowu, et al. 2001. *Report on Reform and Development of the Chinese Social Security System, China,* 212. Renmin University Press.
3. Wang, Yanzhong. 2007. *Commentaries on Development of China Social Security in the Past Ten Years,* 2007 Annual Meeting of China Social Security Forum.
4. Zheng, Gongcheng. 1997. *On the Road of Social Security with Chinese Characteristics,* 497–503. Wuhan University Press.
5. Meng, Zhaoxi. 2009. The organization of social insurance: Based on the insured—The brief development of social insurance business in 30 years after the reform and opening up. *China Social Security* 1:20.
6. Feng, Lanrui. 1994. *Centralization and Decentralization of Social Security Management System, Reform* 4:76–77.
7. Zheng, Gongcheng. 2004. From centralized management of the Government to diverse autonomous management-future development of China's social insurance organization and management model. *Renmin University of China Journal* 5.
8. Yang, Liangchu, Fuchang Zhao, and Fenger Han. 2007. Analysis of power division of social security. *China Social Security* 4:28–29.
9. Wang, Yanzhong. 2009. *Construction of the Basic Integrated Social Pension Security System.* The 7th China Reform Forum.
10. Zhu, Ling. 2010. The research of China's social security system frame work and its evaluation. *Chinese Journal of Population Science* 5:2–12.
11. Hu, Zucai, Yuan Xue, and Yunan Liu. 2009. Social service industry: An important part of demand expansion. *Macro-Economy Management* 9:14–19.
12. Research Group of the Capital Institute of Social Economical Development. 2009. Research on perfecting elderly care mode and system policies in Beijing. *Frontline* 9:43–45.
13. Gui, Shixun. 2001. Reasonably adjust the functional structure of the elderly care institutions. *Journal of East China Normal University (Philosophy and Social Sciences Edition)* 7:98.

[85]See Footnote 84.

14. He, Ping. 2009. *Accelerate the Construction of China's Social Security System, Red Flag Manuscript* 12.
15. Zheng, Gongcheng. 2008. *China's Social Security Reform and Development Strategy—Concept, Objective and Action Plan.* People's Publishing House.
16. International Labor Organization. 1988. *Looking ahead to the 21st Century: The Development of Social Security*, 77–78. The Labor and Personnel Press.
17. Zheng, Gongcheng. 2009. *Science of Social Security: Concept, System, Practice and Thinking*, 384–385. The Commercial Press.
18. Sun, Shuhan. 2008. *Social Insurance*, 96. China Renmin University Press.
19. World Bank. 2003. *World Development Report: Making Services Work for Poor People*, 53. China Financial and Economic Publishing House.
20. Jia, Kang. 2007. Consider the social insurance management system framework from the national supreme interests. *Public Finance Research* 11:2–5.
21. Zheng, Bingwen. 2009. Legal organization forms of social insurance fund: Experience and lessons of EU. *Journal of China University of Political Science and Law* 1:30–45.
22. Zheng, Gongcheng. 2007. *Social Insurance Fund Should Adopt Centralized Supervision, China Labor and Social Security News*, June 11.
23. Li, Wei: *The Main Problems in the Construction of China's Social Elderly Care Service System and Countermeasures.* China Economic Information Network.
24. Zhang, Wenchun. 2009. *International Trend of Social Security Tax Collection, Tax Research Conference,* 4:51.
25. Zheng, Gongcheng. . *Fee Should Not Be Changed into Tax in Social Security, China Economic Times,* February 2, 2000.
26. Zheng, Bingwen. 2007. Based on international development trend, China should not change social security fee into taxes and dues. *Macroeconomics* 3:17–21.
27. Lü, Xuejing. 2006. *Modern Social Security Systems of Different Countries*, 159. China Labor-and-Social Security Publishing House.
28. Yang, Jianmin, and Bingyao Sun. 1998. Comparison and research of social security management system (II). *China Social Security* 6:16–17.
29. National Information Security Engineering Center. 2003. *Overall Design and Technical Achievement of Electronic Government*, 133. Publishing House of Electronics Industry.
30. Zheng, Bingwen. 2007. Risks and solutions of pension insurance investment reform. *China Securities Journal*, November 9.
31. Tang, Xiaoguang. 1999. *Power Shift*, 284. Zhejiang People's Publishing House.
32. Research Team of the Capital Institute of Social Economical Development. 2009. Research on perfecting elderly care mode and system policies in Beijing. *Frontline* 9:43–45.
33. Chen, Jiagui, and Yanzhong Wang. 2010. *China Social Security Development Reform (2010) —Make Fair Social Security Accessible to Everyone.* Social Sciences Academic Press.

Chapter 7
Learning from the International Experience of Pension Security Systems

Yuqi Long

The history of foreign pension security systems, developing from traditional self-support and support from families to the modern pension security system, is an important demonstration of human civilization's progress. Following the development and progress of the economy and society, particularly the acceleration of industrialization and globalization, most countries in the world have established respective pension security systems. As different countries have different situations, political systems, economic development, culture, history, traditions and so forth, they adopt different models, responsibility sharing mechanisms and levels of support for their respective pension security systems. However, all these systems share a common goal of offering security to elderly citizens. Since ancient times, there has been a tradition of respecting and caring for the elderly in China. For centuries, most elderly people in China were supported by their own families while the official pension security system was established at a relatively late stage, and is still undergoing a vital period of reform and improvement. As changes in economic and social development occurred throughout the world, pension security systems were modified and reformed. Via reform and experimentation, various countries formed a variety of unique systems and mechanisms. In reforming and improving China's pension security system, the nation should rationally learn from the international experience of developing pension security systems. Based on the specific conditions in China, the country should establish and improve its pension security system as soon as possible to ensure that the elderly receive security in old age.

Y. Long (✉)
School of Management, Capital Normal University,
#83 West Third Ring Road North, Haidian District,
Beijing 100089, People's Republic of China
e-mail: lyqdy2008@163.com

Y. Wang (ed.), *The Development of Security and Whole Care System for the Aged in China*, https://doi.org/10.1007/978-981-10-6481-4_7

7.1 Historical Characteristics of Pension Security Systems' Development Abroad

7.1.1 Increasing Problem of Population Aging Seriously Challenges Pension Security Systems[1]

The development of a pension security system is closely related to a region's population structure, and must be constantly adjusted and improved as that population structure changes. An important characteristic of current changes in population structure is population aging. Population aging in the world is becoming increasingly serious, and some countries face a rapidly aging population. Population aging first appeared in some developed countries and now in some developing countries. Population aging is an inevitable result of the decrease in death rates and birth rates alongside an increase in life expectancy. The life expectancy of populations in many countries clearly increased between 1960 and 2007, with increases of over 10 years in Australia, Belgium, France, Germany, Japan, Korea, Luxembourg, Mexico, Portugal, Switzerland, Turkey and others over that period. OECD countries enjoyed an increase in average life expectancy of 10.6 years, and some developing countries also saw clear increases.[2] From the proportion of the elderly in the total population, one can see that population aging is becoming more entrenched throughout the world. In 1865, the proportion of the population aged 65 years and over in France exceeded 7%, which made France the first country of the world to become an aging society. The proportion of the population aged 65 or older in 2007 reached an average of 14.3% in OECD countries, with heights of 21.5% in Japan and 20.2% in Germany.[3] Moreover, population aging has been speeding up in countries throughout the world. Japan's aging population is growing the fastest, with the proportion of the population aged 65 or older growing from 7 to 14% in just 26 years, where it took France 115 years, Switzerland 85 years, the United States 65 years and Britain 45 years.[4] Looking at the changes in the respective proportions of the population aged 65 years and over in some countries between 1995 and 2005, it can be seen in the countries listed in Table 7.1 all reached over 10%, and, excluding Ireland, these proportions had all clearly increased. Of these countries, Japan experienced the fastest growth rate, with the proportion of the population aged 65 and over growing from 15% in 1995 to 20% in 2005.

The most direct result of population aging is an increasing growth of the old-age dependency ratio and pressure for elderly support. As to the old-age dependency

---

[1]Long [1].

[2]OECD Factbook 2010: Economic, Environmental and Social Statistics.

[3]OECD figures 2009.

[4]Xiong [2].

Table 7.1 Proportion of the population aged 65 years and over in some countries (1995–2005) (unit: %)

Country	Year		Country	Year	
	1995	2005		1995	2005
Japan	15	20	Denmark	15	15
Italy	16	19	Norway	16	15
Germany	16	19	Luxembourg	14	14
Greece	15	18	Czech Republic	13	14
Sweden	17	17	Netherlands	13	14
Belgium	16	17	OECD	12	14
Portugal	15	17	Poland	12	13
Spain	15	17	Canada	12	13
Austria	15	17	Australia	12	13
France	15	16	United States	13	12
Finland	14	16	New Zealand	12	12
United Kingdom	16	16	Iceland	12	12
Switzerland	15	16	Slovakia	11	12
Hungary	14	16	Ireland	12	11

Source OECD Regions at a Glance: 2009 Edition

ratios in OECD countries[5] in 1950, most OECD countries' old-age dependency ratios were between 0.1 and 0.2, while the figures for countries such as Poland, South Korea, Turkey and Mexico were even lower than 0.1, and only Ireland's ratio exceeded 0.2. In 2005, 26 countries' ratios exceeded 0.2. Of those countries, ratios in Germany, Italy and Japan exceeded 0.3. According to the OECD, by 2030, old-age dependency ratios are projected to exceed 0.4 in 16 countries, with the ratios in Germany, Finland and Japan expected to exceed 0.5. By 2050, old-age dependency ratios are projected to exceed 0.4 in 25 countries, with those in Austria, the Czech Republic, Finland, France and Slovakia expected to exceed 0.5, those in Germany, Spain, Greece, Italy, Poland and Portugal expected to exceed 0.6, that of South Korea expected to exceed 0.75 and Japan's ratio is expected to exceed 0.8. These countries will face a heavy burden in providing elderly support.[6]

It is clear, then, that population aging is becoming entrenched in many countries. Finding solutions to ensure the elderly receive security in old age, providing a better life for the elderly and ensuring medical and cultural services are becoming important tasks and big challenges for the economic and social development of countries throughout the world. Solving these problems has become a common task for all the countries of the world, and one that will require an increase in investment as well as the reform and improvement of pension security systems.

[5]The old-age dependency ratio refers to the ratio of the population aged 65 and over compared to the population aged between 20 and 64.

[6]Pensions at a Glance 2009: Retirement-Income Systems in OECD Countries.

7.1.2 *The Accelerating Economic Globalization Brings Opportunities as Well as Challenges to Pension Security System*

Economic globalization is an unstoppable trend and the exchanges of elements such as personnel, capital, technology, service and goods among countries are free as well as more and more frequent, in the meantime, the exchanges and cooperation among regions are more and more common. Economic globalization promotes the reasonable allocation of essential productive factors, optimization of economic structure in various countries, which accordingly promotes the advance of competitiveness of all countries. The rapid development of modern information technology and transportation, communication technology provide more convenient conditions for economic globalization, with its degree being deeper and its scope being wider. By the impact of financial crisis, the global trade and direct investment flow reduce, but the basic trend of globalization has not been changed.[7] As economic globalization is a "double-edged sword", though it has brought a lot of benefits to the development of many countries, and has promoted the rapid development of national economy and the improvement of the national standard of living, it has also brought a lot of problems to many aspects, for example, problems of the unbalancedness of development among countries and regions, trade protectionism, unfairness of the rules, increase of trade conflict, the widening gap between the rich and poor and so on which are increasingly obvious.[8]

There are also advantages and disadvantages in the influence of economic globalization to pension security systems of various countries, which has not only provided opportunities for the development of pension security systems, but also increased the difficulties of the construction of pension security systems.

From the view of opportunities, it performs that: on the one hand, economic globalization provides opportunities and platform of exchange for the construction of pension security systems, which is beneficial to supporting better technical guidance to the construction of pension security system in various countries. With the scientific and technological advance, international agreements and political liberalization, global networking of information approach and communication media become possible, and the transmission time is shortened and the spatial distance becomes insignificant, so more and more events can be known around the world at the same time, and can have effects in different sites worldwide in shorter and shorter time difference.[9] Economic globalization provides convenience to government officials, experts and scholars, even common people in all countries to exchange information of pension security, and they can discuss the reform of pension security system more freely and swimmingly, exchange reform experience,

[7]Pei [3].

[8]Yan and Long [4].

[9]Franz-Xaver [5].

even provide help to the reform to enhance the cooperation among countries. These are beneficial to guiding the construction of pension security system in various countries from the lays of theory and technology. On the other hand, economic globalization can provide better economic security for the development of pension security system. Economic globalization allocates various kinds of resources by market-oriented approach to raise the efficiency of resource allocation and enhance economic competitiveness of various countries, thus provides better material basis, capital basis and human resource guarantee for the development of pension security system. However, it's worth noting that, it is not certain that, the more the economy develops, the better pension security people obtain, and it also depends on some other factors, such as political factors, institutional factors, policy factors and so on. In addition, economic globalization has provided good technical means for the development of pension security system, and informatization is important means to boost globalization which has provided great convenience for the exchange and communication of people, and has also provided important means for the management and service of pension security system which is beneficial to raising the efficiency of the management and service of pension security system.

From the view of challenges, it performs that: ① Economic globalization brings greater risks to the elderly. For example, some countries blindly pursue competitiveness (reducing product costs), but neglect environmental protection, thus the ever worsening ecological environment increases the disease risk of the elderly; moreover, the industrialization being conducted by some countries brings greater risks to people, who will be faced with greater survival risk and disease risk in the old age; economic globalization has brought various convenience to people, while it also increased people's living costs to a certain degree; economic globalization has raised higher requests to people's qualities and skills, which makes some of the elderly at a disadvantaged position in the process of competition, whose incidence of poverty is higher with respect to that of ordinary workers (see Table 7.2). ② Economic globalization increases the difficulty of the management as well as value preservation and increment of pension security funds. For example, the uncertainty and interconnectedness of the economic development of all countries has brought great challenges to value preservation and increment of some funds under accumulation system. The great upheaval in stock market in the second half of 2008 brought great impacts on pension security system, with pension funds in OECD countries reducing more than 10% averagely in 2008, in which the pension fund of Ireland almost depreciated for more than 35%; the pension funds of USA and Australia almost evaporated for more than 20%. However, there were two exceptions, namely Korea and Turkey, whose pension funds have not significantly depreciated (see Table 7.3).[10] According to the research on the loss sustained by global pension funds in the financial crisis by Professor Zheng Bingwen, on the whole, global pension funds have greatly depreciated in this financial crisis, with the total assets decreasing from 34.77 trillion US dollars at the end of 2007 to 29.88

[10]OECD Observer [6].

Table 7.2 The incidence of poverty of the population and the incidence of poverty of the elderly in some countries in 2005 (unit: %)

Country	The poverty rate of population	The poverty rate of the elderly	Country	The poverty rate of population	The poverty rate of the elderly
Australia	12.4	26.9	Korea	14.6	45.1
Austria	6.6	7.5	Luxembourg	8.1	3.1
Belgium	8.8	12.8	Mexico	18.4	28.0
Canada	12.0	4.4	Netherlands	7.7	2.1
Czech	5.8	2.3	New Zealand	10.8	1.5
Denmark	5.3	10.0	Norway	6.8	9.1
Finland	7.3	12.7	Poland	14.6	4.8
France	7.1	8.8	Portugal	12.9	16.6
Germany	11.0	9.9	Slovakia	8.1	5.9
Greece	12.6	22.7	Spain	14.1	16.6
Hungary	7.1	4.7	Sweden	5.3	7.7
Iceland	7.1	5.0	Switzerland	8.7	17.6
Ireland	14.8	30.6	Turkey	17.5	15.1
Italy	11.4	12.8	UK	8.3	10.3
Japan	14.9	22.0	USA	17.1	23.6

Note The incidence of poverty here refers to the incidence rate of income lower than half of the median of the income of the whole population

Source OECD Income Distribution Database. See OECD (2008), Growing Unequal? Tables 5.1 and 5.3

Table 7.3 Return on investment of pension funds in some countries in 2008 (unit: %)

Country	Rate of return	Country	Rate of return	Country	Rate of return
Mexico	−5.19	Austria	−15.65	Hungary	− 21.04
Czech	−7.18	Denmark	−16.83	Canada	−21.36
Germany	−8.51	Sweden	−16.86	Belgium	−21.55
Slovakia	−10.47	Netherland	−16.87	Iceland	−22.86
Spain	−11.67	UK	−17.39	USA	−26.23
Norway	−11.82	Poland	−17.73	Australia	−26.69
Switzerland	−12.62	Finland	−19.47	Ireland	−37.53
Portugal	−13.81	Japan	−20.10		

Source Statistics of pension funds in OECD

trillion US dollars at the beginning of 2009, making a loss of nearly 5.5 trillion US dollars.[11] ③ With economic globalization going deeper, people go among countries and regions more frequently, which increases the difficulty of the management and

[11]Zheng [7].

service of pension security. ④ The growing right consciousness to pension security by citizens in various countries increases the urgency of the pension security system construction and the complexity of the reform.

7.1.3 Regional Characteristics of Pension Security System Have Reflected, But the Global Model Is Difficult to Form

With the degree of economic globalization deepening, the exchange and cooperation among countries and regions go deeper, in which the cooperation of pension security system is also one of the cooperative fields of various countries, and the cooperation is going deeper. In particular, some international organizations promote the international cooperation of pension security faster and better, among which are the institutions mainly including World Bank, International Labor Organization, and International Social Security Association.

Though the regional cooperation of pension security system is increasingly obvious, and regional common features of the system have reflected, the global model of pension security is still difficult to form due to many conditional restrictions.

At present, with the deepening of exchange and cooperation, transnational and trans-regional flow of labor is increasingly frequent, and countries have signed a series of bilateral cooperation agreements on social security according to equal, mutual beneficial and transferable principles (see Table 7.4). The cooperation in the aspect of social security among countries will certainly be even further in future, and there will be more and more bilateral agreements signed by various countries. Even the income transfers among countries also gradually increase, for example, the mutual aid among countries, aids by various voluntary aid organizations, even personal transnational aids.

Since the cooperation of social security among various countries goes further, under the effect of the elements such as similar politics, economy, culture, region and so on, the cooperation of social security among countries and regions goes

Table 7.4 The situation of bilateral agreements on social security signed by some countries (unit: number)

Country	Number of bilateral agreements	Country	Number of bilateral agreements	Country	Number of bilateral agreements
UK	157	USA	97	Japan	7
Germany	224	Canada	180	South Korea	6
Ireland	20	Australia	66	China	2

Source Relevant websites of social security in various countries and Portability Regimes of Pension and Health Care Benefits for International Migrants: *An Analysis of Issues and Good Practices* by Robert Holzmann, Johannes Koettl, and Taras Chernetsky, May 2005; see Wang [45]

further and the common features of pension security in some countries and regions have already reflected, which have formed different pension security models. According to the division by Espin Andersen, the pension security models in the countries around the world can be divided into Anglo-Saxon model, which mainly includes USA, Canada, Australia and so on; Continental Europe model, which mainly includes Germany, France, Austria, Italy and so on; Scandinavia model, which mainly includes Sweden, Norway, Denmark, Finland and so on.[12] Besides, it can even include Latin American model, East Asian model, East European model and so on. These models are not identical institutional arrangements, but they reflect some common features to a certain degree.

The regional common features of pension security have already been reflected, but it is still comparatively difficult to form a global unified model, and a global unified model of pension security needs many elements, such as socialized mass production, pressure group, basic social value identity, central government and specialized management institutions. However, the national level has already been provided with these elements, but it has not yet been fully provided with on a worldwide scale. ① Also the most core is that a worldwide world government is in absence, and a worldwide transferring payment system is in absence, which is also very difficult to realize. Though with the gradual deepening of economic globalization, the speed of transnational flow of products, capital, service, personnel is accelerating; however, government functions of sovereign states continue being restricted by international political orders, while the social authority is still nationalized, and the functions controlling resources and regulating incomes are still confined within traditional national borders. Even in the regional organization of European Union which is at the highest integration level, its central government located in Brussels is also unable to conduct large-scale social re-distribution by general tax like sovereign states. ② At present, since the universe value standards in common are still in absence, the international politics is still national, regional and geo-relational to a large extent. ③ There also does not exist worldwide strong professionalized network providing social service; the existing professionalized international organizations and non-governmental organizations are expanding working and service objects in the world, meanwhile, they are conducting mutual networking communication. However, their function fields are very limited, which mainly remain in the remedial working scopes of danger relieving, poverty alleviation and neediness help.[13]

Thus it can be seen that, the global model of pension security is still difficult to form for the time being. Under the background of globalization, there exist demands for the internationalization of pension security system, and under the condition that it is difficult to form a global model, it needs to strengthen cooperation in pension security among nations and regions, and relevant international

[12] Costa [8].

[13] Hou [9].

organizations should come into play (e.g. United Nations) with the functions of coordination, communication and guidance.

7.1.4 The Transformation of Labor Market Influences the Development and Reform of Pension Security System

The transformation of labor market is closely linked with the development of pension security system, both of which are mutually effected. The transformation of labor market directly influences the establishment, development and reform of pension security system; similarly, the establishment, development and reform of pension security system also has an important effect on the labor market. Here, we mainly analyze the influence of the transformation of labor market to pension security system.

With the accelerating process of industrialization, urbanization and informatization as well as the deepening of the degree of globalization in various countries, labor markets in many countries have changed dramatically, which has important influences on the development and reform of pension security system, with the main performances as follows:

(1) With the deepening of the degree of population aging, laboring population relatively decreases. This feature of population structure increases the demands for pension security, which brings heavy pension pressure to laboring population. As stated above, the rising dependency ratio of the elderly also brings heavy pressure to the governments of various countries in future, and with the growing population aging, the expenses on old-age pension in some countries are on the rise (see Table 7.5). The development and reform of pension security system must adapt to population structure all the time to better guarantee the material demands, economic demands, service demands and spirit demands of the elderly during the process of the positive interaction with economic and social development.
(2) The diversified forms of employment. With the accelerating process of marketization, industrialization and urbanization, the division of labor becomes increasingly professionalized, and various forms of employment continuously emerge, with not only full-time employment and formal employment, but also a large number of part-time employment and informal employment. In the circumstances that formal employment has not completely been absorbed into the scope of governmental pension security, the pension security of a large number of informal employment is more problematic due to the instability of employment and lack of normalization of labor relations, which is an important task and problem in the development and reform of pension security of various countries in future.

Table 7.5 Expenses on old-age and disability pension in some countries

Country	Proportion in GDP		Proportion in government expenditure		Country	Proportion in GDP		Proportion in government expenditure	
	1990	2005	1990	2005		1990	2005	1990	2005
Australia	3.1	3.5	8.6	9.9	Italy	10.1	14	19.2	29
Austria	11.7	12.6	22.7	25.3	Japan	4.9	8.7		22.7
Czech	6.1	7.3		16.3	South Korea	0.8	1.6	3.8	5.4
Denmark	5.1	5.4	9.2	10.3	Poland	5.1	11.4		26.3
Finland	7.3	8.4	15.1	16.7	Portugal	5	10.2		22
France	10.6	12.4	21.5	23	Spain	7.9	8.1		21
Germany	10	11.4		24.3	Switzerland	5.6	6.8	18.3	19.1
Greece	9.9	11.5		26.6	UK	4.9	5.7	11.9	12.8

Source OECD social expense database, OECD major economic indicator database

(3) The decrease of labor participation rate of the elderly directly influences the need and supply of pension security. From 1960 to 1985, the ratio of elderly male labor force in major developed countries quitting the labor market grew rapidly, resulting in the rapid decrease of labor participation rate of male labor force aged from 55 to 64 years old, which generally drop from 80 to 90% during 1960 to 1962 to 50 to 70%[14]; In the middle of 1990s, the labor participation rate of the labor force of the group aged from 60 to 64 years old in Belgium, Italy, France and Finland is lower than 20%, which has dropped to 35, 40 and 53% in Germany, Spain and USA respectively.[15] The decrease of labor participation rate results in the reduction of pension security supply and the relative growth of demands, which even leads to a number of early retirement or unemployment in some countries with employment recession, and also affects the demand and supply of pension security.

(4) The labor flow increasingly accelerates. Free flow is not only people's rights, but also the need of national development; the acceleration of labor flow is an inevitable phenomenon under the background of marketization and globalization, and not only the labor flow within a country accelerates, but also the labor flow among countries accelerates.[16] The development and reform of pension security system in various countries in future need to adapt to the trend of the acceleration of labor flow, and establish corresponding pension security system.

[14]Zhang [10].

[15]Lin and Chen [11].

[16]Data About Transference of People May Refer to the United Nations Development Programme [12].

7.1.5 *Reform of Pension Security System: Continuous and Formidable Game*

With the change of population as well as economic and social environment in various countries, the reform of pension security system also continues, and especially after 1970s, reforms of pension security system in various countries are more frequent, with reform measures being more diversified. The main target of the reform of pension security system in various countries is to seek financial sustainability of pension security system and dynamic adaptation of pension security system with economic and social development.

The world is constantly changing, and the reform of pension security system is also continuing with no end. Pension security system involves complex relationships among different subjects, and as to the state, it is the duty of the nation to provide appropriate pension security for citizens, guarantee citizens' basic livings, and let citizens share the achievement of development; as to an individual person, pension security is not only rights, but also an obligation, and an individual person can obtain pension security support from the nation according to relevant laws and should also perform appropriate duties. Pension security system is also closely associated with relevant organizations, such as units, labor unions, old-age groups, political parties. Taking political parties for example, as an important group, the elderly has important influence to different political parties and their pension security policies. According to relevant investigations, the older the age, the more ratio of emphasis on pension security system, and the generation at 60 years old is at the highest ratio, which has reaches 57%, while it is 53% of the generation at 50 years old and 27% of the generation at 20 years old. The investigation by Asahi Weekly showed that, in Japanese House of Councilors election in 2004, in the assessment on governmental pension policies, 78% of the voters selecting the option of "higher assessment" voted due to their high attention to old-age pension.[17]

The reform of pension security system means the adjustment of interest models among different subjects, and since reform is of higher sensitivity, it is always quite difficult. With the continual growth of consciousness of civil right and the increasingly development of society and economy, the demands for pension security are also continually increasing, in this circumstance, it should coordinate relations among different subjects even more to formulate perfect reform schemes and reduce resistance to the reform.

[17]Feng [13].

7.2 Main Types and Models of Foreign Pension Security Systems

7.2.1 Main Types and Features of Foreign Pension Security Systems

According to different standards, pension security systems can be divided into different types. According to the ways defining treatment, they can be divided into the two types of fee-payment defining system and fund defining benefit. According to the levels of systems, they can be divided into single-level pension security system and multilevel pension security system. According to the different subjects of responsibility, they can be divided into national statutory pension security system, enterprise complementary pension security system and individual savings pension security system. According to the different voluntary degrees of participating pension security system, they can be divided into mandatory pension security system and voluntary pension security system. According to the fund raising models of pension security fund, they can be divided into the three types of pay-as-you-go system, partly-funded system and fully-funded system. According to the different managing subjects, they can be divided into pension security system managed by the government and pension security system managed by individuals. Thus it can be seen that, according to different standards, pension security systems can be divided into different types, and in the types of pension security systems distinguished by different standards, some are coincident and some are independent. For example, the financing model of pay-as-you-go system always implements fund defining benefit, and partly funded accumulation pension security system or fully funded accumulation pension security system always implement fee-payment defining system. The following analyzes different types of pension security systems mainly according to financing and funding ways of pension security systems.

1. Distinguishing from financing ways

(1) Pay-as-you-go system mainly seeks short-term financial balance, which generally maintains balance between income and expenditure in the year. This type of pension security system are always with no surplus funds or with precious few surplus funds, and the payments by all the insured objects of pension security are used for issuing pension outlays in the current period, that is intergenerational support by the younger generation for the retired generation. The system type implements the principles of social pooling and mutual assistance, which is beneficial to the re-allocation of incomes among different groups. It shows social equity and is simple as well as practical, which is more convenient to operate, with no excessive fund accumulation and no pressure of investment as well as value preservation and increment, and it is unnecessary to worry about the influence brought by economic fluctuation and inflation. However, with the deepening of the degree of population aging, the burden brought by this type of pension security system gets heavier and heavier, and it needs to either increase

the payment by the younger generation to use for pension outlays, or invest by government finance. After the Second World War, the pay-as-you-go system becomes the main model of basic pension security system in industrialized countries, such as UK, France and so on. However, with the emerging of economy stagflation and the deepening of the degree of population aging in 1970s, this type of pension security system sank into financial distress[18] which forced various countries to carry on reforms to it.

(2) Fully-funded system accurately calculates a certain payment level in advance according to the principle of long-run balance between income and expenditure and the treatment level to be provided, and individual persons' payment accumulation for future pension during the period of employment is fully individual account accumulation, the funds accumulated by which will be managed and invested by governments or private institutions. This type of pension security system is conductive to mobilizing personal initiative of payment, which is less affected by the change of population structure, and benefits to coping with the demands of population aging. However, the fully-funded system is lack of mutual assistance, with bad re-allocation effects, and the funds accumulated will be of greater risks under the circumstances of economic fluctuation and inflation, with heavy pressure of value preservation and increment. During all previous financial crisis, countries implementing funded system were severely impacted, part of which had suffered great losses. Countries implementing fully funded accumulation pension security system mainly include the countries and regions in Latin America and Southeast Asia.

(3) Partly-funded system synthesizes the features of the pay-as-you-go system and fully funded accumulation system, and implements the principle of combining social pooling with individual account, while the part of individual payment is used to pay for current pension outlays, and partly used for individual account accumulation, which in fact is the combination of generational balance and horizontal balance during different age groups of individuals of pension security. This type of pension security system is partly constituted of mutual assistance, and it also mobilizes personal initiative of payment at a certain degree, the funds accumulated by which is fewer than those accumulated by fully-funded system, with relatively less risks correspondingly. At present, it is an important trend to implement partly funded accumulation system for the reform of pension security system in various countries in the world, which is also the basic pension security system implemented by China currently.

Refer to Table 7.6 for the comparison on the three types of pension security systems.

[18]Mu [14].

Table 7.6 Comparison on three types of pension security systems

System type	Pay-as-you-go system	Fully-funded system	Partly-funded system
System concept	Emphasis on fairness	Emphasis on efficiency	Combination of fairness with efficiency
Financing features	Defining payment according to outlays	Defining outlays according to payment	Combination of the two
Effect of re-allocation	Better	Poorer	Moderate
Management organization	Managed by the state	Managed by the state or private institutions	Combination of public institutions with private institutions
Income transference	Intergenerational support	Life cycle transference	Combination of the two
Risks faced with	Population risks	Market risks	Management risks
Pressure of value preservation and increment	Smaller	Larger	Moderate

Table 7.7 Models of pension security systems according to financing and funding

Classification of pension security systems		Financing ways	
		Pay-as-you-go system	Fund accumulation system
Funding ways	Fee-payment defined system	Notional defined contribution (NDC), such as Sweden, Poland, Mongolia, Latvia and other countries	Fee-payment defining contribution (FDC), such as Chile, Singapore, Malaysia, Hong Kong of China and other countries and regions
	Fund defining benefit	Defining benefit by pay-as-you-go system, most countries implemented this type before reforms	Fund defining benefit (FDB), such as the enterprise annuity in some developed countries

2. Distinguishing from ways of treatment defining and funds

The treatment level of old-age pension is determined by individual payment level and investment income in fee-payment defining system, which reflects as "defining outlays according to payment". A payment standard is defined in advance, and different subjects pay the fees according to different responsibility sharing mechanisms. The fees paid may enter into social pooling, or individual account also, or the combination of the two. The old-age pension treatment received by individuals when they quit labor market at the old age is closely linked with individual payments. In practice, fee-payment defining pension security system always implements the model of partly funded accumulation or fully funded accumulation. And there is a fixed funding standard in advance in funding defining system, which is not necessarily linked with individual payments, and individual incomes are comparatively fixed, which reflects "defining payment according to outlays". In practice, the comparatively common practice of funding defining system is to implement pay-as-you-go system, and the major systems combining financing ways and funding ways include pay-as-you-go system of fee-payment type, fund accumulation system of fee-payment defining type, pay-as-you-go system of funding defining type, fund accumulation system of funding defining type (see Table 7.7).

7.2.2 Major Controversies on Selections of International Pension Security Systems

In the reform process of global pension security systems, especially in the reform process of the transformation from pay-as-you-go system to fund accumulation system, there arose a lot of controversies. Some international organizations and experts proposed to retain the original pay-as-you-go model, and make adjustment and improvement on this basis; some other organizations and experts supported the transformation to fully funded accumulation system or partly funded accumulation system. There shows some mitigatory and approaching trends on this controversy in recent years, but there still exists comparatively large divergences.

The controversies on the models of pension security systems by international organizations mainly lie in the controversies between World Bank and International Labor Organization. Thereinto, World Bank actively advocates fund accumulation system and multilevel pension security system. World Bank considers that, there exist a lot of problems and deficiencies in pay-as-you-go pension security system. For example, too high payment rate, too heavy burdens on enterprises, too generous treatment cannot comparatively well cope with the challenge of population aging. In 1994, World Bank proposed a three-pillar pension reform scheme: the first pillar was the pay as you go pension security system financed by tax and managed by the government, aiming at relieving poverty of the elderly to realize the re-allocation of incomes; the second pillar was the privately managed occupational pension security system with fully funded accumulation paid by wages; the third pillar was individual voluntary pension security system. In 2005, World Bank proposed a five-pillar model additionally, which added the zero pillar of non-payment type and the fourth pillar of informal form on the basis of three pillars. The reasons why World Bank tended to fund accumulation system are: ① it defined the connections between payment and income, which reduced the distortion of labor markets; ② the income and expenses in the model of fund accumulation was comparatively transparent, which was beneficial to enhancing the incentive of the system; ③ this kind of model was operated according to market mechanism, which was beneficial to improving the efficiency and effect of the system; ④ it is beneficial to promoting economic development by increasing national savings through fund accumulation.

However, International Labor Organization considers that, the major target of pension security system is to protect basic livings of the elderly and maintain social fairness. The five basic principles to promote the reform on pension security system are: ① the system should cover all the population; ② it should prevent the elderly from falling into poverty; ③ predictable guaranteed retirement income should be provided; ④ retirement pension should be adjusted according to inflation; ⑤ representatives of employers and employees are to be absorbed for democratization management.[19] International Labor Organization considers that, fund accumulation pension security system goes against the five principles, and accumulation pension security system is not necessarily beneficial to population aging, on the contrary, it will bring greater risks and uncertainty with comparatively high transition costs. Meanwhile, it considers that, there are advantages of mutual assistance, promoting income re-allocation, reducing management costs and so on in the pay-as-you-go system, and to the problem existing in it, the parameters of pension system should be adjusted and improved by reform, but not transferring to fund accumulation system, for example, measures such as lengthening retirement age, properly raising payment rate, lengthening payment years and so on. International Labor Organization also proposed a multi-pillar reform scheme, whose difference was that, the second pillar was still the pay-as-you-go funding defining system managed by the government, not the fund accumulation system. The view of International Social Security Association is relatively consistent with that of International Labor Organization.

Some experts and scholars have conducted researches on the reform of pension security system, and have presented different opinions. Feldstein advocated implementing fully-funded system in individual accounts and private management. He considered that, transition to individual accounts could resolve the financial distress faced with pension security systems, and increase capital accumulation, which was beneficial to economic growth and national welfare, and he also considered that, the costs of fund accumulation system was lower than that of pay-as-you-go system, which could reduce tax rate distortion and deadweight loss, while pay-as-you-go system generated "crowding-out effect" to savings, which impeded economic growth.[20] He not only required public old-age pension system of pay-as-you-go to transform to individual account system, namely individuals retained the title to old-age pension fund, but also required management right to the accounts by individuals.[21] From the view of plutonomy, Salvador Valdese-Pareto from Catholic University of Santiago, Chile considered that, the pay-as-you-go pension security system was not safe "security", but an asset subject to the moral hazard of politicians.

Stiglitz deeply analyzed ten existing prejudices against old-age pension reform, and considered that, privately managed old-age pension might be conductive to deepening domestic capital market, but it also made the elderly be faced with greater risks, with more resources lavished on administrative costs also. He considered that, the second pillar should be integrated with well-designed treatment

[19]Yang [15].

[20]Martin [16].

[21]Sun and Li [17].

managed by the government to confirm the plan of institutional old-age pension, while the second pillar privately managed was really not the optimal choice. Peter Diamond considered that, the advantages of individual accounts of payment type had been overstated.[22] Diamond and Valdese-Pareto considered that, the decentralized administrative costs of fund accumulation system were too high, with moral hazard and huge implicit debts at the same time. Nicholas-Barr raised doubts about that fund accumulation system could resolve the problem of population aging, and he considered that, fund accumulation system did not necessarily promote the increase of savings, since on the one hand, the fund accumulated needed to be used for payment, on the other hand, the increase of mandatory savings might result in the decrease of voluntary savings; also, even the increase of savings was not necessarily used for investment to increase revenues. Some experts opposing the transformation from pay-as-you-go system to fund accumulation type also considered that, this kind of transition would result in a lot of old-age pension debts, which would bring too much pressure on the younger generation. Nicholas-Barr and Diamond indicated the profiling error existing in the old-age pension reform by World Bank and other relevant organizations, including the too confined and simplex target, which mainly concentrated on the economic aspect; inappropriately using the optimal analysis method; inappropriately using of the static analysis method; incomplete analysis on implicit debts; incomplete analysis on the effect of fund accumulation system; neglect of allocation effects. They considered that, there was no sole and optimal old-age pension system, and different system designs were all possible to increase or reduce national welfare, which concretely depended on the elements such as economic development level, the capabilities of administration and implementation and so on of various countries, and also designed different three-pillar models for low-income countries, middle-income countries and high-income countries respectively.[23]

With the increasingly deepening of reform exploration and theory research in some countries, the controversies on pension security system were somewhat alleviated, and the original standpoints of World Bank and International Labor Organization had somewhat changed. World Bank paid attention to the economic efficiency of fund accumulation system, and started to lay emphasis on its income allocation effect at the same time, meanwhile, it also considered that, the selection of fund accumulation system by various countries were not completely the same, with the different population and economic conditions in various countries, the implementation effect would be comparatively greatly different,[24] and International Labor Organization insisted the effects in the aspects of expanding the coverage of pension security, safeguarding the basic livings of the elderly by pay-as-you-go system, meanwhile, it also considered that, the influence of pension security system to economic efficiency should be taken into account, therefore, while insisting to

[22]Robert and Stiglitz [18].

[23]Barr and Diamond [19].

[24]Holzmann [20].

improve the model of pay-as-you-go system, pay-as-you-go system is proposed to be combined with fund accumulation system.[25] In fact, the five-pillar model by World Bank was also part accumulation system. In fact, the reform trend in recent years also proves that, complete fund accumulation and complete pay-as-you-go system are all difficult to realize sustainable development, especially in the circumstances of the increasingly deepening of population aging and constantly changing of economic developing environment in future, the main features of the two models should be combined, and the two models should be comprehensively applied with emphasis according to national conditions of various countries.

7.2.3 Main Models of Pension Security System in the World[26]

As mentioned above, the establishment and development of the pension security system in various countries are the results of combined effects by various elements, including the aspects of politics, economy, culture and so on. Combining the influences of various elements and the practice of pension security by different countries, with reference to the divisions on welfare models by some scholars, here, we also try to divide the pension security systems in the world. On the whole, they mainly include the following models:

1. Anglo-Saxon model

Anglo-Saxon model originally refers to that, in the countries implementing liberal economic developing model, under the guidance of this economic developing model, the social welfare systems in these countries also abide by Anglo-Saxon principle, and as the core of social welfare system, the pension security systems in these countries are also such model naturally. The features of Anglo-Saxon pension security system model are mainly embodied in the liberalism of ideology; the market-based and nation-auxiliary principle provided by pension security; the principle of safeguarding the basic on treatment funding; comparatively strict in qualification examination; the lowest degree of "de-commodification". The countries implementing this model mainly include USA, Canada, Australia and so on.

Ideologically, Anglo-Saxon model sticks to liberalism, emphasizes on free competition, seeks for efficiency, proposes tax reduction, eases the burden on enterprises, deregulates, encourages privatization and personal wealth accumulation. With connection to liberalism, the features of individualism are quite obvious in this kind of countries, with emphasis on self-reliance and self-independence of individuals, so the pension also mainly depends on themselves, with comparatively little provision by the nation.

---

[25]Colin [21].

[26]Long [22].

The fund procurement of pension security emphasizes individual payment accumulation, and purchasing relevant private pension securities from the market to provide pension securities for individuals. Therefore, the private old-age pension markets in these countries have been relatively well developed. The nations provide comparatively less pension security directly, instead, they encourage the development of private markets by providing certain tax and policy supports to the market. The part with the most rapid increase in the expenses of public pension is the subsidies for taxes and dues of the so-called "private" plan. Certainly, governments also perform appropriate functions in the aspects of legal system construction as well as supervision and administration in addition.

On the treatment payment of pension security, the counties of this model implement remedial or residual way of provision, with the character of helping the poor, and the bits of inclusive transfer payments or limited social security plans provided by the nation are mainly offered to those people in extreme poverty with lower income, namely the vulnerable group in these countries. Though the states provide a certain inclusive salvation treatment, it is still in a very low treatment level, and strict qualification examination, even with some humiliating examination methods needs to be passed. Other groups obtain corresponding treatments or corresponding pension services by the way of market without exception, and these people are always the middle class of the countries, with relatively higher income level, who also comparatively support the market-oriented ways of supply.

From the view of the degree of de-commodification and social rights of the citizens, the degree of de-commodification is the lowest in this kind of model, and the residual type of treatment supply and strict qualification examination greatly weaken the effect of de-commodification. Lower de-commodification degree is strongly restraining the expansion of social rights, which results in the stratification of social groups to a certain degree. It is likely to cause gaps and inequality among different groups. That is to say, the fairness of this model is comparatively low when encouraging to improve market efficiency at the same time, which is likely to turn up with polarization that the rich becomes richer while the poor becomes poorer. Under this model, the superiority of the market is at the cost of sacrificing social security or privileges of civil servants, or both of the two at the same time, with Australia, Canada, Switzerland and USA all to this model. In these countries, the pension security of civil servants also illustrates the responsibilities of the government as the employer, but it is also determined by the market to a great extent.

From the view of class basis, in the liberalism institution, the middle class gradually institutionally combine with the market, so the middle class are the strong supporters of Anglo-Saxon model, who consider they should prepare for their own old-age pension by their own efforts, but not sit back and wait for the provision from the state. For a long time, the middle class mainly obtain pension support from private sectors and the market, and they oppose the inclusive pension treatment provided by the state. In view of the important position of the middle class in political elections, it is quite logical that every step of expansion by the activities of welfare states is all resisted.[27]

2. Continental European model

Continental European model refers to the model of pension security in some Continental European countries, which mainly include Germany, France, Austria, Italy and so on. The countries implementing this type are comparatively greatly influenced by conservatism or corporatism, which lay emphasis on providing pension security for citizens through developing social pension security by the government, and stress the combination of rights with obligations, as well as the connection with employment, while the provision of pension security needs to combine the state with the market. The degree of de-commodification is moderate, which is higher than that of Anglo-Saxon model, but lower than that of Scandinavia model.

Ideologically, Continental European model mainly follows the principle of conservatism or corporatism, with emphasis on mutual cooperation and mutual support among the state, the society, employers and individuals. The government moderately intervenes economic and social development of the country to play a role of coordination and balance, and satisfies different interest demands of different groups by government intervention to achieve the target of coordination and integration. Corporatism is also called as "corporatism" and "corporativism". The formation of corporatism had been comparatively greatly influenced by the church, while the Roman Catholic Church had succeeded to strongly promote social reforms, and *Papal Encyclical* in the late 19th century advocated the combination of statism and corporatism; then, Encyclical in 1931 laid more emphasis on the elements of corporatism. At the same time, Catholic parties also emphasized strict absolutism and authoritarianism. From the view of developing history, conservatism which advocated implementing paternalistic absolute authority had played an important role in Continental European model, and the paternalistic authority rooted in the feudal lords society and the autocratic monarchy system in Europe and Russia, which emphasized strict classism, absolute authoritarianism of the government and absolute obedience to the nation by individuals. In addition, the reason why corporatism is strongly controlling continental European countries is greatly connected with Guild tradition of these countries.

The ideology of corporatism in continental Europe has important influence on the pension security systems in these countries. These countries mainly adopt the social security pension model, however, in the aim of cooperation and integration, different pension security systems need to be adopted for different groups to satisfy the differentiated pension demands of these groups, which results in the serious fragmentization of pension security systems. As Continental European model is

[27]Costa [8].

also influenced by the new middle class, and with influence by the development of conservative political forces, the loyalty of the middle class to the state is institutionalized with the practice of dividing pension security plans according to occupations, which fundamentally safeguards the political elements that make the middle class difficult to form. For example, the large-scale pension reform conducted by Adenauer in 1957 recovered the loyalty of the middle class to the state. Continental European model formulated welfare plans with qualitative differences for the groups in different classes and positions, each of which were all with distinct and characteristic rights and privileges, intending to fix people on appropriate social and living positions; link individual loyalty directly to the authority of the monarch or the central government,[28] so the social class structure was comparatively stable.

The corporatism in the countries of Continental European model has important influence on the pension security systems of civil servants in these countries, and some countries, such as Germany, France, Austria provide civil servants with particularly generous pension treatments, which shows the special position of the civil servant group, aiming at promoting and safeguarding long-term loyalty and reverence by civil servants through special rewards to civil servants and shaping class structure by their control on social policies. In essence, the old-age pension of civil servants possesses the property of occupational pension, which illustrates the responsibilities of the government as the employee; the old-age pension of civil servants is raised and paid by national finance, which is not related to statutory social rights, but related to the special position of civil servants. Therefore, the specificity of the pension security system of civil servants fully reflects the ideology of statism and corporatism in Continental Europe.

Other than the pension security of civil servants, the pension security of other groups implements the principle of insurance and is also closely connected with employment. In a manner of speaking, the pension security systems in Continental European countries strictly follow the principle combining rights with obligations, in which the precondition for individuals to enjoy pension security treatment is the necessity of performing certain payment obligations according to personal income ratio, meanwhile, the state and employers also pay expenses. The treatment receiving also needs to be related to payment years, age, personal physiological status and so on. The payments and treatments among different groups illustrate certain differences. In a manner of speaking, Continental European model not only emphasizes the functions of policy making, financial support, supervision and administration and so on in pension security by the state, but also pays attention to playing the roles of the market and individuals, emphasizing the solidarity and cooperation among different subjects including the state, the society, employees, individuals and so on.

[28]Costa [23].

3. Scandinavia model

Scandinavia model refers to the pension security model in some countries in Scandinavian Peninsula or adjacent to Scandinavian Peninsula, which mainly include the countries of Sweden, Norway, Denmark, Finland and so on. Ideologically, these countries with comparatively high degree of de-commodification implement social democracy, emphasize the principle of equal status, respect social rights of citizens, and implement the principle of generalized system of preferences in pension security systems covering all classes, which is also called as "national welfare" model.

Ideologically, Scandinavia model implements social-democratic regime emphasizing democracy and social rights, which is neither fully capitalism, nor fully socialism. Social democrats cannot tolerate not only the dualization situations between the state and the market, the working class and the middle class, but also the great disparities and oppositions among different groups; they seek for high-level equal target, but not the satisfaction of floor-level basic demands.[29] In other words, the service and funding level of pension security are relatively high, so it can satisfy the pension demands of the low-income group, as well as the pension demands of the middle class at the same time, to ensure that all the people can equally enjoy pension security rights to achieve their respective pension demands. Therefore, Scandinavia model manifests high de-commodification principle and the principle of generalized system of preferences, which conform to the needs of different classes, and ordinary manual workers gradually enjoy equal pension rights to white collar workers or civil servants, with all classes to be incorporated into an inclusive pension security system.[30] Certainly, the generalized system of preferences here is also not fully identical at all, with differences in financing and treatment level. This model with comparatively strong income allocation effect mainly emphasizes the responsibilities of the state and the government, though the function of market cannot be eliminated, it is also quite weak.

Based on the above ideology of democratic socialism, the operation process of pension security system fully illustrates the principles of equality and generalized preferences. The establishment of pension security systems should mainly take the equality of different groups into account, with consideration to the otherness of different groups. The establishment of pension security systems is not separate like Continental European model, but implements relatively unified pension security system, either with identical systems, or partly the same. On the fund procurement of pension security, it is mainly provided by national finance, but individuals must pay relevant taxes and dues for high-level pension security treatments as the material basis of pension security in generalized system of preferences, in other words, these countries reflect the characteristics of high tax and high welfare. On the aspect of treatment level of pension security, the treatment standards are relatively unified, and treatment levels are comparatively high, which also properly give consideration to the otherness of treatment levels among different groups at the

[29]Zheng [24].

[30]Zheng [25].

same time. The treatment examination is not as strict as that of Anglo-Saxon, which always takes residence and citizen qualification as basic conditions. The countries implementing this model are of the highest degree of de-commodification, in which the social rights of all citizens eliminate the influence of positions, privileges and markets, and the countries including Norway, Sweden, and possibly Denmark, Finland as well as Netherlands implement this type of system.

On the whole, the pension security systems in various countries in the world can be divided into three models, but whether there are three models or more models, it is still very difficult to include the pension security systems in all countries in the world, and there does not exist a pension security system in a country with single feature, in most cases, they are versatile, and since the dynamics in the economic and social development in various countries, the pension models of various countries in different periods also change. Scandinavian countries may mainly manifest the features of social democracy, but also not totally eliminate the elements of liberalism. Also, Anglo-Saxon model is also not unmixed, for example, the pension security system of USA possesses the features of re-allocation and enforcement, which is far from actuarial studies, and at least the economic policy at the period of Roosevelt's New Deal is similar to contemporary Scandinavian social democracy. The conservative system in Europe is also constituted by liberalism and social democracy. Over the past few decades, European countries increasingly become neither corporatism, nor totalitarianism.[31]

7.3 Basic Rules of the Development of Foreign Pension Security System

From the view of the practice of development history and reform of foreign pension security systems, the pension security systems in various countries have their own features due to different national conditions, different development histories and development stages, different political systems and cultural traditions. However, we can conclude some basic rules from the development practice of pension security systems in various countries.

7.3.1 It Is the Common Characteristics in Various Countries in the World to Transform from Family Security to Social Security

For a long time, the later lives of the elderly in various countries have been supported by family members, which is called "family security", and the function of

[31]Cheng [26].

family in the aspect of providing for the aged should not be ignored up till now, especially in some eastern countries with the influence by Confucian culture, whose attachment to their families is deep-rooted, with families becoming the destination of people's material life and spiritual life. As an informal and non-institutional pension form, the family has started to shake today, and has gradually started to transform to social and institutional pension forms. The transformation of family pension form is closely bound up with the development of social productive forces and the social changes thus initiated.

Family pension was the main pension form in the agricultural society, when the productive forces had developed to a certain degree but still in an undeveloped stage compared to primitive society. During this period, laborers started to support themselves and their families by their labor when they got personal freedom and means of production. With the growth of age, various skills of laborers became more proficient, and their experience became richer, with more and more disposable resources. Because of this, the elderly possessed higher status in the family and society, who were at the core of the family and were respected by the society and individuals, and it became the due obligations of descendants to support the elderly, otherwise, they would lose family property and be condemned by public opinions.

However, with the continuous development of productive forces in various countries, various types of requirements by people were also increasingly growing, and the original production mode was already unable to meet various types of requirements by people, which had impeded the development of productivity. The operating forms of traditional agriculture and handicraft industry already could not meet the requirements of development. Western countries headed by the UK initiatively started to explore developing modern industry, and the development of modern industry had brought great impact to land as resources for production and the traditional productive mode. People started to go out of families to workshops to obtain sources of income for the existence of individuals and their families. With the emergence of large-scale mechanized industry and the rapid development of modern science and technology, traditional skills possessed by the elderly cannot meet the needs of the development of industry already, and the property, resources and social status possessed by the elderly also gradually declined. With the acceleration of the pace of life, the accompanying and attending at home by the young to the elderly were confined by time and distance. Thus, it was already more and more difficult for the family to be the main form of provision for the aged in industrial society. With the acceleration of industrialization process, various countries started to explore to establish social pension security systems out of the family. The real establishment of modern social pension security system arose in the late 19th century, which was symbolized by the promulgation of *Old-age And Disability Security Law* by Germany in 1889 and implemented old-age and disability securities for all workers and ordinary officials in Germany, with three parties of the state, units and individuals responsible for the financing of security funds together which relied mainly on the payments by units and individuals, with appropriate subsidies by the government. Workers would receive retirement pensions according to their wages on the job after retirement. Those over the age of 71 with premium paid for over 30 years were entitled to receive corresponding retirement treatment.[32] Hereafter, some of the Occident also established different types of social pension security systems in succession.

After the Second World War, pension security systems in various countries continuously improved, and the levels of pension security treatments increasingly raised, with the cooperation mechanisms in the aspect of pension security systems among the state, society, units and individuals constantly improved. Especially after "WWII" (the Second World War), the tide of establishing welfare states in succession in western countries further promoted the development of pension security systems. With the acceleration of globalization, industrialization, marketization, informatization processes, various countries made new requests to the development of pension security systems, and the economic and social development and transformation would further promote the development and improvement of pension security systems in various countries. However, anyway, it is already an irreversible trend in the development process of pension security systems in various countries to transform from family security to social security.

It should be noted that, the objective law of the transformation from family security to social security is not in contradiction to the continuously important function of the family in pension security. In the future, no matter how perfect the social security pension systems will be, and no matter how high the treatment level of pension security will be, it is impossible to shuffle off all the pension responsibilities to the society, and the family will play an irreplaceable role in spiritual consolation for the elderly, which is also very important in the functions of service for the aged and economic support. In the development process of pension securities in various countries in the future, it should properly play the function of the family to effectively combine family security with social security.

7.3.2 The Modes of Government Functions Under Different Pension Security Models Are Different

Pension security systems are institutional ways of old-age care provided by various countries for citizens when entering the industrial society, but the functions by the governments under different pension security system models are different, which actually are different combinations of responsibility sharing mechanisms among the government, market, society, employers and individuals, in which it is an important aspect to illustrate the characteristics of different pension security models by dealing with the relations between the government and market, thus also reflect the different functions of governments under different pension security system models.

The countries of Anglo-Saxon model adopt the ways to attach most importance to the provision by the market, with the provision by the state as auxiliary to

---

[32]Jiang [27].

provide pension security for citizens under the guidance of liberalism. The main duty of the state is to formulate relevant laws, regulations and policies to provide basis and standard for the development of various types of pension securities and enhance qualification examination as well as supervision and administration. The government provides imperative basic pension securities for the vulnerable group of those who are unable to provide self-security, and encourages competent people to obtain securities by the market way. However, they are also not laissez-faire governments that completely let the market alone, but enhance supervision and promote the development and improvement of private pension markets by granting tax or policy support. The countries of Anglo-Saxon model always provide securities through indirect ways, with comparatively less direct provision; through comparatively more legislative, administrative and policy supports, but comparatively less direct financial supports.

The countries of Continental European model implement the pension security system mainly of social insurance, and emphasize the cooperation and equal participation among the state, society, enterprises and individuals due to the tradition of corporatism, which needs to take the differences of pension requirements among different groups into account, not just the basic pension requirements of the vulnerable group. Therefore, the fiscal expenditures on pension securities by the government are obviously more than those of the countries of Anglo-Saxon model. The social securities in the countries implementing this model emphasize the government-leading principle, which not only need to pay most of the expenses for incapable insured objects, but also perform the responsibilities of employers to pay for the pension of civil servants. In addition, the government finance is also the last guarantee for social pension security systems which are faced with payment crisis. In the countries implementing this model, the functions by the government and market in pension security are comparatively balanced with each other. Compared to Anglo-Saxon model, its burden of financial expenditure is relatively heavier; however, compared to Scandinavia model, its burden of financial expenditure by the government is relatively lighter. In other words, Continental European model not only performs the responsibility of proper financial support by the government, but also emphasizes utilizing functions of the market. Other than proper financial support, governments in these countries also shoulder the responsibilities in the aspects of policy making, supervision, administration and so on.

In the countries of Scandinavia model, governments of the states are almost responsible for the overall provision of relatively high-level social welfare for citizens, and in the aspects of burden sharing among the state, society, employers and individuals, governments take most responsibilities for not only providing most capital sources for pension security, but also the specific operation and supervision of pension security systems. And the provision ratio which the market is responsible for is relatively lower. Therefore, in the countries of Scandinavia model, the burdens by governments in social welfare are relatively heavier, which need to be secured by relatively higher tax revenue, and relatively higher tax revenue will affect the competence of enterprises and individual initiatives, even the development of the overall economy in turn. This also becomes the reason for reforms on welfare systems by Scandinavian countries in succession in recent years. However, the reform is faced with more resistance due to the deep impact by the long-developed welfare system and deep-rooted democratic socialism, therefore, it is still very difficult to totally transform the welfare features of these countries in the near future.

7.3.3 The Development of Pension Security Systems in Various Countries Needs to Adapt to Economic Development

The development of pension security systems must adapt to economic development level, which should neither excessively transcend beyond economic development, nor lag much behind economic development. If the development level of pension security systems excessively transcends economic development, it will bring heavy pressure on economic development and restrain economic development, which also affects the development of pension security systems in turn. On the contrary, if the pension security systems lag much behind the economic development level, it will be difficult for citizens to share the achievements brought by economic development, even affect social stability and harmony, which is to the disadvantage of creating a favorable social environment for economic development, as well as the transformation of economic development mode. Specifically, the aspects such as the model design, financing level, financing way, treatment level and so on of pension security systems should adapt to economic development level, thus the development of pension security systems and the economy will stimulate one another for benign interaction.

Throughout the development history of pension security systems in various countries in the world, the economic development has important influences on pension security systems. The quality, level and mode of economic development have specific influences to the aspects such as the security level, security structure, coverage, administration, operation and so on of pension security systems. From the view of the development since the establishment of modern pension security system, in those periods with better economic development, the development of pension security systems is more smooth and rapid owing to good material basis and economic environment. After the end of the Second World War "WWII", various countries carried out national construction to develop national economy in succession, thus the economies of various countries rapidly developed. Meanwhile, the whole social welfare system focusing on pension security systems also rapidly developed, with some European countries declaring to establish welfare counties in succession. Besides, during a long period after the Second World War "WWII", the coverage of pension securities in various countries continuously extended, with the level of pension security continuously advanced and various pension security systems continuously perfected, as well as the administration of pension security being further improved. Citizens in various countries deeply felt the advance of welfare level brought by the development of national economic development, and in a manner of speaking, it was the golden age of the development of pension security systems in various western countries from the end of "WWII" to in the middle 1970s, when the absolute value of financial expenditure as well as its proportion in GDP and financial expenditure for pension securities in various countries increasingly grew.

After the middle of 1970s, with the outbreak of western petroleum crisis, the economies of various countries came out with the phenomenon of low growth, high inflation, high unemployment, and the economic development was severely hindered, while the improvement of social welfare systems in various countries since "WWII" led to the increasing growth of welfare expenditure which brought out many problems by the rigidity of welfare expenditure at that time and aggravated financial burdens on the governments of various countries. Original pension security systems and social welfare systems could already not adapt to the economic development at that time, which were to the disadvantages of enhancing economic competence of various countries and the long-term and sustainable development of pension security system itself was also severely impacted. Therefore, various countries started to conduct reforms on pension security systems, which mainly included increasing income, reducing expenditure, enhancing the financial sustainability of the systems, such as directly reducing treatment level, reducing substitution rate, being strict on the qualification of treatment funding, prolonging retirement age, increasing payment proportion and so on. Some countries even thoroughly reformed original pension security models to transform from pay-as-you-go system to fund accumulation system. After 1980s, some countries boldly operated on welfare systems to strictly control welfare expenditure. From the end of the 20th century to the 21st century, the development of pension security systems in some countries had not gone to the completely opposite extreme, but sought for dynamic balance between economic development and pension security system to achieve healthy development of the two.

The global financial crisis in 2008 strongly proved again, due to the economic crisis brought by financial crisis, the economic development of various countries under the background of globalization had been unprecedentedly impacted. The disadvantageous economic situation not only increased the demands for pension security and social security, but also brought disadvantageous influence to the healthy development of pension security systems, in which an important aspect was the impact to the investment of pension security funds, which were greatly shrunk in various countries by this financial crisis, thus seriously affected the development of pension security systems and the rights of the elderly.

In addition, the economic development mode also had important influence to the selection of pension security systems. Taking Eastern European countries for example, they mainly implemented planning systems before the reform, in which the economy was managed by strict plans of the states, thus the pension security systems also implemented the principle of national security. In the process of transformation from planned economy to market economic system by Eastern

European countries, the pension security model also transformed correspondingly from national security to partly funded accumulation security system.

Also, the development of pension security systems had important influence to economic development, with the functions as the stabilizer and regulator. Pension security and social security systems always generated at the time when the economy was not well developed or with many social problems, for example, the generation of modern social insurance system in Germany in the 19th century was aiming to cope with the various problems came into being since industrialization. The economic crisis in 1930s promoted the establishment of pension security and social security systems in USA, which played comparatively great functions to social stability and economic development recovery at that time. After the middle of 1970s, when the economic developed unsmoothly, the expenditure level of pension securities surpassed the economic development level then, which brought heavy burdens to the economic development of various countries, and stimulated various countries to conduct reforms in consequence. In the global financial crisis in 2007, various countries resolved the problems of living securities of the elderly to relieve laborers' worries about future by improving and perfecting pension security and social security systems, which stimulated the economic development and played the function of economic regulation by pension security systems. Even during normal economic development period, the investment and operation of pension security funds by various countries also directly influenced economic development.

Thus it can be seen that, the development of various countries needs to take the relations between economic development and pension security development into overall consideration all the way to play the normal security functions of pension security systems as well as promote economic development, and further promote the sustainable development of pension security systems. Specifically, the model selection, system design, financing way, security level of pension security systems all need to be closely connected to the economic development of various countries.

7.3.4 When Enhancing Economic Security, More and More Emphasis Is Laid on Service Security and Spiritual Security at the Same Time

In the development process of pension security in various countries, when attaching importance to system security and economic security, more and more emphasis is also laid on pension service security at the same time. Especially under the background of the aging in various countries, the development of pension service and spiritual security have became the important contents of the development of pension security in various countries, which mainly perform in the following aspects:

1. Having formulated relatively complete laws and regulations on pension service and spiritual security

In the process of the development of pension service securities, various countries attach great importance to laws and regulations, and formulate various kinds of laws and regulations about pension service, which play important functions to the development of foreign pension service. Korea formulated *Social Welfare Service Law* and *Elderly Welfare Law* to stipulate the contents of elderly welfare service. Japan promulgated *Elderly Welfare Law* in 1963, *Gerocomy Law* in 1982, which was amended in 1986 and amended again in 1990, "Golden Plan" in 1989 and "Care Insurance"[33] in 2000 to provide strong system guarantee and legal support for community elderly service. Since 1965, the USA have promulgated *Law of the Aged, Elderly Volunteer Work Program, Elderly Nutrition Program, Multi-target Elderly Center Program, Elderly Community Service and Employment Law* and so on in succession to establish the contents of pension service in law, including the aspects of elderly salvation, residence, medical treatment, employment and so on to provide overall pension service for the elderly.

2. Having established the pension service and spiritual security system with multi-agent participation, complete contents and well-equipped facilities

The development of foreign pension service not only pay attention to playing the leading role of the government, but also attach much importance to playing the functions of society and individuals, and integrate resources to establish pension service network with multi-agent participation, complete contents and well-equipped facilities, thus form various forms of pension service systems including family pension, community pension and institution pension support. ① From the view of participating subjects, besides the functions of the government in the aspects of rule making, financial support, policy support and so on, some countries also paid attention to playing the functions of other subjects, such as society, individuals and so on. The USA established Community Volunteer Work Program, and the social services for the elderly were mainly carried out by volunteers. Japanese government encouraged and supported non-governmental institutions to invest in elderly service business with tax preference in addition. Sweden encouraged charities, non-profit organizations to energetically develop elderly public welfare establishment, and encouraged to operate elderly nursing institutions by commercialization with huge tax preference. ② The service contents covered the aspects of daily care, material support, psychological comfort, nursing service, employment service, residence service, education service, social participation and so on of the elderly. Local autonomous organizations in Sweden play important functions in elderly service, which provide welfare residence, transportation preferential services and family door-to-door service for the elderly by formulating elderly service plan, including doing cleaning, cooking food, meal delivery to

[33]Provide relevant living services for the elderly needing help in the community to satisfy the needs of living, medical care, nursing and so on of the elderly by social or individual forces.

home, establishing daytime elderly activity center, organizing the elderly to carrying out activities such as culture, entertainment, physical fitness, travelling, making friends and so on. Korea has enhanced medical security systems for the elderly, meanwhile, it also has established various public facilities to serve for the elderly, including elderly day-care center, home for the aged and so on.[34] The aging service security in Sweden has also accumulated rich experience, which has established strong community elderly service network, with household service to home, house repairs, short-term care, daily activities, community medical treatment and so on as its service contents. Sweden has also established "City Medical Care Service Co., Ltd" in some metropolises to provide more personalized services for the elderly, especial the elderly with higher incomes. The government departments of the UK and volunteers provide services to the elderly such as meal delivery to home, night service, laundry, taking care of children, daytime service, elderly clubs and recreation rooms.[35] ③ From the view of service facilities, many countries have established comparatively complete service facilities. Japan has established various kinds of elderly universities of about more than 400 at present. The elderly who live independently in Japan are equipped with radio emergency call facilities.[36] Besides establishing nursing homes for the elderly, the USA has also established non-profit elderly care center, including health center and social center. The retirement villages in the USA have set up various types of recreation and sports facilities, as well as living facilities. Singapore has established "nursing home for the elderly".

3. Establishing and improving elderly nursing insurance systems

Some countries have established and improved nursing insurance systems in succession to cope with the problems of nursing for the aged brought by population aging. Germany introduced care insurance system in 1995, including the two kinds of in-house nursing and hospital nursing. According to needing intensity, the nursing is divided into three levels, constituted by material objects and currency.[37] Japan formulated the second *Nursing Insurance Law* in the world after Germany, which was duly implemented in 2000. Those who are over 65 years old and those among 40–64 years old who have participated in various kinds of medical insurances may all become the objects of nursing insurance, and the insured can enjoy elderly nursing service in accordance with the regulations after monthly payment of a certain amount of insurance premium.

In 2004, Ministry of Health, Labor and Welfare of Japan submitted *Suggestions on the Modification of Nursing Insurance System* to the government in preparation to modify and improve nursing pension systems, especially to modify the contents and key points of funding. Congress of Japan modified *Nursing Insurance Law* in 2005 to adjust the unit price of service and charging standard of nursing insurance,

---

[34]Zhang and Zhang [28].

[35]Lv [29].

[36]Tong [30].

[37]Lv [31].

which added the contents of establishing comprehensive nursing preventive system, new local nursing service system and so on. The new *Nursing Insurance Law* has come into effect in April 2006.[38] The USA started the long-term nursing insurance quite early, and the business entities of its long-term nursing insurance system are commercial insurance companies, which are in the business scope of commercial insurances, with insurance applicants at different ages voluntarily participating in by purchasing long-term nursing insurance contracts. The financing are mainly from the insurance premium of the insured, the amount of which are relevant to the elements such as the age of the insured, the maximum funding amount that the insured selected, funding period, waiting period and so on.[39]

Some countries pay attention to the fund, policy and facility investment to pension service, meanwhile, they also lay emphasis on the training of professionals in elderly service, including elderly social work personnel, elderly nursing personnel and so on. For example, Japan has established enormous talent teams for elderly professional services, who need strict pre-job examinations and evaluations with corresponding qualification certificates. In the USA, there were 110 thousand medical and public health social workers in the institutions including old people's homes and senior centers in 2004, which made up about 19.6% social workers throughout the USA.[40]

7.3.5 International Organizations Have Played Important Functions in the Reform and Development of Pension Security Systems

With the increasingly growing of the degree of global integration, international organizations have played more and more important functions in the reform of global pension securities. Among them, the functions of the international organizations including World Bank, International Labor Organization, International Social Security Association and so on are quite obvious, which promoted the reforms on pension securities in various countries by publicity, funds, technologies, researches, communications and so on.

Thereinto, World Bank considered that, only by various ways and channels could effectively secure the retirement incomes of the elderly, and it proposed the proposition of three-pillar reforms on pension security system in its research report of *Preventing Aging Crisis* in 1994, namely mandatory public administration pillar, mandatory privately-managed pillar and voluntary pillar. In 2005, World Bank proposed five-pillar pension security system additionally: The first was non-payment "zero-pillar" to provide minimum level of security; the second

[38]Jinwen [32].

[39]Tao [33].

[40]Robert et al. [34].

was payment system linked with incomes, namely the first pillar; the third was mandatory individual savings account, namely the second pillar; the fourth was voluntary pension savings, namely the third pillar; the fifth was non-regular family security pillar.[41] The policy claims by World Bank were responded by many countries, and some countries started the reforms of multi-pillar and multi-level pension security systems. World Bank not only provided some technical programs for the researches of the development of pension security systems in various countries, but also supported the reforms and developments of pension security systems in various countries by pension fund loans. From 1984 to 2004, World Bank provided 204 loans for pension fund reforms to 68 countries in all: Firstly, it directly subsidized specific reforms by technical supports and investment loans; secondly, it indirectly provided funds by projects and structural adjustment loan.[42]

International Labor Organization (ILO) also proposed propositions on the reforms of pension security systems with five basic principles based on the concept of equal security: firstly, pension security systems covering the entire population; secondly, preventing all the people to sink into poverty; thirdly, providing predictable income security to compensate for the decrease of incomes after retirement; fourthly, adjusting retirement pension according to the inflation; fifthly, absorbing representatives of employees and employers to participate in administration.[43] Since 1970s, International Labor Organization has proposed a series of policy claims in succession from the perspective of fully protecting the elderly: provide proper working conditions for the elderly; the elderly should enjoy equal employment opportunities and treatments; establish mandatory security systems to protect the elderly, the capacitated and surviving dependants; establish elastic retirement plan; adopt the way of partly fund accumulation; prolong premium payment years of receiving old-age pension.[44] International Labor Organization also specified and promoted the development of global pension securities by formulating and improving *Social Security (Minimum Standard) Convention.*

In addition, the functions of International Social Security Association (ISSA) in the development of global pension securities also cannot be ignored. International Social Security Association was established in Brussels on October 4, 1927, with the original name of "International Social Insurance Convention", and the targets of ISSA are to enhance inter-state cooperation by developing global social security business, and promote the development of global social security and improve welfare levels of various countries by improving various technical and administrative measures of social securities. The specific modes of activities include: organizing international and regional conferences and conventions, developing international database, developing publications, promoting researches on the issues of social security, organizing training programs for people participating social

[41]Robert et al. [35].

[42]Social Work Website of Ministry of Civil Affairs [36].

[43]Yang [15].

[44]Li [37].

security organizations all around the world, cooperating with other relevant international or regional organizations. The periodical *International Social Security Review* hosted by ISSA provides a good platform for the researches and communications of global pension securities.

Various countries also pay much attention to the global social security conference which is held every four years, with important functions in propagating pension security policies and research achievements, and its 28th global conference was held in Beijing in 2004.

Thus it can be seen that, the functions of different international organizations in the reform and development of global pension securities should not be ignored, which provide beneficial references for various countries to formulate pension security policies. The ways of functions by different international organizations may be different, even with some controversies, but just these controversies made various countries not blindly refer to the suggestions or practice by a certain organization or a certain country in the reform process of pension securities, instead, they deeply studied their own pension security models and formulated pension security systems suitable for them according to their national conditions.

7.3.6 It Is a Trend in the Reforms of Foreign Pension Security Systems to Seek for the Effective Combination of Fairness and Efficiency

Though the ways and measures of pension security reforms in countries with different welfare models are different, they also reflect some converging elements. There exist trends of welfare expansion in many fields simultaneously in the American model which is of the lowest welfare degree, and the direction of expansion is undoubtedly towards the direction of Northern European model. In addition, there still exist indications of expansion in some aspects of the reform towards Northern European model, but there exists strong "re-commercialization" impulsion.[45] On the whole, the combination of fairness and efficiency is the reforming trend and pursuing target in the development of pension security systems in various countries. The development of pension security systems in various countries proved that, full pursuit of absolute fairness and regardless of efficiency would lead to the difficulty in the sustainable development of pension security systems; full pursuit of absolute efficiency of pension security systems and regardless of the goal orientation of fairness go more against the essence of income re-allocation of pension security systems. In the previous processes of developments and reforms, some countries were always unable to attend to all troubles in coping with the relations between the fairness and efficiency of pension security systems, which had brought many problems to the development of systems. In the process of development, many countries absorbed lessons from the past to seek for

[45]Zheng [38].

Table 7.8 Non-payment old-age pension systems in some countries

Type	Country	Coverage (lower limit of age)	Resources of funds	Treatment level (US dollar/month)	Financial burden (proportion of GDP, %)
Household survey type	South Africa	Male 60, female 65	Tax financing	70	1.4
	Brazil	City 67; County male 60, female 55	Tax financing	140	1.3
	Argentina	79	Tax financing	105	0.23
	Chile	70	Tax financing	60	0.38
	Bangladesh	57	Tax financing	2	0.03
	India	65	Tax financing	4	0.01
	Costa Rica	65	Employer contributions, tax financing	30	0.3
	Uruguay	70	Tax financing	90	0.62
Generalized preferences	Lesotho	70	Tax financing	25	2
	Bolivia	60	Bonus of state-owned enterprises, tax financing	26	1.2
	Namibia	60	Tax financing	23	0.9
	Mauritius	60	Tax financing	61 (60–89 years old) 229 (90–99 years old) 262 (over 100 years old)	1.7
	Botswana	65	Tax financing	30	0.5
	Nepal	75	Tax financing	2	0.1
	Samoan Islands	65	Tax financing	33	1.4
	Brunei	60	Tax financing	120	0.4
	Kosovo	65	Tax financing	50	2.7
	Mexico	70	Tax financing	65	0.2

Source Drawing from Yang [46]

the balance between the fairness and efficiency of pension security systems. Specifically, they reflect in the following aspects.

1. Focusing on the effective combination of the functions of the government with the market

The reforms of pension security systems in various countries neither pay too much attention to the functions by the government with neglect of the functions by the market, nor blindly solve problems only by the market, but focus on the effective combinations of the functions the government with the market in the process of reforms. The government is the dominant force in the reform of pension security systems all the time, with important functions in legal construction, policies and systems, fund procurement, supervision and administration, coordinating relations and so on of pension security system reforms, and whether the reform of pension security systems can be smooth or not, successful or not should not be separate from the emphasis and investment by the government, which is also the important force to promote the realization of the target of fairness by pension security.

However, the strength of the government is limited with boundedness after all, and the reforms of pension security systems in various countries should pay attention to playing the proper functions by the market. Thus, it can not only provide more resources for pension securities, but also improve the efficiency of securities, and rational allocation of resources to promote economic development as well. Especially after 1970s, with the transformation of international economic situation, the development of pension security was faced with dilemma, and various countries laid more emphasis to playing the functions of the market, some of which even conducted privatization reforms on pension securities in succession.

2. Combination of basic security with complementary security

There exist obvious differences in the demands of pension security and paying capacity by secured objects due to the otherness in the aspects of personal incomes, family economic status and so on. In the reform processes of pension security systems, some countries adopted the ways to combine basic security with complementary security to cope with the otherness of secured objects. They treat equally to the basic pension security demands of all the citizens without discrimination, or with relatively small difference; the government shoulders relatively more responsibilities for financing and administration in the part of basic security. And complementary security systems or measures are established on the basis of basic security to cope with the discrepant demands of old-age pension by citizens, and the responsibilities for direct financing by the government are relatively small in the complementary security, but it is still responsible for supervision and administration, policy making and so on. And the market and private sectors play more functions in the complementary security. Some countries have established and improved multi-level pension security systems in succession. This practice can not only better satisfy the pension demands by different objects to reduce the financial burdens of the government, but also spread risks to increase the efficiency of securities.

3. Combination of payment security with non-payment security

At present, most countries mainly adopt the payment security model in pension securities, but some countries have introduced non-payment zero-pillar in the process of reform in succession to combine payment security with non-payment security of pension securities, which have been an effective way to achieve the target combining fairness with efficiency in the reform of pension security systems in these countries. To establish and improve non-payment pension security system can expand the coverage area of pension securities and better prevent poverty of the elderly. Up to 2007, there had been 72 countries in the world implementing non-payment pension security systems, 46 of which were low-middle-income countries.[46] According to the practice of non-payment pension security systems in part of the countries, non-payment pension security systems can be divided into two types, namely non-payment pension security system of household investigation and non-payment pension security system of generalized preferences (see Table 7.8). As the name suggests, non-payment pension security system does not need personal payment, most of which is financed by national tax, and fewer of which comes from the bonus of state-owned enterprises and the payment by employers: non-payment pension security system is an important way to secure the achievement of fairness by pension security. However, the non-payment type should not be the main type, or the single type of security model, and it should be effectively combined with payment pension security system to secure both fairness and efficiency.

7.4 Concrete Experience of the Development of Foreign Pension Security Systems

The development of China's pension security system should fully learn from and refer to foreign typical experience with the basis on national conditions. Next, we will analyze the experience of the construction of foreign pension security system from the three aspects of economic security, service security and spiritual security for the elderly.

7.4.1 *International Experience of Economic Security for the Elderly*

There are mainly several aspects of experience can be referred to in the aspect of economic security for the elderly as follows:

1. Establishing and improving multi-level and socialized pension security systems

It is an overall trend of the development of foreign pension security systems to transform from family pension to social pension, during which, various countries have all established and improved different types of pension security systems, therefore, people may obtain institutional economic securities when they are old.

---

[46]Help Age International [39].

After the middle of 1970s, many countries constantly improved pension insurance systems and established multi-level pension insurance systems. They mainly include three levels, namely basic pension insurance system, occupational pension insurance system and individual voluntary saving system, and some countries even implemented four-level or five-level pension insurance systems. In addition to the establishment of basic pension insurance system, the USA has also established complementary pension insurance system and individual saving system, and the UK has established multi-level pension insurance system including national basic pension insurance, national income linked pension insurance, occupational insurance, individual voluntary savings and so on. Japan has also established pension security system and individual voluntary saving system for different occupations on the basis of national pension insurance. Besides, other developed countries and developing countries have also established and improved multi-level pension insurance systems in succession. See the structure of old-age pension systems in some countries in Table 7.9.

2. The investment to pension security is obviously higher than that of other social security programs

From the view of the investment to social security in some countries, the investment to pension security is obviously higher than that of other social security programs. In 2005, Gross Public Social Expenditure in 30 OECD countries accounted for about 24% of net national income.[47] There is great difference in the investment to social security by various countries, among which, the proportions of expenditure by Mexico and Korea are only 8%, but that is up to 33.6% in Sweden. The investments to the elderly and surviving dependants account for the largest proportion in the programs of social security, which are more than 1/3 of the overall investment in social security in many countries (Table 7.10).[48]

3. Having established the sharing financing mechanism of responsibilities for pension securities

Various countries strengthen the financial input to pension security by the government, and meanwhile, they have established sharing financing mechanism of responsibilities. The establishment of rational sharing mechanism of responsibilities combines rights with obligations to promote the efficiency of pension security and the sustainability of systems. Different countries are different in the aspect of the proportion of sharing responsibilities. Some countries attach most importance to the payment by employers, with the payment by employees as auxiliary; while in some countries, employers and employees equally share the payment of pension insurance; some countries attach most importance to the payment by employees, with the

[47]The differences between gross social security expenditure and net social security expenditure are: some funding by social security should pay individual income tax, and the former is the expenditure without tax deduction, while the latter is the expenditure with tax deduction.
[48]Zheng [40].

Table 7.9 Structure of old-age pension systems in some countries

Country	First level: full coverage, re-allocation			Second level: mandatory insurance	
	Targeted	Basic	Minimum	Public (type)	Private (type)
OECD countries with high income					
Australia	√				DC
Austria	√			DB	
Belgium	√		√	DB	
Canada	√	√		DB	
Denmark	√	√		DB + DC	DC
Finland	√			DB	
France	√		√	DB + points	
Germany	√			points	
Greece	√		√	DB	
Iceland	√				DB
Ireland	√	√			
Italy	√			n.acs.	
Japan		√		DB	
Korea		√		DB	
Luxembourg	√	√	√	DB	
Netherlands	√	√			DB
New Zealand		√			
Norway	√	√		points	
Portugal	√		√	DB	
Spain			√	DB	
Sweden	√			n.acs.	DB + DC
Switzerland	√		√	DB	Defined credit
UK	√	√	√	DB	
USA	√			DB	
Central Europe and central Asia					
Bulgaria	√		√	DB	DC
Croatia		√		points	DC
Czech	√	√	√	DB	
Estonia	√	√		points	DC
Hungary			√	DB	DC
Latvia			√	n.acs.	DC
Lithuania		√		DB	DC
Poland			√	n.acs.	DC
Slovakia			√	points	
Turkey	√		√	DB	

(continued)

Table 7.9 (continued)

Country	First level: full coverage, re-allocation			Second level: mandatory insurance	
	Targeted	Basic	Minimum	Public (type)	Private (type)
Latin America and the Caribbean					
Argentina		√			DC
Chile	√				DC
Colombia	√				DC
Costa Rica	√			DB	DC
Dominica	√				DC
Salvador	√				DC
Mexico	√				DC
Peru	√				DC
Uruguay	√			DB	DC
The Middle East and North Africa					
Algeria			√	DB	
Bahrain			√	DB	
Djibouti			√	DB	
Egypt			√	DB	
Iran			√	DB	
Jordan			√	DB	
Libya			√	DB	
Morocco			√	DB	
Tunisia			√	DB	
Yemen				DB	

Note n.acs. is nominal account
Source OECD, Pension at Glance 2009

payment by employers as auxiliary. However, on the whole, most countries attach most importance to the payment by employers (see Table 7.11).

4. Having strengthened the reserve accumulation of pension security funds

Some countries have established pension reserve funds to cope with the growing population aging, with the funds as strategic reserve for pension mainly financed by the government. In the aspect of the ways of reserve, some countries regard pension as the components of comprehensive social security reserve funds, while some countries have established special pension reserve funds (see Table 7.12). The amount of pension reserve funds becomes larger and larger by the accumulation for several years, which will play important functions in coping with population aging.

Table 7.10 The proportion of public social security expenditure in the national net income (NNI) of OECD countries in 2005 (unit: %)

Country or region	Old-age pension of the elderly and dependants	Income support for people at job age	Medical and health services	Services other than medical and health services	Positive employment program	Total amount
Sweden	8.8	7.8	7.7	7.8	1.5	33.6
France	14.0	5.8	8.9	3.3	1.0	33.2
Luxembourg	10.0	9.3	9.7	2.5	0.7	32.2
Austria	14.9	6.8	8.1	1.6	0.7	32.1
Germany	13.2	5.2	8.9	2.6	1.1	31.1
Belgium	10.6	8.5	8.6	2.0	1.3	31.0
Finland	9.9	8.0	7.3	4.4	1.0	30.5
Italy	16.6	3.2	8.1	1.1	0.7	29.7
Hungary	10.7	6.4	7.6	3.3	0.4	28.4
Portugal	12.5	5.3	8.8	0.8	0.8	28.2
Spain	9.7	6.0	7.0	1.8	0.9	25.5
Czech	9.5	5.4	8.2	2.0	0.3	25.5
Poland	13.6	5.1	5.1	0.7	0.5	25.1
Average of 30 OECD countries	8.5	5.3	7.3	2.6	0.7	24.4
Greece	13.2	2.2	6.5	1.6	0.1	23.6
UK	6.2	5.0	7.7	3.8	0.6	23.3
Japan	10.7	1.9	7.8	2.2	0.3	22.9
Switzerland	7.4	5.5	6.6	1.8	0.8	22.2
Slovakia	8.0	5.1	6.8	1.0	0.4	21.3
Canada	4.8	3.1	8.0	3.0	0.3	19.3
USA	6.8	2.2	7.9	1.0	0.1	18.1
Turkey	6.3	0.2	4.4	0.1	0.0	11.0
Korea	1.8	1.6	3.7	0.8	0.1	8.0
Mexico	1.5	0.8	3.3	2.3	0.0	7.9

Source OECD, Social Expenditure database (www.oecd.org/els/social/expenditure)

5. Emphasis on the adaptability of economic security for the elderly with economic development level

Some countries lay emphasis on the adaptability to economic development level in the process of improving pension security system and strengthening economic security for the elderly with constant adjustment and reforms. From the end of "WWII" to the middle of 1970s, with the rapid economic and social development and under the guidance of Keynesianism theory, the pension security systems in various countries experienced a golden period of rapid development. The input to

Table 7.11 Payment rate of public pension in some countries in 2009 (unit: %)

Country	Employees	Employers	Country	Employees	Employers
Austria	10.3	12.6	Poland	9.8	9.8
Belgium	7.5	8.9	Slovakia	4.0	14.0
Canada	5.0	5.0	Slovenia	15.5	8.9
Chile	28.8	1.0	Spain	4.7	23.6
Czech	6.5	21.5	Sweden	7.0	11.9
Estonia	2.0	20.0	Switzerland	4.9	4.9
Finland	4.5	17.1	Turkey	9.0	11.0
France	6.8	9.9	USA	6.2	6.2
Germany	10.0	10.0	OECD34	8.4	11.2
Greece	6.7	13.3	Argentina	11.0	12.7
Hungary	9.5	24.0	Brazil	11.0	20.0
Israel	3.9	3.1	China	8.0	20.0
Italy	9.2	23.8	India	12.0	12.0
Japan	7.7	7.7	Indonesia	2.0	4.0
South Korea	4.5	4.5	Saudi Arabia	9.0	9.0
Luxembourg	8.0	8.0	EU27	7.9	14.0

Source OECD (various years), Taxing Wages; OECD (2008), Revenue Statistics; Social Security Administration, United States (various years), Social Security Programs throughout the World; OECD Pension and Tax Models

Table 7.12 Pension reserve funds in some OECD countries (unit million US dollars)

Country	Name of pension reserve fund	The year of establishment	Asset accumulation
USA	Supporting funds of social security	1940	2,048,112
Japan	National reserve fund	1959	1,217,551
Norway	Government pension funds	1990	278,124
Korea	National pension funds	1988	190,842
Sweden	National pension funds	2000	117,468
Canada	Pension planned reserve funds of Canada	1962	86,392
Spain	Social security reserve fund	1997	44,875
France	Pension reserve fund	1999	39,140
Ireland	National pension reserve fund	2001	23,710
Australia	Future fund	2006	13,678
Portugal	Financial stability fund	1989	8330
Mexico	IMSS reserve fund		7392
New Zealand	Annuity fund	2001	6666
Poland	Population reserve fund	2002	1760
Denmark	Social security fund	1964	650

Source OECD (2007)

pension security systems by the governments of various countries continuously increased, and the coverage continuously expanded, the treatment level continuously raised, the institutional systems continuously improved, the management system continuously perfected, the operating mechanism continuously matured. After the middle of 1970s, with the outburst of petroleum crisis, there emerged malignant situation of "high inflation rate, high unemployment rate, low economic growth rate" in some countries, and the economic development of the west was faced with serious dilemmas. The expansion of the expenditure on social security and welfare after "WWII" laid heavy burdens of financial expenditure on some countries. The deterioration of economic situation, the acceleration of the process of population aging, the rigidity of welfare expenditure compelled various countries to reform and adjust their original pension insurance systems to promote its sustainable development and adapt to economic and social development. With the slowing down of economic development and the increasing of the financial burdens by pension insurance, some countries thoroughly reformed the original pension insurance system model to transform from pay-as-you-go system to partly-funded system or fully-funded system, and it becomes an important trend to lay emphasis on fund accumulation in the reform of pension insurance systems by various countries. The privatization reforms in Latin American countries are comparatively prominent. Some countries started to explore the implementation of the model of "notional defined contribution", which include Sweden, Poland, Italy, Mongolia, Latvia and so on. Most countries conducted parameter reforms on pension security systems, namely the adjustment on the parameter of systems, to achieve the target of increasing income and reducing expenditure as well as promote the financial sustainability of pension insurance systems to cope with the population aging in various countries. This includes increasing payment rate, reducing old-age pension treatment, being strict with funding conditions, prolonging retirement age and so on.

7.4.2 International Experience in Service Security for the Elderly

Some foreign countries, especially some western developed countries have developed and improved pension security systems and enhanced economic security for the elderly, meanwhile, they have also paid great emphasis on the development of service security for the elderly, and have accumulated some experience worthy of reference, which mainly include the following aspects:

1. Having established the service security system for the elderly to satisfy the diversified and multi-level demands

Some countries have established service systems to satisfy the different demands of the elderly with the participation of various forces. The governments of the UK,

Japan and some other countries actively encourage and support social forces and capitals to establish commercial age service institutions, and grant preferential policy supports such as tax reduction and so on to strongly support the input to charitable organizations, non-governmental organizations, private funds and the contributions by all orders of society at the same time. White Paper Of Community Care In the UK in 1989 proposed that: "Community Care should form a spectrum of care to provide from household support and care to daytime care for those who need deep care, up to provide hospital care and long-term nursing service and so on for those with more need." The community-based senior care services in Japan structures relatively perfect policy and law systems, which form diversified organizational forms including governments, enterprises, non-profit organizations, with rich service contents suitable for the elderly of different levels. Japan provides different health care and medical treatment service programs for the elderly with different physical conditions, including regular health examination, door-to-door medical treatment, home nursing, institution nursing or rehabilitation training and so on; Different service programs should also be designed for the elderly with different educational levels. The community-based senior care services in Singapore and some other countries not only provide physical and spiritual help for the elderly, but also satisfy their demands to devote their remaining energy, which provide manageable operating posts for them, therefore, the elderly with capacities can participate in community services. The service systems for the elderly in Singapore cover the four levels of individuals, families, communities and the government with functional complementation, which form harmonious service systems for the elderly to satisfy multi-level demands of age services.

2. Having established complete systems of laws and regulations for senior care services

In the process of developing age service securities in some countries, legislation goes ahead, which lays emphasis on the establishment and improvement of laws and regulations for age services to strongly promote the development of age services. In the USA, *Retirement Pension Law Of Old-age Employees* was issued and implemented in Alaska in 1915, and there had already been 28 state continents in the USA approving *Retirement Pension Law Of Old-age Employees* up to 1933; In 1965, the USA issued a series of laws on the aspect of pension, including *Elder Law*, *Elder Volunteer Work Program*, *Elder Nutrition Program*, *Multi-target Elder Center Program*, *Elder Community Service Employment Law* and so on. *Elder Welfare Law* issued by Japan in 1963 was called as "Elder Charter", which stipulated the contents of community-based senior care services for the first time and was the beginning of socialized pension by Japan; The subsequent *Elder Health Care Law* further stipulated medical treatment and health services for the elderly again; *Ten-year Strategic Planning to Promote Health Care Welfare for the Elderly in Advanced Ages* implemented in 1989 defined the dominant position of the community in age services again; *Nursing Care Insurance Law* implemented in April, 2000 stipulated that, all the citizens over the age of 40 should pay care insurance fees to solve the problems of care when they were old. *Nursing Insurance Law* issued in 1995 in Germany promoted the nursing insurance for the elderly to formal development. Korea formulated *Social Welfare Service Law* and *Elderly Welfare Law* to stipulate the contents of elderly welfare service.

3. Emphasis on the functions of different subjects in age services

The age service systems in western developed countries all take community pension as the subject, with multi-level and multi-form pension institutions as beneficial supplement, and as to most of the elderly, to live in their familiar communities not only accord with the psychological needs of "home love" of the elderly, but also benefit to improving their living qualities. It is convenient for the community to carry out diversified services such as daily care, housekeeping, health care and medical treatment, psychological counseling, amusement and so on for the elderly, which also gives full play to socialized market operating mechanism in the aspect of nursing. The long-term nursing insurance system in the USA started from 1970s, with commercial insurance companies as the underwriting subjects, and relevant nursing fees generated from various personal care services at home or in pension institutions by the insured were undertaken by insurance companies. The USA fully utilized social funds and organized forces to participate in age service business with the government as the leading role, which granted policy and tax preferences, and community service centers as well as pension institutions are generally operated by social funds to guarantee service qualities by full competition and good supervision systems. Sweden has established strong community service network for the elderly, which includes service contents of door-to-door housework service, house repairs, short-term care, daily activity, community health care, and the government encourages commercial operation of nursing institutions for the elderly, with enormous tax preferences. In the UK, apartments for the elderly have been built by local governments, and nursing institutions for the elderly have been organized and established by individuals or volunteers, therefore, the elderly may make choices according to their own situations; the elderly in the UK may enjoy socialized medicine in community health centers, and governments have specially set up aged hospitals for the elderly with specialized invigilation interviewers to regularly visit the elderly in their homes according to the suggestions and instructions by doctors, and provide suggestions on the aspects of treatment, recovery and so on. There are complete community-based senior care service networks in all parts of the country.

4. Emphasis on the professionalization of age service

From the experience by Japan, high-quality and specialized community-based senior care service working personnel is the important guarantee for high-quality age services, and different from Japan, there is no enormous housewife groups in communities of China, however, there are a large number of undergraduates, middle school students, pupils and many healthy elders in China, whose participation will bring new vitality and vigor to community-based senior care services and will be potential forces of community-based senior care services. China should draw on the experience of Japan to strengthen educations and training for specialized age services, set up and perfect relevant specialized systems to guarantee the constant provision of specialized personnel for communities and secure the economic and social positions of this group. In addition, education and training on family members should also be strengthened. Family members are those who are the most in touch with the elderly, and it is of great significance for community-based senior care services to popularize nursing knowledge and improve their nursing levels, meanwhile, it should also strengthen education and training on volunteers so that every volunteer can be equipped with certain knowledge levels before providing services, only this can they scientifically serve the elderly, at the same time, they should be fully acquainted with the importance of volunteer service and more people will participate in the service teams for the elderly by encouragement and propaganda. In the UK, the community-based senior care service plays a very important role in social work and service; therefore, the government should make great efforts on its system construction.

The work system is constituted of administrators, key staffs and caregivers, all of whom are generally professional social workers or semi-professional support staffs.[49]

5. Establishing and reforming nursing insurance systems

Some countries have established nursing insurance systems in succession and gradually reformed on them to satisfy the increasing demands of age services brought by the population aging. Among them, Germany was the first to introduce the nursing care insurance in 1995, which included the two types of home nursing and hospital nursing. According to needing intensity, the nursing is divided into three levels, constituted by material objects and currency. Japan formulated the second *Nursing Insurance Law* in the world after Germany, which was duly implemented in 2000. In 2004, Ministry of Health, Labor and Welfare of Japan submitted *Suggestions on the Modification of Nursing Insurance System* to the government. In 2005, Congress of Japan modified *Nursing Insurance Law*. The USA started the long-term nursing insurance quite early, and the nursing insurance was operated by commercial insurance companies, with insurance applicants at different ages voluntarily purchasing insurances.

7.4.3 International Experience in Spiritual Security for the Elderly

1. Governments in some countries lay comparatively more emphasis on spiritual security for the elderly

At the beginning of 1980s, some of the welfare countries which entered into aging society earlier, such as Denmark, Sweden and so on, had reminded people against

[49] Chen and Sun [41].

the facility-only doctrine in age services based on their own lessons. When providing basic living securities for the elderly, the satisfaction of their spiritual demands should be paid more attention to at the same time.[50] In the process of developing security systems for the elderly, governments in some countries not only laid emphasis on the economic security, health care security and service security for the elderly, but also attached great importance to the spiritual security for the elderly. The USA adopted *Social Security Act* in 1935, after then, it also issued *The Older Americans Act, Law On Prohibition Against Discrimination In Employment For The Elderly*, which secured the legal interests of the elderly, and modified *Law On Prohibition Against Discrimination In Employment For The Elderly* with countermand of legal provisions of mandatory retirement of employees under 70 years old. Over the past 20 years, Singaporean government enhanced the construction of spiritual civilization, with the launch of courtesy activities and "week of respecting the elderly" every year since 1979, and encouraged non-governmental organizations and other organizations to care for the welfare issues of the elderly as well as encouraged the elderly to participate in social activities through various activities. The newspapers in Singapore frequently carry out the education of "loving other's elders as your elders", and propose to promote the oriental ethical traditions by families and schools to cultivate the new generation of Singaporean.[51] Japan has also attracted social attention to the aging issues by consensus propaganda, for example, the government has established September 15th as the day of respecting the elderly to promote national care for the elderly.

2. The family is of very important functions in the spiritual security for the elderly

Some countries pay attention to playing the functions of governments, as well as the functions of families in the spiritual security for the elderly. Especially in some Eastern Asian countries, with the deep influence by Confucian culture, the functions of families are more obvious in the spiritual security for the elderly. Singaporean government always strongly advocates the principle of three generations under one roof, and Housing Development Board of Singapore has adopted measures to coordinate with the principle of three generations under one roof advocated by the government, for example, to allow those who plan to live neighborly with their immediate families to participate in drawing lots for housing allotment in advance and so on. With reference to western developed countries, Japan also lays emphasis on its national tradition of respect to the elderly and advocates the elderly and their families living together. As to living models, Japan has developed and built a batch of new residences for the elderly and their families to live together with moderate separation in daily life of the two generations to adapt to the tendency of kernelization of families in Japan.

[50]Zhou [42].

[51]Shi [43].

3. Taking the spiritual life of the elderly into consideration at residence design

Some countries have taken the issues of the spiritual life of the elderly into consideration when conducting design and construction of residences, at the same time, they have also thought about the characteristics and demands of the young, for example, Singapore has developed various types of multi-generation HDB blocks, with basically the same spatial relationships, which are divided into main rooms and one-room apartments connected by the living room, and the two families are separate as well as integrate, thus can adapt to the harmonious coexistence of the two generations. With reference to western developed countries, Japan also lays emphasis on its national tradition of respect to the elderly and advocates the elderly and their families living together. As to living models, Japan has developed and built a batch of new residences for the elderly and their families to live together with moderate separation in daily life of the two generations to adapt to the tendency of kernelization of families in Japan. The spatial relationships of this type of "two generations under one roof" form of adolescent families can approximately be divided into: ① Cohabitation and boarding type: the same portal, the same kitchen and living room, and the habitable room of the elderly is only attached with a toilet or simple cooking equipment; ② Cohabitation and separate-living type: the same portal, and the kitchen, toilet and living room are all separate with respective supporting facilities; ③ Neighboring living with shared occupancy type: separate portals, the same living room, separate toilets and kitchens; ④ Completely neighboring type: separate portals connected by the living room, separate toilets and kitchens.

4. Emphasis on the participation of social forces including non-profit organizations and so on

American government cultivates non-profit organizations to undertake a large number of specific social affairs. For example, American "Community Developing Cooperative Organization" started in 1966 adopted the model of community resident autonomy with the leadership of prestigious and influential community residents to strive for supports from governments, banks, religious circles, charitable organizations and the cooperation with them, which had carried out a series of community-based senior care services, such as caring for the elderly, skill training for the re-employment of the elderly, mediating family disputes, developing health clinics, services for enriching cultural lives of residents and so on. The business circles in Japan have given enormous support to the development of old-age undertakings and subsidized academic conferences related to the elderly and aging. After entering aging society, all sectors of society in Korea, from the theoretical cycle to the decision-making level, have all paid extensive attention to it.[52]

[52]Liu [44].

5. Encouraging social participation by the elderly

In addition to the supports of governments and families, some countries also lay emphasis on the own functions of the elderly to encourage their social participation. Japan has encouraged the elderly to take active parts in various social activities to bring their experience and intelligence into play, and on the one hand, Japanese government has established "Elderly Club" to strengthen the interaction and communication among the elderly. On the other hand, it has established "Longevity Social Development Center" in the center, and "Bright Longevity Social Promotion Institution" in prefecture to secure the constant development of the undertakings of "Elderly Club", enhance the cognition of social participation of the elderly by citizens, cultivate leaders to revitalize social activities of the elderly, and establish model municipalities comprehensively developing physical activities and other health care activities for the elderly.[53]

From the said foreign experience of the aspects of economic security, service security and spiritual security for the elderly, there is something in common: ① the emphasis by the government is the key element determining whether or not the security system of the elderly can develop well; ② the emphasis on the legal construction of the development of security system for the elderly and legislation first are important features; ③ the emphasis on socialization development of security for the elderly; ④ the mechanism of multi-agent participation has been established. This experience deserves further study and use for reference by China.

References

1. Long, Yuqi. *Study on Development and Reform of Pension Security Systems for Foreign Public Servants*, a doctoral dissertation at Renmin University of China in 2011.
2. Xiong Bijun. 2009. *Economics of Aging*, 15. Beijing: China Social Press.
3. Changhong, Pei. 2010. The trend of economic globalization in post-crisis era and its new characteristics and new situation. *International Economic Review* 4: 27.
4. Yan, Zhong, and Yuqi, Long. 2010. *China's Social Security System under the Current Economic Environment, China Europe Social Security Forum.*
5. Franz-Xaver, Kaufman. 2004. *Challenges Faced with Social Welfare States*, 101. Trans. Xuedong Wang. Beijing: The Commercial Press [Germany].
6. OECD Observer. 2009. OECD in Figures, 89.
7. Zheng Bingwen. 2009. The Impact of Financial Crisis to Global Pension Assets and Its Challenges to the Investment System of Pension Assets in China. *International Economic Review* 5–29.
8. Costa, Espin Andersen. 2003. *The Three Worlds of Welfare Capitalism* (trans: Zheng Bingwen). Law Press [Denmark].
9. Hou, Hong. 2002. Whether Social Security System Can Be Globalized, World Economy, the 8th edition, 73–80.

[53]Chen and Sun [41].

10. Zhang, Shibin. 2010. *Transformation of Labor Market and Reform of Social Pension Security in China- based on Historical Review to Pension Security System, Comparative Economic & Social System*, the 2nd edition, 123–130.
11. Lin, Yi, and Zhiguo, Chen. 2010. *International Comparison and Revelation of Retirement System Reform*, recorded in *International Comparison of Social Pension Security* with Zhou Hong as the chief editor. China Labor and Social Security Publishing House, 103.
12. Data About Transference of People May Refer to the United Nations Development Programme. 2009. *Human Development Report in 2009*, Chinese Financial & Economic Publishing House.
13. Feng, Gan. 2007. Trend of Japanese Old-age Pension Crisis and Political Parties-Taking Congressional Elections from 2003 to 2007 as the Perspective. *Northeast Asia Forum* 6: 64–65.
14. Mu, Huaizhong. 2002. *International Comparison on Social Security*, 170. China Labor & Social Security Publishing House.
15. Yang, Jae-jin. 2000. *The 1999 Pension Reform and a New Social Contract in South Korea.* Ph.D. Dissertation. New Brunswick, New Jersey.
16. Martin, Feldstein. 1974. Social Security, Induced Retirement and Aggregate Capital Accumulation. *Journal of Political Economy* 82:905.
17. Sun, Yongyong, and Zhen, Li. 2005. *Does It Need To Select Either Individual Account Or Pay-as-you-go Of Old-age Pension, Macroeconomic Study*, the 6th edition, 34–36.
18. Robert, Holtzman, and Joseph E Stiglitz. 2004. *Old-age Pension System of Sustainable Development in 21st Century*, 15–36. China Labor & Social Security Publishing House.
19. Barr, Nicholas, and Peter Diamond. 2009. Reforming Pensions: Principles, Analytical Errors and Policy Directions. *International Social Security Review* 2: 5–29.
20. Holzmann, Rorbert. 2000. The World Bank Approach to Pension Reform. *International Social Security Review* 1: 11–34.
21. Colin, Gillion. 2000. *The Development and Reform of Social Security Pensions: The Approach of the International Labor Office. International Social Security Review* 53(l):35–63.
22. Long, Yuqi. 2011. *Comparative Study on the Development and Reform of Foreign Pension Security System of Civil Servants.* A doctoral dissertation of Renmin University of China.
23. Costa, Espin Anderson. 2003. *The Three Worlds of Welfare Capitalism*, 26 (trans: Zheng Bingwen). Law Press [Denmark].
24. Zheng, Bingwen. 2005. *Comparative Study on "Welfare Models" and Empirical Analysis on Welfare Reform, Academics*, the 3rd edition.
25. Zheng, Bingwen. 2005. Social Rights: Annotation on the Origin of Modern Welfare State Model. *Journal of Shandong University, the 2nd edition.*
26. Costa, Espin Anderson. 2003. *The Three Worlds of Welfare Capitalism*, 31 (trans: Zheng, Bingwen). Law Press [Denmark].
27. Jiang, Xiangqun. 2005. *Old-age Social Security System-History And Revolution*, 23. China Remin University Press.
28. Zhang, Baoren, and Huizhi, Zhang. 2000. Analysis on Korean Social Security System. *Population Journal*, the 4th edition.
29. Lv, Xuejing. 2006. *Modern Social Security Systems in Various Countries*, 103. China Labor Security Publishing House.
30. Tong, Baogui. 2009. *Retirement Pension Systems at all Times and in all over the World*, 85–90.
31. Lv, Xuejing. 2006. *Modern Social Security Systems in Various Countries*, 7. China Labor Security Publishing House.
32. Jinwen, Song. 2010. Analysis On Japanese Nursing Insurance Reform and Trend. *Japanese Studies* 4: 107–121.
33. Tao, Jing. 2010. Establish Long-term Nursing Insurance System Model Suitable for China. *Insurance Studies* 4: 62–77.

34. Robert, Holtzman, and Hienz, Richard, et al. 2006. *Income Security For The Elderly In 21st Century-Comparison On Reforms Of Pension Systems*, 1–2 (trans. Zheng, Bingwen). China Labor Security Publishing House.
35. Robert, Holtzman, and Hienz, Richard, et al. *Income Security For The Elderly In 21st Century-Comparison On Reforms Of Pension Systems*, 71 (trans Zheng, Bingwen). China Labor Security Publishing House.
36. Social Work Website of Ministry of Civil Affairs. http://sw.mca.gov.on/article/tszs/200710/20071000002712.shtml.
37. Li, Zhen. 2001. *Social Security Theory*, 306. China Labor Security Publishing House.
38. Zheng, Bingwen. 2005. *Reform Trend and Revelation of Foreign Social Security Systems*, published in *China Social Security Green Paper (2004)* by Chen Jiagui, Wang Yanzhong, Social Sciences Academic Press, 261–287.
39. Help Age International. 2007. *Feasibility: Counties with a Social Pension*. http://www.helpage.org.
40. Zheng, Gongcheng. 2011. supervised the publication of: *Reform and Development Strategy of Social Security in China* (Volume of Pension Security), 270. People's Publishing House.
41. Chen, Chengwen, and Xiulan Sun. 2010. Community-based Senior Care Services: Practice Models and Revelation of UK, USA and Japan. *Socialism Study* 1: 116–120.
42. Zhou, Shaobin. 2005. The Spiritual Demands of the Elderly and the Social Policy Significance. *Market & Demographic Analysis* 6: 68–72.
43. Shi, Jinqun. 2013. A Tentative Study on the Construction of Spiritual Security System for the Elderly. *Social Security Studies* 2: 1–15.
44. Liu, Qingzhi. 2009. Experience and Revelation Coping with Population Aging by USA, Japan Korea. *Northwest Population Journal* 4: 73–75.
45. Wang, Yanzhong, and Anan, Wei. 2010. *The Internationalization Development Issue on the Social Security in China*, recorded in *Green Paper of Social Security in China, No. 4*, with Chen Jiagui and Wang Yanzhong as chief editors, Social Sciences Academic Press, 2010, 255.
46. Yang, Juan. 2010. *Research and Review on Non-payment Old-age Pension Systems, Economic Perspectives,* the 4th edition, 121–125.

Chapter 8
Building a Trinity Pension Security System in China: A Preliminary Framework

Yanzhong Wang

China's pension security system has evolved from the family-supporting stage to the one witnessing the weakening family support and the budding social pension system. At present it is still in the early stage of rapid development. Overall, the Chinese pension security system has made remarkable achievements and played an important role in securing the basic livelihood of the elderly, maintaining social stability and promoting economic development. However, this system, far from perfect, cannot meet the need of an aging population, confronts many problems, and sees prominent contradictions between the various needs of elderly people and insufficient supply. "Getting old before getting rich", changes in family structure, acceleration of urbanization and changes in labor market all bring new challenges to the development of the pension security system.

8.1 The History and Status Quo of China's Pension Security System

8.1.1 The History of China's Pension Security System

The history of China's pension security system after the founding of People's Republic of China can be roughly divided into the following phases:

The first phase (1949 to early 1980s): Family support played the major role with supplemented government, unit and collective security. Due to economic underdevelopment and tradition, family support was in a dominant place in this

Y. Wang (✉)
Institute of Ethnology and Anthropology, Chinese Academy of Social Sciences, Building 6, #27 Zhongguancunnan Street, Haidian District, Beijing 100081, People's Republic of China
e-mail: wangyanzhong01@163.com

Y. Wang (ed.), *The Development of Security and Whole Care System for the Aged in China*, https://doi.org/10.1007/978-981-10-6481-4_8

phase, and especially in the vast rural areas, the vast majority of elderly people relied on their family in their old age. The family provided financial support, life care, health care, mental consolation and other more comprehensive protection for the elderly but on lower levels, so it was difficult to meet the needs of the elderly. Due to the impact of the planned economic system, there were also a number of other forms of pension security provided by the government, unit and collective in addition to the dominant family support. For example, the government and unit provided urban workers with pensions and certain old-age services, and in rural areas, the collective provided financial and in-kind assistance to the elderly.

The second phase (mid-1980s to the late 1990s): Family support was constantly weakened, while social pension system began the exploration. The acceleration of marketization, industrialization, and urbanization on the one hand promoted the rapid development of national economy and provided a certain economic base for the elderly, and on the other hand, it also increased the difficulty in supporting the elderly to a certain extent. Especially against the background of quick population flow, family structure changes and weakening soil security, the traditional family support got continuously weakened. In this period, the country was in the process of rapid economic and social transformation. Although socialized pension began its exploration, it was still in its infancy. The state began to explore and establish the basic pension insurance system for urban workers, but the system for the rest of the population was not fully established. The vast majority of people except government department staff and some urban workers were not integrated into the pension security system, especially in the vast rural area, there was extremely lack of security for the elderly, and rural social pension insurance system stalled. In the context of the growing aging population, how to support the old has become an important topic of great concern to the whole society.

The third phase (after the late 1990s): social support for the old develop into a new era. In the twentyfirst century, the Party and the state pay more attention to the cause of people's livelihood including pension security, take a series of pension security measures, really begin to explore social pension security, and constantly improve the social pension security initiatives against the background of weakening household support and accelerating population aging. In 1997, it established a unified basic pension insurance system for urban employees. To accelerate the development of undertakings for the aged, the National Committee on Aging was established in 1999, serving as the coordinating body of the State Council in charge of the cause of aging, including overall planning and guidance. On August 19, 2000, the CPC Central Committee and the State Council promulgated the "Decision of the CPC Central Committee and State Council on Strengthening the Work for the Aged"; in 2001 and 2006, it promulgated "Planning for the Development of Undertakings for the Aged" for the 10th and 11th Five-Year period; the office of China Committee on Aging and other departments issued "Opinions on Accelerating the Development of Pension Service Industries" (February 2006), "Opinions on Comprehensively Promoting Home Care Services" (January 2008); in November 2010, Ministry of Civil Affairs held a national old-age care service conference in Wuxi in Jiangsu Province. All these initiatives play an important role in promoting pension security services, especially the cause of social pension. During this period, the various pension and security systems including urban basic pension insurance system, new rural social pension insurance system, and pension allowance system were established and improved one after another.

8.1.2 Achievements of Chinese Pension Security System

First, pension security regulatory system continues to improve. The "Law of the People's Republic of China on Protection of the Rights and Interests of the Elderly" passed in 1996 listed comprehensive, specific provisions on pension security and the protection of interests of the elderly, and had a significant impact on the development of pension security. It provides that "the state and society should take measures to improve the social security system for the elderly, gradually improve their life, health and conditions for social development participation, in order to ensure the support, medical care, learning and enjoyment for the elderly"[1] In the security system, the laws and regulations from "Labor Insurance Rules" in the 1951 to "Social Insurance Law" in 2010 indicate the gradual improvement in pension security system. In this process, "Basic Pension Insurance System for Urban Workers" continues to improve; rural social pension insurance system is also constantly exploring and developing; urban and rural pension allowance system will be generally established, and the regulations on pension security services are also constantly improving (see Table 8.1). The coverage of pension security gradually expands along with the gradual improvement of the system. At the end of 2009, the population participating in urban basic pension insurance was 218.91 million, an increase of 17.54 million over the previous year; the rural population participating in basic pension insurance was 24.16 million, an increase of 5.7 million over the previous year; the corporate employees participating in basic pension insurance was 199.51 million, an increase of 17.16 million over the previous year; the population participating in rural pension insurance was 55.95 million, an increase of 4.24 million over the previous year; a total of 5.12 million farmers received pensions that year, 1.2 million more than last year.[2] By the end of 2012, the population participating in urban basic pension insurance was 304.27 million, including 229.81 insured in-service workers and 74.46 insured retirees. The number of migrant workers participating in basic pension insurance was 45.43 million. All county-level administrative regions carried out comprehensive national

[1]See the "Law of the People's Republic of China on Protection of the Rights and Interests of the Elderly" (passed in 1996).

[2]See Ministry of Human Resources and Social Security: "Statistical Communique on Human Resources and Social Security Development in 2009".

Table 8.1 Relevant laws, regulations and planning

Time	Name of laws and regulations (planning)	Promulgated by
December 14, 1994	"The Aging Cause in China Program for Seven Years (1994–2000)"	State Education Commission, Ministry of Civil Affairs, Ministry of Finance, State Development Planning Commission, the Ministry of Labor, Federation of Trade Unions, Ministry of Personnel, Ministry of Health
August 29, 1996	"Law of the People's Republic of China on Protection of the Rights and Interests of the Elderly"	NPC Standing Committee
February 27, 2000	"Opinions on Speeding up the Realization of Social Welfare"	Ministry of Civil Affairs, the State Planning Commission, State Economic and Trade Commission, Ministry of Education, Ministry of Finance, Ministry of Labor and Social Security, the Ministry of Construction, Ministry of Foreign Trade, Ministry of Health, State Administration of Taxation
August 19, 2000	"Opinion of CPC Central Committee and State Council on Strengthening the Work for the Elderly"	CPC Central Committee and the State Council
July 22, 2001	"The 10th-Five Year Planning for The Aging Cause in China (2001–2005)"	State council
February 20, 2006	"Opinions on Accelerating the Development of Pension Security Service Industries"	Office of the National Committee on Aging, Development and Reform Commission, Ministry of Education, Ministry of Civil Affairs, the Ministry of Labor and Social Security, Ministry of Finance, the Ministry of Construction, Ministry of Health, Population and Family Planning Commission, State Administration of Taxation
August 16, 2006	"The 11th-Five Year Planning for the Aging Cause in China"	National Committee on Aging
January 29, 2008	"Opinions on Comprehensively Promoting Home Services for the Aged"	Office of the National Committee on Aging, National Development and Reform Commission, Ministry of Education, Ministry of Civil Affairs, Ministry of Labor and Social Security, Ministry of Finance, the Ministry of Construction, Ministry of Health, Population and Family Planning Commission, State Administration of Taxation
September 1, 2009	"Opinions on Carrying Out New Rural Social Pension Insurance Pilot"	State Council

(continued)

Table 8.1 (continued)

Time	Name of laws and regulations (planning)	Promulgated by
February 5, 2009	On Soliciting the Public Comments on "Rules on Migrant Workers' Participation in Basic Pension Insurance" and "Interim Measures on Transferring and Continuing Pension Insurance for Urban Enterprise Employees"	Ministry of Human Resources and Social Security
October 2010	"Social Insurance Law of People's Republic of China"	NPC Standing Committee
June 7, 2011	"Guidance on Conducting Social Pension Insurance Pilot for Urban Residents"	State council
September 17, 2011	"The 12th-Five Year Planning for The Aging Cause in China"	State council
December 28, 2012	"Law of the People's Republic of China on Protection of the Rights and Interests of the Elderly"	NPC Standing Committee
July 30, 2013	"Guiding Opinions on Promoting Old-Age Services Assessment Work"	Ministry of Civil Affairs
December 23, 2013	"Notice on Carrying Out Comprehensive Reform in Old-Age Service Industry"	Ministry of Civil Affairs, the NDRC
February 21, 2014	"Opinions on Establishing A Unified Basic Pension Insurance System for Urban and Rural Residents"	State Council

Source Summarized by the author based on publicly available data

social pension insurance work for urban and rural residents. The insured population was 483.70 million, of which the actual population receiving pension was 130.75 million.[3]

Second, the social pension system has been basically set up. With national economic and social progress, social security system gradually evolves from the traditional state-unit support to social security. In terms of the pension security, the security concept of the urban and rural elderly began to show a diversified trend along with the changes in economic and social environments, breaking the dominance of traditional concept of family support,[4] and the concept of social pension security gradually comes into shape. In addition to the traditional family support, the socialized pension security system has been basically completed. First, the coverage of pension security has expanded. By the end of 2012, the population

[3]See Ministry of Human Resources and Social Security: "Statistical Communique on Human Resources and Social Security Development in 2012".

[4]Baoan [12].

participating in urban basic pension insurance was 304.27 million, an increase of 20.36 million over the previous year. Among them, there were 229.81 insured in-service workers and 74.46 insured retirees, an increase of 14.16 million and 6.19 million over the previous year respectively. The number of migrant workers participating in basic pension insurance was 45.43 million, an increase of 4.03 million over the previous year. Second, the urban and rural basic pension insurance system has achieved full coverage institutionally. All county-level administrative regions carry out national social pension insurance for urban and rural residents in a comprehensive way. At the year end, the insured population was 483.70 million, an increase of 151.87 million over the previous year, and the actual population receiving pensions was 130.75 million.[5] Third, market-oriented enterprise annuity system and commercial pension insurance system saw the rapid development. By the end of 2012, there were 54,700 enterprises having established enterprise annuity, an increase of 21.8% over the previous year; the number of participating workers was 18.47 million, an increase of 17.1% over the previous year; the accumulated enterprise annuity fund balance at the end was 482.1 billion yuan.[6] Commercial annuity insurance has become one of the fastest growing commercial insurance products, and further enhances the status of insurance funds in the capital market. Finally, the elderly welfare system with old age pension and allowance as representative continues to expand, effectively improving the living standards and quality of life of the elderly. Some wealthy seniors are considering institutional or community care. The concept of social pension helps reduce the burden on families and also provides better security for the elderly.

Third, old-age service system undergoes rapid development. In the process of developing pension security, China has begun to focus on the development of old-age service system to meet various service needs of the elderly. In 2000, the State Council promulgated the "Opinions on Speeding up the Socialization of Social Welfare" which included relevant provisions on the socialization of pension benefits. In November 2010, Ministry of Civil Affairs held a national social elderly care service conference in Wuxi, Jiangsu Province, which is of great significance for promoting the construction of social care service system. In pension services, Beijing, Chengdu, Guangzhou, Shanghai and other cities have carried out several explorations and gained valuable experience.[7] By the end of 2009, the number of various types of elderly welfare institution reached 38,060, an increase of 908 over the previous year. Among them, 5291 urban pension service adopted 323,000 old people at year end; 31,286 rural pension service institutions adopted 1.73 million old people at year end; 1401 homes for disabled veterans adopted 46,000 old people at year end; 47 honorable military rehabilitation hospitals adopted 4000 old people

[5]See Ministry of Human Resources and Social Security: "Statistical Communique on Human Resources and Social Security Development in 2012".

[6]Ministry of Human Resources and Social Security: "Statistical Communique on Human Resources and Social Security Development in 2012", website of Ministry of Human Resources and Social Security.

[7]See Wang Lianxin: "*Report on Pension Security Service System Research*".

at year end; 35 sanatoriums for demobilized soldiers adopted 4000 old people at year end.[8]

Fourth, financial investment in pension security is increasing. The state financial investment in pension security is also rising with the growing elderly population. There is a substantial increase in the expenditures for the retired elderly after the 1990s. The expenditure for the retired elderly was only 39.62 billion yuan in 1990, and it rose to 525.32 billion yuan in 2005, an increase of 12 times over 15 years.[9] Financial subsidies to social insurance fund rose from 2.155 billion yuan in 1998 to 177.67 billion yuan in 2009, an increase of 81.44 times, and the subsidies to social insurance fund in 1998–2009 totaled 1.09158 trillion yuan.[10] Most of state financial subsidies to social insurance fund expenditure went to pension expenses.[11] At the same time, a variety of social pension insurance funds have become an important part of the expenditure of public funds. The total revenue of urban basic pension insurance fund was 2.0001 trillion yuan in 2012, an increase by 18.4% over the previous year, of which the collection contributed 1.6467 trillion yuan, an increase by 18.0% over the previous year. The financial subsidies to basic pension insurance funds at all levels totaled 264.8 billion yuan. The total annual expenditure was 1.5562 trillion yuan, an increase of 21.9% over the previous year. Pension security expenditure and pension expenditure accounted for about 20% of the expenditure of public funds.[12] Meanwhile, the benefits from pension security and a variety of social insurance also become an important source of income for rural and urban residents. According to our household survey, 14.5% of urban and rural residents' incomes were from social security in 2012, and most of the incomes for the retired elderly come from pensions.[13] The increasing financial investment in pension security plays an important role in expanding pension coverage and improving the level of pension security.

8.1.3 Major Problems in Chinese Pension Security System

Chinese pension security system has made some achievements after several years of development and played an important role in guaranteeing and improving the quality of life of elderly people. However, there are still many problems and limitations in Chinese pension security system, and the current supply of pension

[8]Ministry of Civil Affairs: "Statistical Communique on Civil Affairs in 2009".

[9]Zhu et al. [13].

[10]Wang and Long [14].

[11]Take the data in 2009 for an example, the national fiscal subsidies to social insurance fund were 177.673 billion yuan, of which 143.7 billion went to basic pension insurance fund, accounting for 81%.

[12]Ministry of Human Resources and Social Security: "Statistical Communique on Human Resources and Social Security Development in 2012", website of Ministry of Human Resources and Social Security.

[13]Wang et al. [15].

security can hardly meet the growing demand. The following problems need to be solved in its future development.

First, there are still many difficulties in economic security for the elderly. It is the need and the right of the elderly to get adequate income, maintain basic living and improve the quality of life in later years. Economic security is directly related to the survival and well-being of the elderly. Although the rapid economic growth and constantly growing national income create relevant conditions, economic security for the elderly is still not satisfying. There are differences in incomes of the elderly in rural and urban areas. The annual per capita income of the urban elderly was 12,805 yuan in 2005, slightly higher than the national per capita income of urban residents (11,321 yuan), but significantly lower than the average wage (18,200 yuan) at that point; the per capita annual income of rural elderly people was 2936 yuan in 2005, significantly lower than the per capita income of rural residents (3255 yuan) of the same year.[14] The income sources of the elderly include market income, pension income, public transfer income, household transfer income and other incomes; with the rapid loss of ability to work and innovation of science and technology, the elderly's market income relying labor will continue to get reduced, and they have to rely on other sources of income. According to the survey, less than 20% of those in advanced ages (over 80 years old) enjoy retirement payments, and only about 7% rely on social assistance. However, the remaining 71.4% rely on their children and grandchildren as financial resources, and their socio-economic security rate is 17% lower than the average, with less than 500 yuan per month which is just enough to maintain basic living.[15] Due to the limitations in pension security system, the narrow coverage, the changes in family structure and the growing family burden, it is difficult for the elderly to get incomes from family and government sources; in particular, the majority of the rural elderly find it even more difficult to obtain pension security income from the state. By the end of 2009, the population participating in urban basic pension insurance was 23,550, and the employed population was 77,995. The insured population accounted for only 30.2% of all the employed population, and only less than a third of employees participated in urban basic pension insurance.[16] In rural areas, the exploration of new rural social pension insurance pilot just begins. Only 86.91 million people participate in rural social pension insurance, and the vast majority of rural residents do not. If 52.93 million retirees and 16.414 million people living on minimum subsistence allowances are subtracted from 159 million nationwide aging population, 80 million people are not covered by institutional arrangements of pension insurances. According to the population sampling survey in 2007, the elderly aged 70–79 accounted for 4.57% of national population, amounting to 50 million or

[14]Data is from the tracking survey on the status of urban and rural elderly population in 2006, see Zhang and Guo [16].

[15]Liu and Chang [17].

[16]The data is from the Ministry of Human Resources and Social Security: "Statistical Communique on Human Resources and Social Security Development in 2009".

more; 1.43% are over 80 years old, amounting to 18 million; even if 20 million or more retirees and people living on minimum subsistence allowances were excluded, there were still about 40 million over-70 people out of the institutional arrangements of pension insurances.[17] The limitation in financial security for the elderly affects the quality of their life and even puts many elderly people into poverty, let alone enable the elderly to share the fruits of economic and social development.

Second, elderly health care is far from satisfying. As mentioned earlier, compared with adults, elderly people face a much higher risk of disease. While the elderly suffers physical and mental pains, the state and government should establish appropriate security mechanisms to reduce the elderly's fear for disease. When current urban and rural medical security system is not well developed nowadays, the health care of the elderly is far from satisfying, and some will plump into poverty in the case of low incomes, which tends to increase the family burden. Currently, the imperfect health care system makes the health care for the elderly far from satisfying, and there is still a long way to go. At present, although China's health care system has achieved remarkable results in coverage, yet many problems still need to be solved before arriving at its real effects. The current health insurance system is not fair enough, lacking appropriate policy support for the elderly and other weak groups. The health care system design, especially compensation mechanism design, still exposes many deficiencies, such as lack of mutual aid role of individual accounts, unreasonable compensation ways, narrow coverage, low reimbursement levels, and excessive reimbursement restrictions.[18] In terms of medical services, the problem is mainly reflected in that primary care services are still weak, medical staff are lack of service awareness and medical prices are too high, making it prominently "expensive and difficult to receive medical treatment". In terms of medical expenses, according to the survey, only 13.8% of senior citizens (over 80 years old) enjoy free medical care, and the rest have to rely on themselves and their families. This situation is more serious in rural areas. Only 4.1% of senior citizens enjoy free medical care, and 95.5% have to rely on themselves and their families. Medical expenses bring great pressures to the elderly and their families and constitute an important reason for the decline in the quality of life of the elderly.[19] In addition, there are many restrictions and inconveniences when the elderly receive medical treatment in other places; more efforts are still needed in the education and chronic disease prevention for the elderly; there is a lack of health promotion and intervention mechanisms for the elderly. How to strengthen future health care for the elderly and improve the quality of their life remains an important task in the development of China's endowment insurance system.

Third, elderly service security still is lagging far behind. Despite the development and progress in elderly service system, there are still many problems and it is

[17]See Ministry of Civil Affairs: "Report on Strategic Planning for China's Basic Old-Age Service System Construction" (draft).

[18]Wang and Long [18].

[19]Liu and Chang [17].

difficult to meet the service needs of the elderly. Especially in the case of growing demand for elderly services, the contradiction between elderly service security supply and demand has become increasingly prominent. Currently, the problems in elderly service security are mainly reflected in these aspects: First, the development of pension system lacks a systematic planning. Although the state authorities have introduced a number of planning and opinions on this regard in recent years, yet so far, there is no systematic planning for pension security system construction, making the development of pension security system lack of macro guidance. Second, the investment in elderly services is inadequate and there is a lack of fund security mechanism. In the current shortage of social security investment, limited financial fund mostly goes for the construction of pension insurance system, and only a little is used for elderly services. Some planning and opinions just strengthen the improvement of financial investment in principle, but fail to propose any clear financial investment guarantee mechanism. Third, the elderly service institutions are insufficient, causing prominent contradictions between supply and demand. In December 2009, Dou Yupei pointed on the 5th National Nursing Home President Forum that China would be faced with a grim aging population problem for a long period in the future, and the social demand for elderly service institutions and professional care personnel would increase dramatically. But now, China faces a serious shortage of elderly service institutions and personnel, far from meeting the needs. In accordance with the international average 50 beds per thousand old people, it is estimated that a total of 8 million elderly service beds are needed. However, there are only 2.5 million beds nowadays, a shortfall of 5.5 million. There are still many problems in the management and operation of old-age service institutions, such as immature system, unprofessional staff and fund shortage, etc.[20] Fourth, there is a serious lack of elderly service personnel. At present, China's urban disabled and semi-disabled elderly accounts for 14.6% of the total, while it is more than 20% in rural areas. This people require professional care and attention. According to the ratio of 3 old persons to 1 caregiver, China needs about 10 million elderly service nurses. However, there are only 220,000 workers at nursing homes in China, of which only 20,000 have obtained vocational qualifications. The current supply is too little compared with the potentially huge demand of tens of millions disabled elderly, and even cannot meet needs of 2 million people in nursing institutions.[21] The pension security backwardness is more prominent in rural areas. It becomes extremely urgent to strengthen elderly service security system construction in the context of the growing aging population and the weakening family support.

Fourth, there is also an extreme shortage of elderly mental security. Compared with the economic security and service security, mental security for the elderly is in greater need. The elderly needs more spiritual solace compared with adults, and the

---

[20]Wu [1].

[21]See Ministry of Civil Affairs: "Report on Strategic Planning for China's Basic Elderly Service System Construction" (draft).

mental security problem of the elderly is even more important along with economic development and constantly improved living standards; if not handled properly, it will affect the quality of their life, and there will be more and more aging-society problems. The absence of elderly mental security is mainly reflected as follows: First is the understanding of the society on mental security. It is believed that the pension security is mainly on material and economy levels, at most, with some life care and service (you can pay others to do this), and that this will meet their needs; it is also believed that spiritual solace is the high-level need and a luxury for the elderly, and it is dispensable based on economic conditions. The mental security of the elderly is likely to be overlooked under the influence of such beliefs. Second, the reality weakens the family's spiritual support for the elderly. With the acceleration of urbanization, the mobility and pace of life also accelerates, and young people are busy all year round and likely to overlook the care and mental comfort for the elderly. Especially some elderly people separate and cannot get care from their children for many years, and will feel very lonely,[22] and even catch some psychological problems. This problem is more prominent especially in the core and small families. Third, the government fails to take effective measures to solve elderly mental security problems. Despite the emphasis on mental security of the elderly in a number of laws and regulations, the government does not take forceful measures to ensure the mental security of the elderly in reality; it does not pay sufficient attention to mental security problems of the elderly and offers no guidance in terms of family and social mental support for the elderly. Thus, some old people have to resort to the court to force their children to go home and see them. The government also fails to fulfill its obligation in directly providing mental security for the elderly, such as constructing entertainment facilities and organizing social activities for the elderly. Fourth, there is a lack of specific standards for elderly mental security and specialized personnel providing mental security services. In addition, the role of social organizations and relevant relative groups in pension security is also worth strengthening.

8.2 Preliminary Thoughts on the Construction of Trinity Pension Security System

8.2.1 Major Objectives of Building Chinese Pension Security System

A clear, scientific and rational development objective is the premise of developing Chinese security system for the aged. In response to the challenges of an aging population, a number of international organizations propose the related concept of aging, including "healthy aging" and "active aging". As for "healthy aging",

[22]Luo and Peng [2].

the World Health Organization put forward this concept in 1990 and its three criteria as well: First, physical health; second, mental health; and third, a good state to adapt to the society. On "active aging", the Political Declaration of United Nations Second World Assembly on Aging gave a definition of active aging on April 12, 2002: the process where the old people maximize the benefits of health, participation and security opportunities in order to improve the quality of life. Active aging is based on the recognition of human rights of the elderly and the United Nations' principle of independence, participation, dignity, care and self-realization. "Positive" stresses the continuing participation in social, economic, cultural and public affairs.[23] These two concepts mainly emphasize the physical and mental health, social participation and rights of the elderly. On this basis, we can also take "happy aging" as the policy target of an aging society, which is a target on a higher level than "active aging" and "healthy aging". It emphasizes the overall improvement of the elderly's life quality, so that they could enjoy the joy of the later life based on physical and mental health, social participation and rights. In fact, the "Law of People's Republic of China on Protecting the Rights and Interest of the Elderly" put more clear and more specific objectives for the pension security system, namely, "they are provided for, have access to necessary medical care, have opportunities for their own pursuits and studies and enjoy themselves". Based on these, this paper puts forward the four main objectives in developing Chinese future pension security system.

First, relieve the economic worries of the elderly. This is for economic security and serves as the basis in developing the pension security system. Economic security is the fundamental need of the elderly, and the security system should provide basic necessities of life for the elderly to protect their basic economic needs. To relieve the economic worries of the elderly is not just a matter when they really get old, but should be appropriately prepared before and after getting old. Only in this way can it cut the elderly's worries before getting old and access to economic security after it.

Second, relieve the physical and psychological pains of the elderly. This is mainly for physical and mental security. The elderly is more vulnerable to the risk of mental disorders and diseases. To strengthen the security system for the aged, we should strengthen health education and prevention of chronic diseases for the elderly through the appropriate institutional arrangements and capital investment, and enable them to access adequate health care after catching diseases, thus relieving their sufferings. We should also establish and improve elderly mental security mechanisms to provide effective mental security for the elderly. Only when the physical and psychological pains of the elderly are relieved or eliminated can they have a better mental state.

Third, ensure sufficient service supply for the elderly. This is mainly for elderly service security. Due to the declining body function and self-care ability, the elderly needs more care services. Currently, the elderly service system is still in great

[23] Wang [3].

shortage in the development of the pension security system, and can hardly meet the growing demand for services among the elderly. The lack of services for the elderly has a serious impact on their lives. The government should take the lead in the future to double the efforts in system planning, personnel and funding, so that the elderly people can receive timely, effective and adequate elderly services.

Fourth, ensure the happiness of the elderly. This is a high-level target in developing the pension security system. The security system should not be just confined to basic livelihood security, but it ought to take into account their economy, health, service, mental state and community involvement, thereby giving the elderly a happy life in the old age. Particularly along with constant economic and social development, pension security should also be constantly improved.

8.2.2 *Principles for Building a Trinity Pension Security System*

China should follow the following basic principles in building the pension security system.

First, combine fairness and efficiency. Fairness, justice and sharing are the basic values of the social security system,[24] and the pension security system as an important part of the whole social security system should follow the principle of fairness. The pension security system should cover all the elderly regardless of their census register, gender, ethnicity, region, professional background and other factors, and constantly improve the level of pension security while ensuring their basic economic security, health security, service security and mental security, so that the elderly can share the achievements of reform and development in a fair way with other groups. Efficiency should also be emphasized on the basis of fairness principle. There is no real fairness without efficiency. The development of pension security system needs to introduce market forces to mobilize more resources to improve the efficiency of pension security. In addition, it should also pay attention to the differences and diversity of the needs of the elderly. The pension security system needs always to stick to the fairness and efficiency principal throughout the entire process, and any bias is not desirable.

Second, combine government guidance and social participation. To achieve "a sense of security" and develop a pension security system is the bounden duty of governments at all levels. In developing the pension security system, the government must play a major role in legal improvement, policy development, investment increase, personnel training, supervision and management, organization and coordination and other aspects. The development of pension security system cannot achieve good results and the corresponding results cannot be realized without a strong government leadership. But the total reliance on the government is also not

[24]Zheng [4].

desirable, for the development of the pension security system needs the participation of a variety of social forces. It needs to take advantage of market mechanisms and social resources and guide the participation of businesses, nonprofit organizations, volunteers and other forces, which will further increase the supply of pension security and improve the quality of pension security. Of course, even when the market and social forces play their parts, we still need the government to play a role in guidance, policy support, supervision and management, etc.

Third, be compatible with the level of economic development. The development of s pension security system must be compatible with the current level of economic development, and the security system excessively lagging behind or ahead of the level of economic development is unsustainable. If it lags far behind economic development, the pension security system will be incomplete and on low levels, and it is hard to achieve the target and difficult for the elderly to share the fruits of economic and social progress, which is contrary to the purpose of economic and social development. At present, the Chinese pension security system is lagging far behind economic development, and the government needs to play the major role in increasing the investment to improve it as soon as possible. Conversely, if the pension security system is over ahead of economic development, it will bring burdens to economic development, which will not only hinder economic development but also affect the sustainable development of the pension security system. Nowadays, when China gets "old before getting rich", the pension security system still has a long way to go, which requires more investment. But, this should be done appropriately so as to keep pace with economic development.

Fourth, focus on the diversified needs of the elderly. Older age groups show differences, and therefore the demands for pension security also vary. The elderly of different ages (the oldest and younger elderly), genders, education levels, incomes, health statuses, regions and nationalities have different needs for pension security. The pension security system should not be the same for different people, and we should make the pension security more targeted, more efficient, more multi-levelled and more diversified. For the younger elderly, they should be guided to participate in social welfare activities; for some senior citizens, their health care and geriatric services should be strengthened; for the elderly with good health, they can be provided with opportunities for employment; and for the elderly with low incomes, economic security should be strengthened; for empty-nest elderly, economic security, service security and mental security should all be enhanced; for the high-income elderly, they should be encouraged to purchase services available in the market. While paying attention to the different needs, we need to provide a consistent, universal, basic pension security for the elderly, such as establishing a fairly shared pension system and health care system and improving community care service facilities.

Fifth, respect the participating role of the elderly. The elderly is the object of pension security and also one of participants in pension security. While grasping their different needs and providing targeted pension security for them, we should also respect their participating role and reduce their reliance on the state and family. We should encourage some able-bodied younger elderly to get some economic incomes through appropriate labor; encourage some older people to engage in community welfare activities and cultivate old-age volunteer team; encourage the "younger old" to take care of the "advanced old"; establish and improve older people's associations to increase their participation in social and sports activities. The respect for the participating role of the elderly is also the inevitable requirement to offer them "opportunities for their own pursuits and enjoy themselves".

Sixth, carry forward the traditional endowment culture. Old saying goes, "children should not travel afar when living with parents" and "respect others' parents just as my own parents and love others' children just like my own children". Respecting and caring for the elderly is a fine tradition of the Chinese people and has a long history. The concept of filial piety can be traced back to the word on bones or tortoise shells of the Shang Dynasty. According to *Shuo Wen Jie Zi*, "filial piety" is being kind to the elderly.[25] In *Book of Filial Piety*, Confucius regarded the respect for the elderly as the nature of human beings, and he said that "human being is the most important in the world, and the most valuable behavior of human being is filial piety," "filial piety is the unalterable truth in the cosmos and the foundation of common people."[26] According to Confucius, the respect for the elderly falls on three levels: "filial piety falls into three types: the greatest is to respect them, followed by not to shame them, and then by just able to support them." That is, to respect parents is the greatest filial piety, not to let them humiliated is the moderate "filial piety", and only to support them is "small filial piety." According to Mencius, "the greatest deed of Yao and Shun is nothing but filial piety,"[27] which reflect the importance of filial piety and respect for the old. The traditional filial piety and endowment culture plays an important role in the economic, service, mental security of the elderly. However, with the development of productive forces and the changes in economic and social environment, especially with the accelerated process of industrialization, the traditional endowment culture begins to weaken,[28] bringing new issues to the development of pension security. When developing pension security system in the future, we still need to stick to our traditional culture of respecting, caring, loving and supporting the elderly to provide them with better security.

8.2.3 Specific Contents of Chinese Pension Security System

Based on the previous analysis, this paper explains some ideas on building Chinese pension security system. Overall, the pension security system should cover

[25]Xu [5].

[26]*Book of Filial Piety*, translated and annotated by Wang Shoukuan, Shanghai Ancient Books Publishing House, 2007, pp. 28–40.

[27]Li and Lv [6].

[28]Yao [7].

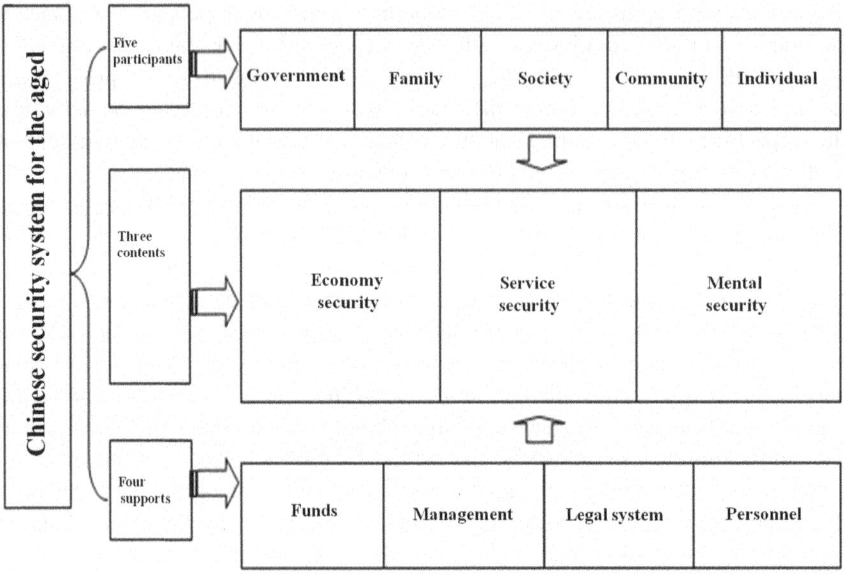

Fig. 8.1 Chinese pension security system

economic security, service security and mental security, which requires funds, institutions, management and talent as support and demands the participation of the government, family, society, communities and individuals (Fig. 8.1). The following will discuss the specific content of building Chinese security system for the aged.

8.2.4 *Three Contents*

Chinese pension security system includes the following three aspects:

One is economic security. Economic demand is the basic needs of the elderly, and lack of income (poverty) is one of the risks for the elderly. Only by ensuring the elderly has no economic worries will they have a happy life in later years. Thus, adequate and reasonable economic security is a prerequisite for a happy life of the elderly. Economic security should be the priority in the process of building the pension security system. We need to establish a scientific economic security mechanism to ensure that the elderly will have a sufficient and stable source of income to sustain life in their later years. The provision of basic economic security for the elderly is one of the government's responsibilities, and the government must establish and improve various pension security systems and elderly benefits as soon as possible. At the same time, the government also takes some mandatory measures to ensure that the children will provide certain economic security for the elderly. The economic security should be compatible with economic development.

A too low security level can hardly achieve desired effects, while a too high security level will be burden on the government and the family and not conducive to its sustainable development. Economic security should show certain urban-rural, regional and occupational (before the old age) differences on the basis of the basic economic security.

Second, service security. Service security is essential to helping the elderly get through the old age with improved quality of life. We should, under the guidance of the government, establish appropriate regulatory systems and service agencies, to provide comprehensive, specialized, multi-level service security for the elderly with the participation of governments, families and communities. Service security needs to focus on the life care and medical care for the elderly in advanced ages. It should combine paid and unpaid patterns. For the “five guarantees” or the elderly with no family or in poverty, we should offer some basic free services; for a number of relatively high-income elderly, we encourage market purchases (especially some non-essential care services), but the government needs to strengthen market supervision and provide some policy support. Compared to economic security, service security has higher requirements, is more specialized, and needs certain services techniques and a large number of professional service personnel.

Third, mental security. Mental need is one of the essential needs of each normal person. Compared with adults, the elderly needs more mental security, and the mental emptiness and loneliness easily lead to senile mental illness and social problems. But in reality, people tend to focus on economic security while ignoring the mental security of the elderly, and believe that mental security is a high-level need and a luxury that can be ignored rather than the basic need of the elderly. This is a wrong perception. Just as the other securities, mental security is also important for the elderly and directly related to the quality of life of the elderly. Mental security may come from formal and informal channels. The formal mental security channel mainly refers to the government. The government at all levels should play a role in the mental security for the elderly by introducing appropriate laws and regulations, strengthening mental security investment and personnel training, and vigorously promoting the attention to mental security for the elderly. Meanwhile, mental security from family, relatives, society and other informal channels is also very important, of which family is a particularly important source of mental security and its role in mental security needs to be particularly strengthened.

Economic security, service security and mental security are indispensable to the pension security system. They interact with and influence each other and also show a certain hierarchy. Economic security is the foundation and has an impact on service security and mental security. Economic security has a direct impact on the quality of service security. A better economic security can enhance the availability and quality of service security; economic security is also directly related to the quality of mental security, and the elderly without economic security can hardly be regarded as happy. Service security is the carrier of pension security and closely linked with the lives of the elderly; if there is just economic security without necessary service security, the life of the elderly will meet some trouble, affecting their quality of life; service security is also directly related to the mental life of the elderly. Mental security is on a relatively high level but also an integral part of pension security. Mental security includes a part of economic security and service security, and the quality of economic security and service security is directly related to that of mental security.

8.2.5 Four Supports

Whether it is economic security, service security or mental security, the following four supports are needed in order to truly meet the needs of the elderly and guarantee a satisfying effect.

First, appropriate funds. You can never make bricks without straw. Adequate funding is a fundamental prerequisite for building pension security system. Insufficient investment is an important bottleneck restricting our pension security system, and the future development of the pension security system needs to raise funds from multiple channels. The government plays the major role and needs to increase financial investment and take pension security system building as an important part of people's livelihood and economic transformation. Meanwhile, it needs to absorb social investment and encourages social forces to participate in the construction of pension security system. It also needs to establish reasonable burden-sharing mechanism, and the elderly pays certain fees at work. It should be noted that fund raising and investment should always be compatible with economic development, and that a too low or too high investment is not conducive to the sustainable development of the security system.

Second, an improved legal system. An improved legal system is an important guarantee for the sustainable development of the pension security system. Despite some progress, the legal system for China's security system is still lagging behind, which calls for further strengthening. Legal construction for pension security includes: establishing and improving universally-covered and fairly-shared pension system; establishing and improving old-age benefits system to enable the elderly to share the results of economic and social development; establishing and improving elderly services system to ensure the fairness, efficiency and standardization of the services for the elderly; establishing elderly mental security system, which is a new requirement of economic and social development for security system building. If conditions permit, we can consider developing a comprehensive "Pension Security Law" to specify economic security, health security, service security, mental security for the elderly.

Third, efficient management. That is, to establish a scientific, rational, and efficient pension security management system and mechanism so as to ensure the good organization and management of the pension security system and improve its fairness and efficiency. Currently, pension security management confronts certain problems such as department separation, vaguely defined responsibilities, inadequate operational mechanisms and inefficient management. In the case of growing pension security demand and constantly improved market economy system, we need to further strengthen pension security management system construction in the future, further clarify the specific responsibilities of administrative management, financial management, operational management and service management, and establish a pension security management system with clear responsibilities, scientific management, efficient operation and strict supervision.

Fourth, professional personnel. The fund-raising and management of pension security as well as the provision of service security and mental security all need professional talents. It can be said that the quality and efficiency of the pension security is directly related to the quality of people. From a practical point of view, there is still a shortage of all kinds of professional talents for pension security, and there is an urgent need to strengthen the training of professional pension security personnel. The security personnel mainly include agency personnel, health care personnel, social workers, service institution management personnel. The institutions of higher education need to offer relevant majors to strengthen the cultivation of pension security professionals, or carry out short-term training on a regular basis to combine theory with practice and enhance the practical abilities of various professionals.

8.2.6 Five Participants

The development of pension security system needs the participation of government, family, society, community and individual.

Government. The government is the most important participant in the pension security system, and its attention paid to the security system directly influences the development of the latter; the government's overall planning directly determines the direction of the latter; the government's financial investment directly determines the level of the latter; the government's supervision and management directly determines the sustainable operation of the latter. Government is the planner, organizer, manager and supervisor of the security system for the aged. The development of security system always needs the government to play a leading role and guide other forces to provide adequate human, material and financial resources for its development.

Family. Family support has been the dominant way to support the elderly, and it is still an important participant in modern pension social security system. Family is where the elderly spends their life and the best place for solace, which can hardly be substituted by other participants. The family still remains an important force to provide financial security for the elderly, and especially in rural areas where the pension security system is not mature, the elderly's (especially those in advanced ages) income mainly comes from their children. Family is also an important service provider for the elderly, and the elderly service provided by families will have better results than that by other institutions or personnel, prone to a sense of intimacy and spiritual solace. Therefore, the role of the family should be consolidated while socializing the pension security system.

Society. As a member of the society, the elderly should get corresponding social support and help. The development of pension security system needs social forces under the guidance of the government which can mobilize social resources to provide more and better security, including financial support, elderly service, and mental solace. We need to nurture and mobilize some non-profit organizations and voluntary organizations to offer their help, encourage members of society to provide assistance for the elderly, and can even encourage some for-profit organizations to provide services for the elderly thus increasing the diversity of pension security offers.

Community. The community refers to the living community in a certain geographical area for a number of inherently interacted residents. The community is closely related to people's lives and offers a variety of convenience for their lives. Similarly, community is an important provider of services and mental solace for the elderly. Community can provide for the elderly various convenient and friendly services, and it is easier for older people to participate in interactive activities in the community and thus enhance their sense of participation in society. To enable the elderly to participate in community outreach activities and sports activities within their capacity is also an important way to give them "opportunities for their own pursuits and enjoy themselves".

Individual. The individual here mainly refers to the elderly themselves. Although the elderly is the object of old-age security, but they are not entirely negative and passive, but should take the initiative to do something in their own ability in order to reduce the burden on the government, family and society as far as possible. In particular, some "young elderly" still having certain ability to work should work appropriately to obtain some incomes. This is the requirement of old-age security and also the need to make full use of human resources.

8.3 Giving Full Play to the Role of Government in Building a Trinity Pension Security System

To develop security system for the aged and provide comprehensive, all-round and high-quality pension security for the elderly is an important task in China's social security reform and developing people's livelihood, and also the requirement of building of a harmonious society and transforming economic development patterns. Government is the most important participant in developing the pension security system, and both the property of pension security system and the function of the government demand that the government should play its role in the pension security system.

The pension security system is of prominent public nature. If seen from the definition of public goods, the US economist Paul Samuelson issued in 1954 "The Pure Theory of Public Expenditure" where the pure public goods or services are defined as: any person' consumption of such goods or services will not reduce the other's consumption of such goods or services.[29,30] Thus, it is known that public goods have two characteristics, namely the "non-exclusive" and "non-competitive." But Samuelson's definition mainly refers to pure public goods, and in reality, there are still a large number of products ranging between pure public goods and private goods. In *Public Finance in Democratic Process*, Buchanan argued that "any goods or services that the group or social group decides to be provided by collective organization for any reason are defined as public"; "this broad category includes the so-called 'pure public' goods by Samuelson and others, and also other goods and services with public degree ranging from 0 to 100%.[31] In other words, there are a number of non-exclusive and non-competitive products in varying degrees.

As can be seen from the above definition, the vast majority of contents of pension security system are of prominent public nature, and are pure public goods or quasi-public goods. Due to the complexity of the security system for the aged, it is difficult to make a precise definition. If unbundled, the basic pension security is public goods or quasi-public goods, such as basic elderly economic security, basic old-age health care, basic elderly services and basic mental security. However, non-basic pension security shows quasi-public or non-public nature (private nature), such as supplementary pension insurance, annuity, commercial pension insurance, and home elderly services. On the whole, the public nature of pension security system is greater than its non-public nature, and the development of pension security system is of great social and public nature. Even the contents of private nature in the pension security system also need government support and supervision in order to ensure security quality and efficiency. Especially when the current market economy in China is not well developed nowadays, the government is more needed to play a role in the pension security system.

Not only does the nature of pension security require the government to play a role, but the function of the government its own demands it should establish and improve the pension security system.

In the studies of government functions, Keynes elaborates his state intervention ideas in *The General Theory of Employment Interest and Money*. He believes that the market alone cannot solve the problem of lack of effective demand, and thus the active state intervention is demanded to stimulate effective needs in order to achieve full employment and promote economic growth. In the specific policy, he advocates high progressive tax policy to help the unemployed and the poor through fiscal transfer payments and to redistribute the income to increase the propensity to consume; he advocates the socialization of investment and direct investment by the state; the state should create jobs and solve the unemployment problem through public projects to increase consumption.[32] Keynes' theory of government function is a direct reflection of the role of government in old-age security system, and it

---

29 Paul [8].

30 Samuelson [9].

31 James [10].

32 John [11].

plays an important role in establishing and improving pension security systems in many countries.

A market economy needs to give play to the basic role of the market in allocating resources and also needs the government to strengthen market regulation and macro-control to prevent "market failure". Especially for some products that the market is unwilling to provide, have low economic efficiency but are closely related to the interests of people, the state should directly provide or purchase them. In improving the socialist market economic system, China puts forward the goal of transforming government functions to establish a service-oriented government and give full play to the role of government in economic regulation, market supervision, social management and public services. Those functions of government are closely related to the development of the pension security system, and it is an important task of the government' economic regulation and transformation to establish and improve the security system for the aged; the market provision of pension security and allocation of pension security resources need the government to strengthen market supervision; it is an important part of strengthening social management and building a harmonious society to establish and improve the pension security system; it is an important part of government's public service function to establish a fairly shared pension security system. Thus, the establishment and improvement of pension security system is an important duty of the government, and we should give full play to the role of government in this process.

Currently, there are still some limitations in the government's development of pension security system, which is an important cause of inadequate security system development and supply-demand contradiction. The limitations are as follows: vague positioning and responsibilities; absence of role in overall planning and coordination; absence of role in legal system construction and policy development; absence of role in financial investment; absence of role in organization, implementation and supervision.

The future development of pension security system must give correct and full play to the leading role of the government in order to provide quality, efficient and available security for the elderly. The government needs to strengthen the publicity, guidance and overall planning of the security system, further improve related regulations and policies, improve its financial input mechanisms, and give full play to the role in implementation, management and supervision.

References

1. Wu, Yang. 2007. Several issues to be noticed in building the endowment system of a harmonious society—Analysis of the building, operation and management of Elderly Care Institutions in Pudong New District. *Shanghai Economic Research* 11: 71–78.
2. Luo, Fang, and Daiyan Peng. 2007. Empirical analysis of the impact of migrant children on elderly care in rural empty-nest families. *Chinese Rural Economy* 6: 21–27.
3. Wang, Shuxin. 2003. Population aging and active aging in Beijing. *Population and Economics* 4: 1–8.

4. Zheng, Gongcheng. 2008. *The Strategy of China's Social Security Reform: Principles, Goals and Action Plans*. People's Publishing House. p. 17.
5. Xu, Shen. 1981. S*huo Wen Jie Zi*, Shanghai Ancient Books Publishing House, 1981, p. 398.
6. Li, Zhengang, and Hongping, Lv. 2009. On the relation between culture of respecting, revering the old and providing for the old in China. *Population Journal* 5: 27–41.
7. Yao, Yuan. 1998. A cultural interpretation of the weakening family support for the old. *Population Studies* 5: 48–50.
8. Samuelson, Paul and William, Nordhause. 2003. *Economics (16th ed.)*, Huaxia Publishing House. p. 78.
9. Samuelson, P.A. 1954. The pure theory of public expenditures. *The Review of Economics and Statistics* 36: 387–389.
10. James Buchanan. 1992. *Public Finance in Democratic Process*. Translated by Tang Shouning and Shanghai Sanlian. Bookstore Publishing House. p. 13.
11. John Maynard Keynes. 1999. *The General Theory of Employment Interest and Money*. Translated by Gao Hongye. The Commercial Press. pp. 10–13.
12. Baoan, Song. 2006. A sociological analysis of endowment of the elderly. *Jilin University Journal (Social Sciences Edition)* 4: 90–97.
13. Zhu, Qing, et al. 2010. *Studies on Improving China's Social Security System and Optimizing the Structure of Fiscal Expenditure*. China Renmin University Press. p. 185.
14. Wang, Yanzhong, and Yuqi, Long. 2011. Analysis of Chinese Government's expenditure on social security since the reform and opening Up. *Finance and Trade Economics* 1: 17.
15. Wang, Yanzhong, et al. 2013. *General Report on "Social Security Income Reallocation Report"*. Social Sciences Academic Press.
16. Zhang, Kaiti and Guo, Ping. 2010. *Blue Book of China's Population Aging and Elderly Population*. China Social Sciences Publishing House. pp. 80–90.
17. Liu, Yuzhi, and Yuan, Chang. 2003. Economic situation and health care of senior citizens: Status quo, problems and countermeasures. *Population and Economics* 1: 12–16.
18. Wang, Yanzhong, and Yuqi, Long. 2010. Characteristics, problems and recommendations of the mid- and low-income group's medical service demand: Based on the survey of 1642 families. *Chinese Health Policy Research* 3: 9–15.

The manufacturer's authorised representative in the EU is Springer
Nature Customer Service Centre GmbH, Europaplatz 3, 69115 Heidelberg,
Germany. If you have any concerns regarding our products, please
contact ProductSafety@springernature.com

Printed and bound by CPI Group (UK) Ltd, Croydon, CR0 4YY
25/09/2026
02222898-0016